GNU/Linux Application Programming

LIMITED WARRANTY AND DISCLAIMER OF LIABILITY

THE CD-ROM THAT ACCOMPANIES THE BOOK MAY BE USED ON A SINGLE PC ONLY. THE LICENSE DOES NOT PERMIT THE USE ON A NETWORK (OF ANY KIND). YOU FURTHER AGREE THAT THIS LICENSE GRANTS PERMISSION TO USE THE PRODUCTS CONTAINED HEREIN, BUT DOES NOT GIVE YOU RIGHT OF OWNERSHIP TO ANY OF THE CONTENT OR PRODUCT CONTAINED ON THIS CD-ROM. USE OF THIRD-PARTY SOFTWARE CONTAINED ON THIS CD-ROM IS LIMITED TO AND SUBJECT TO LICENSING TERMS FOR THE RESPECTIVE PRODUCTS.

CHARLES RIVER MEDIA, INC. (“CRM”) AND/OR ANYONE WHO HAS BEEN INVOLVED IN THE WRITING, CREATION, OR PRODUCTION OF THE ACCOMPANYING CODE (“THE SOFTWARE”) OR THE THIRD-PARTY PRODUCTS CONTAINED ON THE CD-ROM OR TEXTUAL MATERIAL IN THE BOOK, CANNOT AND DO NOT WARRANT THE PERFORMANCE OR RESULTS THAT MAY BE OBTAINED BY USING THE SOFTWARE OR CONTENTS OF THE BOOK. THE AUTHOR AND PUBLISHER HAVE USED THEIR BEST EFFORTS TO ENSURE THE ACCURACY AND FUNCTIONALITY OF THE TEXTUAL MATERIAL AND PROGRAMS CONTAINED HEREIN. WE HOWEVER, MAKE NO WARRANTY OF ANY KIND, EXPRESS OR IMPLIED, REGARDING THE PERFORMANCE OF THESE PROGRAMS OR CONTENTS. THE SOFTWARE IS SOLD “AS IS” WITHOUT WARRANTY (EXCEPT FOR DEFECTIVE MATERIALS USED IN MANUFACTURING THE DISK OR DUE TO FAULTY WORKMANSHIP).

THE AUTHOR, THE PUBLISHER, DEVELOPERS OF THIRD-PARTY SOFTWARE, AND ANYONE INVOLVED IN THE PRODUCTION AND MANUFACTURING OF THIS WORK SHALL NOT BE LIABLE FOR DAMAGES OF ANY KIND ARISING OUT OF THE USE OF (OR THE INABILITY TO USE) THE PROGRAMS, SOURCE CODE, OR TEXTUAL MATERIAL CONTAINED IN THIS PUBLICATION. THIS INCLUDES, BUT IS NOT LIMITED TO, LOSS OF REVENUE OR PROFIT, OR OTHER INCIDENTAL OR CONSEQUENTIAL DAMAGES ARISING OUT OF THE USE OF THE PRODUCT.

THE SOLE REMEDY IN THE EVENT OF A CLAIM OF ANY KIND IS EXPRESSLY LIMITED TO REPLACEMENT OF THE BOOK AND/OR CD-ROM, AND ONLY AT THE DISCRETION OF CRM.

THE USE OF “IMPLIED WARRANTY” AND CERTAIN “EXCLUSIONS” VARIES FROM STATE TO STATE, AND MAY NOT APPLY TO THE PURCHASER OF THIS PRODUCT.

GNU/Linux Application Programming

M. Tim Jones

CHARLES RIVER MEDIA, INC.
Hingham, Massachusetts

Acquisitions Editor: James Walsh
Cover Design: The Printed Image

CHARLES RIVER MEDIA, INC.
10 Downer Avenue
Hingham, Massachusetts 02043
781-740-0400
781-740-8816 (FAX)
info@charlesriver.com
www.charlesriver.com

This book is printed on acid-free paper.

M. Tim Jones. *GNU/Linux Application Programming.*
ISBN: 1-58450-371-8

Library of Congress Cataloging-in-Publication Data
Jones, M. Tim.
GNU/Linux application programming / M. Tim Jones.
p. cm.
Includes bibliographical references and index.
ISBN 1-58450-371-8 (pbk. with cd-rom : alk. paper)
1. Linux. 2. Operating systems (Computers) I. Title.
QA76.76.O63J665 2004
005.4'32—dc22
2004024882

Printed in the United States of America
05 7 6 5 4 3 2 First Edition

CHARLES RIVER MEDIA titles are available for site license or bulk purchase by institutions, user groups, corporations, etc. For additional information, please contact the Special Sales Department at 781-740-0400.

Requests for replacement of a defective CD-ROM must be accompanied by the original disc, your mailing address, telephone number, date of purchase, and purchase price. Please state the nature of the problem, and send the information to CHARLES RIVER MEDIA, INC., 10 Downer Avenue, Hingham, Massachusetts 02043. CRM's sole obligation to the purchaser is to replace the disc, based on defective materials or faulty workmanship, but not on the operation or functionality of the product.

This book is dedicated to my wife, Jill, and my children, Megan, Elise, and Marc—especially Elise, who always looks for what's most important in my books.

Contents

Reader's Guide

This book was written with GNU/Linux application developers in mind. You'll note that topics such as the Linux kernel and device drivers are absent. This was intentional, and while they're fascinating topics in their own right, they are rarely necessary to develop applications and tools in the GNU/Linux environment.

This book is split into five parts, each focusing on different aspects of GNU/Linux programming. Part I, "Introduction," introduces GNU/Linux for the beginner. It addresses the GNU/Linux architecture, a short introduction to the process model, and also licenses and a brief introduction to open source development and licenses.

Part II, "GNU Tools," concentrates on the necessary tools for GNU/Linux programming. The de facto standard GNU compiler toolchain is explored, along with the GNU `make` automated build system. Building and using libraries (both static and dynamic) are then investigated. Coverage testing and profiling are then explored, using the `gcov` and `gprof` utilities. Finally, the topic of application bundling and distribution is discussed with `automake` and `autoconf`.

With an introduction to the GNU/Linux architecture and necessary tools for application development, we focus next in Part III, "Application Development Topics," on the most useful of the services available within GNU/Linux. This includes pipes, Sockets programming, dealing with files, both traditional processes and POSIX threads, message queues, semaphores, and finally shared memory management.

In Part IV, "GNU/Linux Shells and Scripting," we move up to application development using shells and scripting languages. Some of the most useful GNU/Linux commands that you'll encounter in programming on GNU/Linux are covered, and there is a tutorial of the Bourne-Again Shell (bash). Text processing is explored using two of the most popular string processing languages (awk and `sed`). Finally, we explore the topic of parser generation using GNU `flex` and `bison` utilities (lex and yacc-compatible parser generator).

Finally, in Part V of the book, "Debugging and Testing," debugging is addressed from a variety of perspectives. We investigate some of the unit-testing frameworks that can help in automated regression. The GNU Debugger is introduced, with treatment of the most common commands and techniques. Finally, the topic of code hardening is explored, along with a variety of debugging tools and techniques to assist in the development of reliable and secure GNU/Linux applications.

While the book was written with an implicit order in mind, each chapter can be read in isolation, depending upon your needs. Where applicable, references to other chapters are provided if more information is needed on a related topic.

THREADS IN THIS BOOK

This book can be written part by part and chapter by chapter, but a number of threads run through it that can be followed independently. A reader interested in pursuing a particular aspect of the GNU/Linux operating system can concentrate on the following sets of chapters for a given topic thread:

GNU/Linux Interprocess Communication Methods: Chapters 11, 12, 15, 16, and 17.

Scripting and Text Processing: Chapters 20, 21, 22, and 23.

Building Efficient and Reliable GNU/Linux Applications: Chapters 4, 8, 24, and 26.

Multiprocess and Multithreaded Applications: Chapters 13 and 14.

GNU/Linux Testing and Profiling: Chapters 7, 8, 24, and 25.

GNU Tools for Application Development: Chapters 4, 5, 9, and 25.

GNU Tools for Packaging and Distribution: Chapters 5, 9, and 19.

Acknowledgments

My first exposure to open source was in the summer of 1994. I had just come off of a project building an operating system kernel for a large geosynchronous communication spacecraft in the Ada language on the MIL-STD-1750A microprocessor. The Ada language was technically very nice, safe, and easily readable. The MIL-STD-1750A processor was old, even by early 1990 standards (it was a 1970s instruction set architecture designed for military avionics), but was still very elegant in its simplicity.

I moved on to working on a research satellite to study gamma ray bursts and, on the side, supported the validation of a project called "1750GALS." This project, managed by Oliver Kellogg, consisted of a GCC compiler, assembler, linker, and simulator for the Ada language, targeted to the 1750A processor family. Since I had some background in Ada and the 1750A and the gamma ray burst project was just ramping up, I loaned some time to its validation. Some months later, I saw a post in the comp.compilers usenet group, of which a snippet is provided here:

```
'1750GALS', the MIL-STD-1750 Gcc/Assembler/Linker/
Simulator, now has a European FTP home, and an American
FTP mirror.

[snip]

Kudos to Pekka Ruuska of VTT Inc. (Pekka.Ruuska@vtt.fi),
and M. Tim Jones of MIT Space Research (mtj@space.mit.
edu), whose bugreports made the toolkit as useable as
it now is. Further, Tim Jones kindly set up the U.S.
FTP mirror. [1]
```

I was automatically world famous, and my 15 minutes of fame had begun. This is of course an exaggeration, but my time devoted to helping this project was both interesting and worthwhile and introduced me to the growing world of Free Software (which was already 10 years old) and open source (whose name would not be coined for another three years).

This book is the result not only of many months of hard work but of many decades of tireless work by UNIX and GNU tool developers around the world. Since an entire book could be written about the countless number of developers who created and advanced these efforts, I'll whittle it down to four people who (in my opinion) made the largest contributions toward the GNU/Linux operating system:

Dennis Ritchie and Ken Thompson of AT&T Bell Labs designed and built the first UNIX operating system (and subsequent variants) and also the C programming language.

Richard Stallman (father of GNU and the Free Software Foundation) motivated and brought together other free thinkers around the world to build the world-class GNU/Linux operating system.

Linus Torvalds introduced the Linux kernel and remains the gatekeeper of the kernel source and a major contributor.

I'm extremely grateful to Jim Lieb, whose wealth of UNIX knowledge and comprehensive review of this text improved it in innumerable ways. I'm also appreciative for the hard work of Curtis Nottberg, who submitted the chapters on GNU `make` (Chapter 5) and `automake/autoconf` (Chapter 9) and otherwise lent his GNU/Linux expertise whenever asked.

---

[1] "[announce] 1750GALS Now Have an FTP Home" at *http://compilers.iecc.com/comparch/article/94-11-043*

FIGURE I.1 Copyright © 1999, Free Software Foundation, Inc. Permission is granted to copy, distribute, and/or modify this image under the terms in the GNU General Public License or GNU Free Documentation License.

Part I Introduction

Chapter 1: "GNU/Linux History"
Chapter 2: "GNU/Linux Architecture"
Chapter 3: "Free Software Development"

In this first part of the book, we'll explore a few introductory topics of the GNU/Linux operating system and its development paradigm. This includes a short history of UNIX, GNU, and the GNU/Linux operating system, a quick review of the GNU/Linux architecture, and finally a discussion of the Free Software (and open source) development paradigm.

Chapter 1, "GNU/Linux History"

The history of GNU/Linux actually started in 1969 with the development of the first UNIX operating system. This chapter discusses the UNIX development history and the motivations (and frustrations) of key developers that led to the release of the GNU/Linux operating system.

Chapter 2, "GNU/Linux Architecture"

The composition of the GNU/Linux operating system is the topic of the second chapter. We identify the major elements of the GNU/Linux operating system and then break them down to illustrate how the operating system works at a high level.

Chapter 3, "Free Software Development"

The free software development paradigms are detailed in this chapter, including some of the licenses that are available for free software. The two major types of open development—called free software and open source—are discussed, as well as some of the problems that exist within the domains.

1 GNU/Linux History

In This Chapter

- UNIX History
- Richard Stallman and the GNU Movement
- Linus Torvalds and the Linux Kernel

INTRODUCTION

Before we jump into the technical aspects of GNU/Linux, let's invest a little time in the history of the GNU/Linux operating system (and why we use the term GNU/Linux). We'll review the beginnings of the GNU/Linux operating system by looking at its two primary sources and the two individuals who made it happen.

HISTORY OF THE UNIX OPERATING SYSTEM

To understand GNU/Linux, let's first step back to 1969 to look at the history of the UNIX operating system. Although UNIX has existed for over 30 years, it is one of the most flexible and powerful operating systems to have ever been created. A timeline is shown in Figure 1.1.

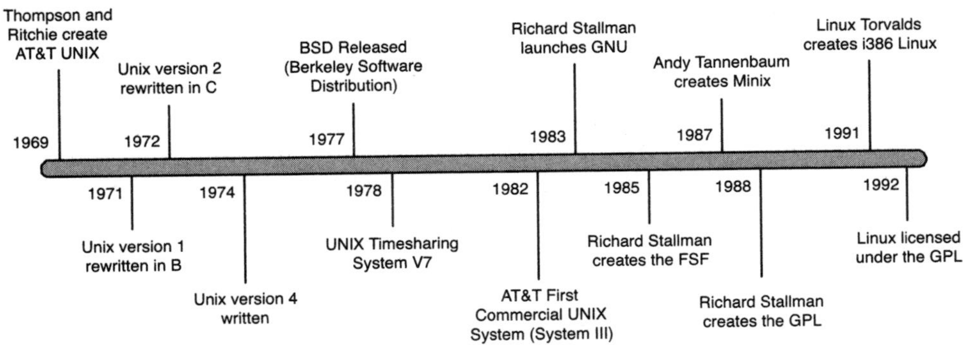

FIGURE 1.1 Timeline of UNIX/Linux and the GNU. [RobotWisdom02]

The goals for UNIX were to provide a multitasking and multiuser operating system that supported application portability. This tradition has continued in all UNIX variants and, given the new perspective of operating system portability (runs on many platforms), UNIX continues to evolve and grow.

AT&T UNIX

UNIX began as a small research project at AT&T Bell Labs in 1969 for the DEC PDP-7. Dennis Ritchie and Ken Thompson designed and built UNIX as a way to replace the current Multics operating system already in use.

Once Multics was withdrawn as the operating system at AT&T, Thompson and Ritchie developed UNIX on the PDP-7 in order to play a popular game at the time called Space Travel [Unix/Linux History04].

The first useful version of UNIX (version 1) was introduced in late 1971. This version of UNIX was written in the B language (precursor of the C language). It hosted a small number of commands, many of which are still available in UNIX and Linux systems today (such as `cat`, `cp`, `ls` and `who`). In 1972, UNIX was rewritten in the newly created C language. In the next three years, UNIX continued to evolve, with four new versions produced. In 1979, the Bourne shell was introduced. Its ancestor, the bash shell, is the topic of Chapter 20, "Bourne-Again Shell (bash)" [Unix History94].

BSD

The BSD (Berkeley Software Distribution) operating system was created as a fork of UNIX at the University of California at Berkeley in 1976. BSD remains not only a strong competitor to GNU/Linux, but in some ways is superior. Many innovations were created in the BSD, including the Sockets network programming paradigm and the variety of IPC mechanisms (addressed in Part III of this book, "Application Development Topics"). Many of the useful applications that we find in GNU/Linux today have their roots in BSD. For example, the vi editor and termcap (which allows programs to deal with displays in a display-agnostic manner) were created by Bill Joy at Berkeley in 1978 [Byte94].

One of the primary differences between BSD and GNU/Linux is in licensing. We'll address this disparity in Chapter 3, "Free Software Development."

GNU/LINUX HISTORY

The history of GNU/Linux is actually two separate stories that came together to produce a world-class operating system. Richard Stallman created an organization to build a UNIX-like operating system. He had tools, a compiler, and a variety of applications, but he lacked a kernel. Linus Torvalds had a kernel, but no tools or applications for which to make it useful.

A controversial question about GNU/Linux is why it's called GNU/Linux, compared to the commonly used name Linux. The answer is very simple. Linux refers to the kernel (or the core of the operating system), which was initially developed by Linus Torvalds. The remaining software—the shells, compiler tool chain, utilities and tools, and plethora of applications—operate above the kernel. Much of this software is GNU software. In fact, the source code that makes up the GNU/Linux operating system dwarfs that of the kernel. Therefore, to call the entire operating system Linux is a misnomer, to say the least.

Richard Stallman provides an interesting perspective on this controversy, which is covered in his article, "Linux and the GNU Project" [Linux/GNU04].

GNU and the Free Software Foundation

Richard Stallman, the father of open source, began the movement in 1983 with a post to the *net.unix-wizards* Usenet group soliciting help in the development of a free UNIX-compatible operating system [Stallman83]. Stallman's vision was the development of a free (as in freedom) UNIX-like operating system whose source was open and available to anyone.

Even in the 1970s, Stallman was no stranger to open source. He wrote the Emacs editor (1976) and gave the source away to anyone who would send a tape (on which to copy the source) and a return envelope.

The impetus for Stallman to create a free operating system was the fact that a modern computer required a proprietary operating system to do anything useful. These operating systems were closed and not modifiable by end users. In fact, until very recently, it was impossible to buy a PC from a major supplier without having to buy the Windows operating system on it. But through the Free Software Foundation (FSF), Stallman collected hundreds of programmers around the world to help take on the task.

By 1991, Stallman had pulled together many of the elements of a useful operating system. This included a compiler, a shell, and a variety of tools and applications. Work was underway in 1986 to migrate MIT's TRIX kernel, but divisions existed on whether to use TRIX or CMU's Mach microkernel. It was not until 1990 that work began on the official GNU Project kernel [Stallman02].

The Linux Kernel

Our story left off with the development of an operating system by the FSF, but development issues existed with a kernel that would make it complete. In an odd twist of fate, a young programmer by the name of Linus Torvalds announced the development of a "hobby" operating system for i386-based computers. Torvalds wanted to improve on the Minix operating system (which was widely used in the day) and thought a monolithic kernel would be much faster than the microkernel that Minix used. (While this is commonly believed to be true, operating systems such as Carnegie Mellon's Mach and the commercial QNX and Neutrino microkernels provide evidence to the contrary [Montague03].)

Torvalds released his first version of Linux (0.01) in 1991, and then later in the year he released version 0.11, which was a self-hosted release (see Figure 1.2). Torvalds used the freely available GNU tools such as the compiler and the bash shell for this effort. Much like Thompson and Ritchie's first UNIX more than 20 years earlier, it was minimal and not entirely useful. In 1992, Linux 0.96, which supported the X windowing system, was released. That year also marked Linux as a GNU software component.

Linux, much like the GNU movement, encompassed not just one person but hundreds (and today thousands) of developers. While Torvalds remains one of the top maintainers of Linux, the scope of this monolithic kernel has grown well beyond the scope of one person.

From Figure 1.2, it's important to note why the released minor version numbers are all even. The even minor number represents a stable release, and odd minors represent development versions. Since development releases are usually unstable, it's a good idea to avoid them for production use.

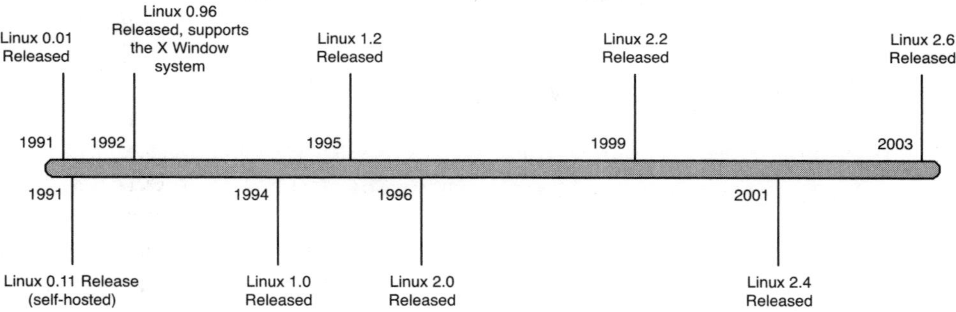

FIGURE 1.2 Linux development timeline [Wikipedia04].

Bringing It Together

The rest, as they say, is history. GNU/Linux moved from an i386 single-CPU operating system to a multiprocessor operating system supporting many processor architectures. Today, GNU/Linux can be found in large supercomputers and small handheld devices. It runs on the x86 family, ARM, PowerPC, Hitachi SuperH, 68K, and many others. But even with this achievement, BSD still garners the most architectures supported.

GNU/Linux has evolved from its humble beginnings to be one of the most scalable, secure, reliable, and highest performing operating systems available. GNU/Linux, when compared to Windows, is less likely to be exploited by hackers [NewsForge04]. Considering Web servers, the open source Apache HTTP server is far less likely to be hacked than Microsoft's IIS [Wheeler04].

LINUX DISTRIBUTIONS

In the early days, running a GNU/Linux system was anything but simple. Users sometimes had to modify the kernel and drivers in order to get the operating system to boot. Today, GNU/Linux distributions provide a simple way to load the operating system and selectively load the plethora of tools and applications. Given the dynamic nature of the kernel with loadable modules, it's simple to configure the operating system dynamically and automatically to take advantage of the peripherals that are available. Projects such as Debian [Debian04] and companies such as Red Hat [RedHat04] and Suse [Suse04] introduced distributions that contained the GNU/Linux operating system and precompiled programs on which to use it. In fact, most distributions typically include over 10,000 packages (applications) with the kernel, making it easy to get what you need.

SUMMARY

The history of GNU/Linux is an interesting one because at three levels, it's a story of frustration. Thompson and Ritchie designed the original UNIX as a way to replace the existing Multics operating system. Richard Stallman created the GNU and FSF as a way to create a free operating system that anyone could use, free of proprietary licenses. Linus Torvalds created Linux out of frustration with the Minix [Minix04] operating system that was used primarily as an educational tool at the time. Whatever their motivations, they and countless others around the world succeeded in ways that no one at the time would have ever believed. GNU/Linux today competes with commercial operating systems and offers a real and useful alternative. Even in the embedded systems domain, Linux has begun to dominate and operates in the smallest devices.

REFERENCES

[Byte94] "Unix at 25" at *http://www.byte.com/art/9410/sec8/art3.htm*

[Debian04] Debian Linux at *http://www.debian.org*

[Linux/GNU04] "Linux and the GNU Project" at *http://www.gnu.org/gnu/linux-and-gnu.html*

[Minix04] Minix Operating System at *http://www.cs.vu.nl/~ast/minix.html*

[Montague03] "Why You Should Use a BSD-Style License," Bruce R. Montague at *http://63.249.85.132/open_source_license.htm*

[NewsForge04] "Linux and Windows Security Compared," Stacey Quandt at *http://os.newsforge.com/os/04/05/18/1715247.shtml*

[RedHat04] Red Hat at *http://www.redhat.com and http://fedora.redhat.com*

[RobotWisdom02] "Timeline of GNU/Linux and UNIX" at *http://www.robotwisdom.com/linux/timeline.html*

[Stallman83] "Initial GNU Announcement" at *http://www.gnu.org/gnu/initial-announcement.html*

[Stallman02] "Free as in Freedom," Richard Stallman, O'Reilly & Associates, Inc. 2002.

[Suse04] Suse Linux at *http://www.suse.com*

[Unix/Linux History04] "History of UNIX and Linux" at *http://www.computerhope.com/history/unix.htm*

[Unix History94] "Unix History" at *http://www.english.uga.edu/hc/unixhistoryrev.html*

[Wheeler04] "Why Open Source Software / Free Software," David A. Wheeler at *http://www.dwheeler.com/oss_fs_why.html*

[Wikipedia04] Timeline of Linux Development at *http://en.wikipedia.org/wiki/Timeline_of_Linux_development*

2 GNU/Linux Architecture

In This Chapter

- High-Level Architecture
- Architectural Breakdown of Major Kernel Components

INTRODUCTION

The GNU/Linux operating system is organized into a number of layers. While understanding the internals of the kernel isn't necessary for application development, knowing how the operating system is organized is important. In this chapter we'll look at the composition of GNU/Linux starting at a very high level and then work our way through the layers.

HIGH-LEVEL ARCHITECTURE

Let's take a high-level look at the GNU/Linux architecture. Figure 2.1 shows the 20,000-foot view of the organization of the GNU/Linux operating system. At the core is the Linux kernel, which mediates access to the underlying hardware resources such as memory, the CPU via the scheduler, and peripherals. The shell (of which there are many different types) provides user access to the kernel. The shell provides command interpretation and the means to load user applications and execute them. Finally, applications are shown that make up the bulk of the GNU/Linux operating system. These applications provide the useful functions for the operating system, such as windowing systems, Web browsers, e-mail programs, language interpreters, and of course, programming and development tools.

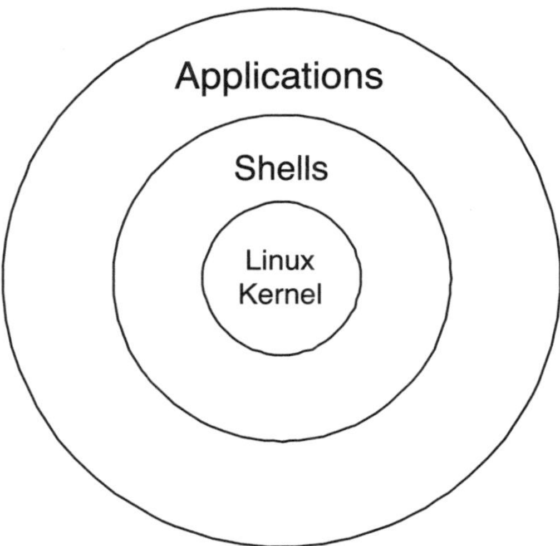

FIGURE 2.1 High-level view of the GNU/Linux operating system.

Within the kernel, we also place the variety of hardware drivers that simplify access to the peripherals (such as the CPU for configuration). Drivers to access the peripherals such as the serial port, display adapter, and network adapter are also found here.

This is a simplistic view, but next we'll dig in a little deeper to understand the makeup of the Linux kernel.

LINUX KERNEL ARCHITECTURE

The GNU/Linux operating system is a layered architecture. The Linux kernel is monolithic and layered also, but with fewer restrictions (dependencies can exist between noncontiguous layers). Figure 2.2 provides one perspective of the GNU/Linux operating system with emphasis on the Linux kernel.

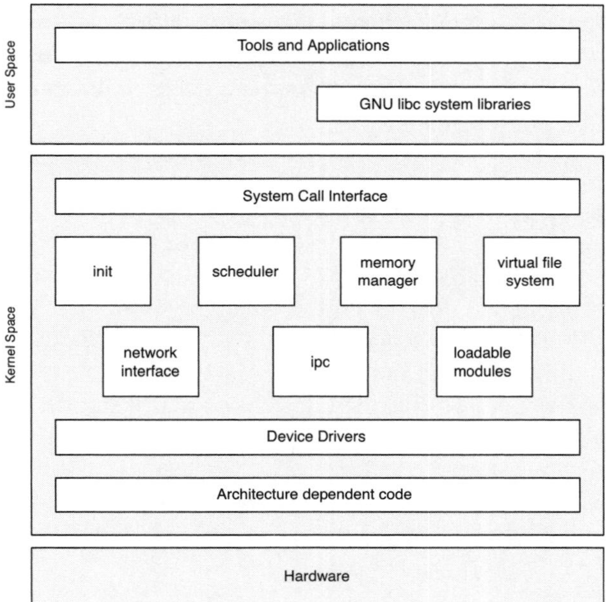

FIGURE 2.2 GNU/Linux operating system architecture.

Note here that the operating system has been split into two software sections. At the top is the user space (where we find the tools and applications as well as the GNU C library), and at the bottom is the kernel space where we find the various kernel components. This division also represents address space differences that are important to note. Each process in the user space has its own independent memory region that is not shared. The kernel operates in its own address space, but all elements of the kernel share the space. Therefore, if a component of the kernel makes a bad memory reference, the entire kernel crashes (also known as a *kernel panic*). Finally, the hardware element at the bottom operates in the physical address space (which is mapped to virtual addresses in the kernel).

Let's now look at each of the elements of the Linux kernel to identify what they do and what capabilities they provide to us as application developers.

GNU/Linux is primarily a monolithic operating system in that the kernel is a single entity. This differs from microkernel operating systems that run a tiny kernel with separate processes (usually running outside of the kernel) providing the capabilities such as networking, filesystem, and memory management. Many microkernel operating systems exist today, including CMU's Mach, Apple's Darwin, Minix, BeOS, Next, QNX/Neutrino, and many others. Which is better is certainly hotly debated, but microkernel architectures have shown themselves to be dynamic and flexible. In fact, GNU/Linux has adopted some microkernel-like features with its loadable kernel module feature.

GNU System Libraries (glibc)

The glibc is a portable library that implements the standard C library functions, including the top half of system calls. An application links with the GNU C library to access common functions in addition to accessing the internals of the Linux kernel. The glibc implements a number of interfaces that are specified in header files. For example, the `stdio.h` header file defines many standard I/O functions (such as `fopen` and `printf`) and also the standard streams that all processes are given (`stdin`, `stdout`, and `stderr`).

When building applications, the GNU compiler will automatically resolve symbols to the GNU libc (if possible), which are resolved at runtime using dynamic linking of the libc shared object.

In embedded systems development, use of the standard C libraries can sometimes be problematic. The GCC permits disabling the behavior of automatically resolving symbols to the standard C library by using `-nostdlib`*. This permits a developer to rewrite the functions that were used in the standard C library to his own versions.*

When a system call is made, a special set of actions occurs to transfer control between the user space (where the application runs) and the kernel space (where the system call is implemented).

System Call Interface

When an application calls a function like `fopen`, it is calling a privileged system call that is implemented in the kernel. The standard C library (glibc) provides a hook to go from the user space call to the kernel where the function is provided. Since this is a useful element to know, let's dig into it further.

A typical system call will result in the call of a macro in user space. The arguments for the system call are loaded into registers, and a system trap is performed. This interrupt causes control to pass from the user space to kernel space where the actual system call is available (vectored through a table called `sys_call_table`).

Once the call has been performed in the kernel, return to user space is provided by a function called `_ret_from_sys_call`. Registers are loaded properly for a proper stack frame in user space.

In cases where more than scalar arguments are used (such as pointers to storage), copies are performed to migrate the data from user space to kernel space.

The source code for the system calls can be found in the kernel source at `./linux/kernel/sys.c`.

Kernel Components

The kernel mediates access to the system resources (such as interfaces, the CPU, and so on). It also enforces the security of the system and protects users from one another. The kernel is made up of a number of major components, which we'll discuss here.

`init`

The `init` component is performed upon boot of the Linux kernel. It provides the primary entry point for the kernel in a function called `start_kernel`. This function is very architecture dependent because different processor architectures have different `init` requirements. The `init` also parses and acts upon any options that are passed to the kernel.

After performing hardware and kernel component initialization, the `init` component opens the initial console (`/dev/console`) and starts up the `init` process. This process is the mother of all processes within GNU/Linux and has no parent (unlike all other processes, which have a parent process). Once `init` has started, the control over system initialization is performed outside of the kernel proper.

The kernel `init` *component can be found in* `linux/init` *in the Linux kernel source distribution.*

Process Scheduler

The Linux kernel provides a preemptible scheduler to manage the processes running in a system. This means that the scheduler permits a process to execute for some duration (an epoch), and if the process has not given up the CPU (by making a system call or calling a function that awaits some resource), then the scheduler will temporarily halt the process and schedule another one.

The scheduler can be controlled, for example by manipulating process priority or chaining the scheduling policy (such as FIFO or Round-Robin scheduling). The time quantum (or epoch) assigned to processes for their execution can also be manipulated. The timeout used for process scheduling is based upon a variable called jiffies. A *jiffy* is a packet of kernel time that is calculated at `init` based upon the speed of the CPU.

The source for the scheduler (and other core kernel modules such as process control and kernel module support) can be found in `linux/kernel` *in the Linux kernel source distribution.*

Memory Manager

The memory manager within Linux is one of the most important core parts of the kernel. It provides physical to virtual memory mapping functions (and vice-versa) as well as paging and swapping to a physical disk. Since the memory management aspects of Linux are processor dependent, the memory manager works with architecture-dependent code to access the machine's physical memory.

While the kernel maintains its own virtual address space, each process in user space has its own virtual address space that is individual and unique.

The memory manager also provides a swap daemon that implements a demand paging system with a least-recently-used replacement policy.

The memory manager component can be found in `linux/mm` *of the Linux kernel source distribution.*

Elements of user-space memory management are discussed in Chapter 17, "Shared Memory Programming," and Chapter 18, "Other Application Development Topics," of this book.

Virtual File System

The Virtual File System (VFS) is an abstract layer within the Linux kernel that presents a common view of differing filesystems to upper-layer software. Linux supports a large number of individual filesystems, such as ext2, Minix, NFS, and Reiser. Rather than present each of these as a unique filesystem, Linux provides a layer into which filesystems can plug their common functions (such as `open`, `close`, `read`, `write`, `select`, and so on). Therefore, if we needed to open a file on a Reiser journaling filesystem, we could use the same common function `open`, as we would on any other filesystem.

The VFS also interfaces to the device drivers to mediate how the data is written to the media. The abstraction here is also useful because it doesn't matter what kind of hard disk (or other media) is present, the VFS presents a common view and therefore simplifies the development of new filesystems. Figure 2.3 illustrates this concept. In fact, multiple filesystems can be present (*mounted*) simultaneously.

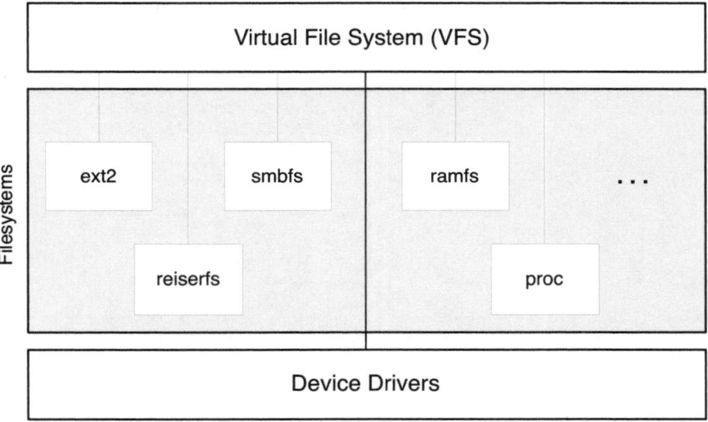

FIGURE 2.3 Abstraction provided by the Virtual File System.

The Virtual File System component can be found in `linux/fs` *in the Linux kernel source distribution. Also present there are a number of subdirectories representing the individual filesystems. For example,* `linux/fs/ext3` *provides the source for the third extended filesystem.*

While GNU/Linux provides a variety of filesystems, each provides characteristics that can be used in different scenarios. For example, xfs is very good for streaming very large files (such as audio and video), and Reiser is good at handling large numbers of very small files (<1KB). Filesystem characteristics influence performance, and therefore it's important to select the filesystem that makes the most sense for your particular application.

The topic of file I/O is discussed in Chapter 10 of this book, "File Handling in GNU/Linux."

Network Interface

The Linux network interface offers a very similar architecture to what we saw with the VFS. The network interface component is made up of three layers that work to abstract the details of networking to higher layers, while presenting a common interface regardless of the underlying protocol or physical medium (see Figure 2.4).

Common interfaces are presented to network protocols and network devices so that the protocol and physical device can be interchanged based upon the actual configuration of the system. Like the VFS, flexibility was a design key.

The network component also provides packet scheduler services for quality of service requirements.

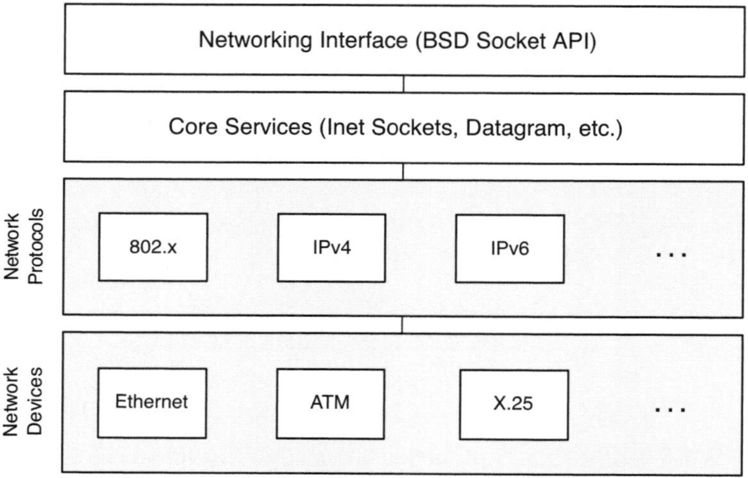

FIGURE 2.4 Network subsystem hierarchy.

NOTE

The network interface component can be found in `linux/net` *of the Linux kernel source distribution.*

The topic of network programming using the BSD Sockets API is discussed in Chapter 10 of this book.

Inter-Process Communication (IPC)

The IPC component provides the standard System V IPC facilities. This includes semaphores, message queues, and shared memory. Like VFS and the network component, the IPC elements all share a common interface.

The IPC component can be found in `linux/ipc` *of the Linux kernel source distribution.*

The topic of IPC is detailed within this book. In Chapter 16, "Synchronization with Semaphores," semaphores and the semaphore API are discussed. Chapter 15, "IPC with Message Queues," discusses the message queue API, and Chapter 17 details the shared memory API.

Loadable Modules

Loadable kernel modules are an important element of GNU/Linux as they provide the means to change the kernel dynamically. The footprint for the kernel can therefore be very small, with required modules dynamically loaded as needed. Outside of new drivers, the kernel module component can also be used to extend the Linux kernel with new functionality.

Linux kernel modules are specially compiled with functions for module `init` and cleanup. When installed into the kernel (using the `insmod` tool), the necessary symbols are resolved at runtime in the kernel's address space and connected to the running kernel. Modules can also be removed from the kernel using the `rmmod` tool (and listed using the `lsmod` tool).

The source for the kernel side of loadable modules is provided in `linux/kernel` *in the Linux kernel source distribution.*

Device Drivers

The device drivers component provides the plethora of device drivers that are available. In fact, almost half of the Linux kernel source files are devoted to device drivers. This isn't surprising, given the large number of hardware devices out there, but it does give you a good indication of how much Linux supports.

The source code for the device drivers is provided in `linux/drivers` *in the Linux kernel source distribution.*

Architecture-Dependent Code

At the lowest layer in the kernel stack is architecture-dependent code. Given the variety of hardware platforms that are supported by the Linux kernel, source for the architecture families and processors can be found here. Within the architecture family are common boot support files and other elements that are specific to the given processor family (for instance, hardware interfaces such as DMA, memory interfaces for MMU setup, interrupt handling, and so on). The architecture code also provides board-specific source for popular board vendors.

Hardware

While not part of the Linux kernel, the hardware element (shown at the bottom of our original high-level view of GNU/Linux, Figure 2.1) is important to discuss given the number of processors and processor families that are supported. Today we can find Linux kernels that run on single-address space architectures and those that support an MMU. Processor families including Arm, PowerPC, Intel ×86 (including AMD, Cyrix, and VIA variants), MIPS, Motorola 68K, Sparc, and many others. The Linux kernel can also be found on Microsoft's Xbox (Pentium III) and

Sega's Dreamcast (Hitachi SuperH). The kernel source provides information on these supported elements.

The source for the processor and vendor board variants is provided in `linux/arch` *in the Linux kernel source distribution.*

SUMMARY

The Linux kernel is the core of the GNU/Linux operating system. It is monolithic in nature and defined in a number of layers segregating its necessary elements. The kernel has been designed in such a way that adding new device drivers or protocols is simple, given the uniform interfaces that are available. This chapter provided a very high-level look at the architecture, with discussion of the major elements. References to Linux kernel source were provided where applicable.

RESOURCES

Linux Kernel Archive at *http://www.kernel.org*
Linux Kernel Cross-Reference at *http://lxr.linux.no/*

3 Free Software Development

In This Chapter

- The Free Software/Open Source Development Paradigm
- Open Source versus Free Software
- Free and Open Software Licenses
- Problems with Free Software

INTRODUCTION

While many consider the concept of free software (or open source) something that surfaced with the GNU/Linux operating system, the source can be traced back to early use in universities and research labs where source was released for others to use, modify, and hopefully improve. The goal of open source is to make source code available to others so that they can identify bugs, create new features, and generally evolve the software. Free software can promote greater reliability and quality through increased use of the software, in addition to greater visibility into how it works.

In this chapter, we'll discuss the free software development models and introduce some of the licenses that are used to release open source. To keep it fair and balanced, we'll also discuss some of the common problems with free software.

Open Source versus Free Software

Before we discuss free software, let's first cover one of the many religious debates that exist. *Open source* was a term coined by Eric Raymond with the creation of the Open Source Initiative (OSI) in 1997. The term *free software* was coined by Richard Stallman with the release of the GNU Project in 1984 and the founding of the Free Software Foundation in 1985.

While the two terms appear to be similar, and most open source software is released under the GPL (also created by Stallman), the debate is over the motivation for the release of open software. Richard Stallman defines it best: "Open source is a development methodology; free software is a social movement" [Stallman04]. Raymond has also been criticized by many for hijacking the free software movement for his own self-promotion (13 years after it was originally created) [Raymond04].

Anatomy of a Free Software Project

Free software (including open source) is simply a set of useful source code that is licensed under a free software license such as the GNU General Public License (or GPL). Popular Web sites such as SourceForge and Freshmeat provide a means to make free software available to the Internet community. In addition to providing a means for others to find free software, these sites also serve as a meeting place for free software developers. Developers can create new free software projects or join existing projects. This is the essence of free software: developers coming together to build software that is both useful and free to the wider community.

The fact that source code is available means that if something doesn't work the way it should, it can be modified to suit the needs of others. The availability of source also solves the myth of proprietary software, called "security through obscurity." Companies believe that since their software isn't provided in source form, it's more secure because it can't be opened to identify exploits. In fact, what happens in free software is that since it's open, exploits are found and fixed faster, making them less likely to be exploited as the distribution of the software widens. Proprietary software is quickly proving that obscurity does not provide security.

As free software gains in popularity, so does the desire of others who want to help. Free software gains not only in development support but also in documentation, testing, and advertising (typically word-of-mouth).

Free software has gained so much popularity that even large companies contribute source code to the community. In 2003, IBM donated source code under the Common Public License (in addition to $40 million) to the Eclipse Consortium to help in the development of the Visual Editor Project [*zdnet.com*03]. IBM has been very supportive of open source and has stated that it is one of the key factors fueling software discovery and innovation around the world [*ibm.com*04].

OPEN SOURCE LICENSES

Now let's take a quick look at the free and open source licenses. We'll look at a few different examples that provide different aspects of licensing (from preventing commercial distribution to supporting it).

GPL

The GNU General Public License is one of the most popular licenses used in free software. The GPL provides the user with three basic "rights":

- The right to copy the software and give it away
- The right to change the software
- The right to access the source code

Within these rights is the catch with GPL software. Any changes that are made to GPL software are covered by the GPL and must be made available to others in source form. This makes the GPL unattractive from a commercial perspective because it means that if a company changes GPL software, that source code must be made available to everyone (including competitors). This is what's called the "tainting" effect of GPL software. What "touches" GPL software becomes GPL software (otherwise known as a derivative work, something derived from the GPL).

There also exists a variation of the GPL called the LGPL (Library GPL, or what is now called the Lesser GPL to indicate the loss of freedom). A library released under the LGPL makes it possible for proprietary software to be linked with the libraries without the tainting effect. For example, the GNU C library is released under the LGPL, allowing proprietary software to be developed on GNU/Linux systems.

Software built into the Linux kernel is automatically GPL, though differences of opinion exist in the case of kernel modules (which can be viewed both ways). The issue of kernel modules has yet to be challenged in court.

Qt Public License

The Qt Public License (QPL) is an oddity in the open source community because it breaks the openness created by other public licenses. The QPL permits two types of licenses: a free license and a commercial license. In the free version, any software that links to the Qt framework must be opened using either the QPL or GPL. A developer could instead purchase a commercial license for Qt, which allows him to build an application using the Qt framework and keep it closed (it's not required to be released to the open source community).

From the perspective of openness, "buying-out" of the license makes the Qt framework less useful.

BSD

If one could identify a spectrum of licenses with the GPL on the left, the BSD license would exist on the right. The BSD license offers a more commercial friendly license because a program can be built with BSD licensed software and not be required to then be BSD licensed itself. The BSD community encourages returning modified source code, but it's not required. Despite this, the BSD UNIX operating system is as advanced, if not more so, as the GNU/Linux operating system.

The issue of the BSD license is what's called *forking* (two or more versions of source code existing from a single source distribution). A commercial incentive exists to fork rather than make your hard work available to your competitors. The BSD UNIX operating system has itself been forked into a number of variants, including FreeBSD, NetBSD, and OpenBSD.

The lack of distribution restrictions defines the primary difference between the BSD and GPL. GPL specifies that if one uses the GPL in a program, then that program becomes a derivative work and therefore is GPL itself. BSD has no concept of a derivative work, and developers are free to modify and keep their changes.

License Summary

Many licenses exist and can be viewed at the Open Source Initiative or in reference [*gnu.org*04], which also identifies their relation to the GPL and free software.

How one defines freedom determines how one views these licenses. If freedom means access to source code and access to source code that uses the original code, then the GPL is the license to use. If freedom is viewed from a commercial perspective (the freedom to build a product without distributing any changes to the source base), then BSD is a fantastic choice. From our short list, the QPL tries to straddle both extremes. If a commercial license is purchased, then the application using the Qt can be distributed without source. Otherwise, without a commercial license (using the so-called free license), source code must be made available.

The reader is encouraged to read the available references and license resources discussed at the end of this section. Like anything legal, nothing is really black and white, and a careful review is suggested.

Problems with Open Source Development

It wouldn't be fair to discuss the open source development paradigm without mentioning any of the problems that exist. Open source is a wonderful development paradigm, but it does suffer from many of the same problems as proprietary software development.

Usability/Reliability Ramp

The early days of the GNU/Linux operating system were not for the faint of heart. Installing GNU/Linux was not a simple task, and commonly, source changes were necessary to get a system working. The operating system, after all, was simply a hobby in the early days and did not have the plethora of drivers and hardware support that exists today. New open source projects mirror some of these same issues. Early adoption can be problematic if one is not willing to get one's hands dirty. But if an application is truly of interest, just wait a few releases, and someone will make it more usable.

Documentation

Documentation on any software project is one of the last elements to be done. Open source is no different in that respect. Some have claimed that the only real way to make money from open source is to sell documentation (such as what the Free Software Foundation does today, though it's also freely downloadable).

Ego

Like proprietary software, ego plays a large part in the architecture and development direction. Ego is a key reason for the failure of many open source projects, probably more than technical failings, but this is a personal opinion. Related to ego are the conflicts that can arise in open source development. One developer sees an application moving in one direction, while another sees a different path. A common result is forking of an application, which in itself can be beneficial if viewed from the perspective of natural selection.

Fanaticism

The open source movement is filled with fanatically committed people. Phrases such as "GNU Zealot" and "Linux Zealot" are not uncommon from those on the "outside." Arguments over, for example, which operating system is better mirror many of the political debates to which we've become accustomed. The danger of fanaticism is that we don't see the real issues and focus on what's really important. One can use both Windows and Linux and lead a full and productive life. It's not an either/or argument, it's more about the best tool for the job.

Many of the deeply fervent debates on open source result in arguments such as "open source is better, just because it is." The argument becomes a disjunctive syllogism, such as "Windows or GNU/Linux; definitely not Windows, therefore GNU/Linux." One could argue the merits of the vi editor over Emacs (or vice versa), but ultimately what's most important is that the editor does what one needs it to do. Can one operate efficiently using it? From personal experience, engineers are aghast that someone would use such an editor as vi. But if one can edit as efficiently in

vi as in any other editor, why not? Personal preference definitely plays a part in the use of open source software.

SUMMARY

The Free Software movement and open source community have changed the way that people look at pro bono software development. The GNU/Linux operating system, the Apache Web server, and the Python object-oriented scripting language (just to name a few) have resulted from distributed and sometimes ad hoc development around the world. Free and open software licenses have been created to protect free software and maintain it as free, but differences exist depending upon the goal. Free software development isn't a panacea to today's software development issues, as it does suffer from the same issues of proprietary software development.

REFERENCES

[*ibm.com*04] "Innovation Thriving, Depends on Openness to Continue" at *http://www.ibm.com/news/us/2004/08/042.html*

[*gnu.com*04] "Various Licenses and Comments About Them" at *http://www.gnu.org/philosophy/license-list.html*

[Raymond04] Eric S. Raymond Wikipedia at *http://www.campusprogram.com/reference/en/wikipedia/e/er/eric_s__raymond.html*

[Stallman04] "Why 'Free Software' Is Better Than 'Open Source'" at *http://www.gnu.org/philosophy/free-software-for-freedom.html*

[*zdnet.com*03] "IBM Donates Code to Open Source Project" at *http://zdnet.com.com/2100-1104_2-5108886.html*

RESOURCES

Open Source Initiative (OSI) at *http://www.opensource.org/*

Developer.com at *http://www.developer.com/open/*

Open Source Technology Group at *http://www.ostg.com/*

GNU Project License List at *http://www.gnu.org/licenses/license-list.html*

Freshmeat at *http://www.freshmeat.net*

SourceForge at *http://www.sourceforge.net*

"Why Open Source Software/Free Software (OSS/FS)? Look at the Numbers!" at *http://www.dwheeler.com/oss_fs_why.html*

Part II GNU Tools

In this part of the book, we'll focus on GNU tools. Since there's a plethora of tools available, we'll focus primarily on those that are necessary to build, segment, test and profile and finally distribute applications.

Chapter 4, "The GNU Compiler Toolchain"

The GNU compiler toolchain (known as GCC) is the standard compiler on GNU/Linux systems (it is, after all, an acronym for the GNU Compiler Collection). We'll address compiling C programs for native systems, but GCC provides front end for a number of different languages and back ends for almost any processor architecture you can think of.

Chapter 5, "Building Software with GNU `make`"

The GNU `make` utility provides a way to automatically build software based upon defined source files, source paths, and dependencies. But that's not all! The `make` utility is a very general utility that can be used for a variety of tasks that have ordered dependencies. We'll look at the typical—and some not so typical—uses.

Chapter 6, "Building and Using Libraries"

Software libraries allow us to collect and compile software (objects) into a single entity. We'll explore the methods for creating both static and dynamic libraries as well as the API functions that allow applications to build and use dynamic (shared) libraries.

Chapter 7, "Coverage Testing with GNU `gcov`"

The topic of testing will be analyzed from the perspective of coverage testing using the `gcov` utility. The `gcov` utility provides a way to identify path execution of an application. This tool can be very useful in determining full test path coverage of an application (where all paths are taken for a given regression of an application). It can also be useful in identifying how often a given path was taken and therefore is a useful performance tool.

Chapter 8, "Profiling with GNU `gprof`"

Profiling tools can be useful in identifying where the majority of time is taken for a given application. In this chapter, we'll investigate the `gprof` utility, which can be used to better focus optimization efforts in an application by profiling the application to see where time is spent.

Chapter 9, "Building Packages with `automake/autoconf`"

Finally, we'll go back to the topic of application building with a look at `automake` and `autoconf`. These tools can be used to automatically create build files for `make` based upon the given architecture and available tools. In this process, `autoconf` and `automake` can determine if the given system has the necessary elements (such as tools or libraries) to build an application correctly.

4 The GNU Compiler Toolchain

In This Chapter

- A Review of the Compilation Process
- Introduction to Common GCC Patterns
- Using the GCC Optimizer
- Using the GCC Warning Options
- Architectural Specification to GCC
- Related Tools Such as `size` and `objdump`

INTRODUCTION

The GNU Compiler Collection (otherwise known as GCC) is a compiler and set of utilities to build binaries from high-level source code. GCC is not only the de facto standard compiler on GNU/Linux, but it's also the standard for embedded systems development. This is because GCC supports so many different target architectures. For example, our use here will concentrate on host-based development (building software for the platform on which we're compiling), but if we were cross compiling (building for a different target) then GCC provides for 40 different architecture families. Examples include ×86, RS6000, Arm, PowerPC, and many others. GCC can also be used on over 40 different host systems (such as Linux, Solaris, Windows, or the Next operating system).

GCC also supports a number of other languages, outside of standard C. We could compile for C++, Ada, Java, Objective-C, FORTRAN, Pascal, and three dialects of the C language.

In this chapter, we'll look at some of the basic features of GCC and some of the more advanced ones (including optimization). We'll also look at some of the related tools within GCC that are useful in image construction (such as `size`, `objcopy`, and others).

We'll address the 3.2.2 version of GCC in this chapter. This is the default version for Red Hat 9.0. Newer versions of GCC now exist, but the details explored here remain compatible.

INTRODUCTION TO COMPILATION

The GNU compiler takes a number of different stages in the process of building an object. These stages can be filtered down to four: preprocessing, compiling, assembling, and linking (see Figure 4.1).

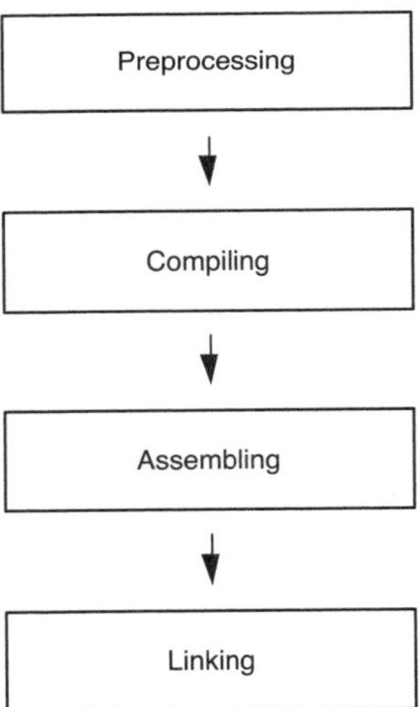

FIGURE 4.1 The stages of compilation.

The preprocessing, compiling, and assembling stages are commonly collected together into one phase, but they're shown as independent here to illustrate some of the capabilities of GCC. Table 4.1 identifies the input files and files that result.

In the preprocessing stage, the source file (`*.c`) is preprocessed with the include files (`.h` headers). At this stage, directives such as `#ifdef`, `#include`, and `#define` are resolved. The result is an intermediate file. Usually, this file isn't externally gener-

TABLE 4.1 Compilation Stages with Inputs and Outputs

Stage	Input	Output	GCC Example
Preprocessing	`*.c`	`*.i`	`gcc -E test.c -o test.i`
Compiling	`*.i`	`*.s`	`GCC -S test.i -o test.s`
Assembling	`*.s`	`*.o`	`GCC -c test.s -o test.o`
Linking	`*.o`	`*`	`GCC test.o -o test`

ated at all, but we show it here for completeness. With the source file now preprocessed, it can be compiled into assembly in the compiling stage (`*.s`). The assembly file is then converted into machine instructions in the assembling stage, resulting in an object file (`*.o`). Finally, the machine code is linked together (potentially with other machine code objects or object libraries) into an executable binary.

That's enough preliminaries. Let's now dig into GCC and see the variety of ways it can be used. We'll first look at a number of patterns that illustrate GCC in use, and then we'll explore some of the most useful options of GCC. This includes options for debugging, enabling various warnings, and optimizations. We'll then investigate a number of GNU tools that are related to GCC.

Patterns for GCC (Compile, Compile and Link)

The simplest example from which to begin is the compilation of a C source file to an image. In this example, the entire source necessary is contained within the single file, so we use GCC as follows:

```
$ GCC test.c -o test
```

Here we compile the `test.c` file and place the resulting executable image in a file called `test` (using the `-o` output option). If instead we wanted just the object file for the source, we'd use the `-c` flag, as follows:

```
$ GCC -c test.c
```

By default, the resulting object is named `test.o`, but we could force the output of the object to `newtest.o`, as:

```
$ GCC -c test.c -o newtest.o
```

Most programs we'll develop will involve more than one file. GCC handles this easily on the command line as:

```
$ GCC -o image first.c second.c third.c
```

Here we compile three source files (`first.c`, `second.c`, and `third.c`) and link them together into the executable named `image`.

In all examples where an executable will result, all C programs require a function called `main`. *This is the main entry point for the program and should appear once in all the files to be compiled and linked together. When simply compiling a source file to an object, the link phase is not yet performed, and therefore the* `main` *function is not necessary.*

Useful Options

In many cases, we'll keep our header files in a directory that's separate from where we keep our source files. Consider an example where our source is kept in a subdirectory called `./src`, and at the same level is a directory where our include files are kept, `./inc`. We can tell GCC that the headers are provided here (while compiling within the `./src` subdirectory as:

```
$ gcc test.c -I../inc -o test
```

We could specify numerous include subdirectories using multiple `-I` specs:

```
$ gcc test.c -I../inc -I../../inc2 -o test
```

Here we specify another include subdirectory called `inc2` that is two directories up from our current directory.

For configuration of software, we can specify symbolic constants on the compile line. For example, defining a symbolic constant in source or header as

```
#define TEST_CONFIGURATION
```

could be just as easily defined on the command line using the `-D option` as:

```
$ gcc -DTEST_CONFIGURATION test.c -o test
```

The advantage to specifying this on the command line is that we need not modify any source to change its behavior (as specified by the symbolic constant).

One final useful option provides us with the means to emit a source and assembly interspersed listing. Consider the following command line:

```
$ gcc -c -g -Wa,-ahl,-L test.c
```

Most interesting in this command is the `-Wa` option, which passes the subsequent options to the assembler stage to intersperse the C source with assembly.

Compiler Warnings

While the GCC compiler will abort the compilation process if an error is detected, the discovery of warnings indicates potential problems that should be fixed, though it might still result in a working executable. GCC provides a very rich warning system, but it must be enabled to take advantage of the full spectrum of warnings that can be detected.

The most common use of GCC for finding common warnings is the `-Wall` option. This turns on "all" warnings of a given type, which consists of the most generally encountered issues in applications. Its use is this:

```
$ gcc -Wall test.c -o test
```

A synonym for `-Wall is –all-warnings`. Table 4.2 lists the plethora of warning options that are enabled within `-Wall`.

TABLE 4.2 Warning Options Enabled in `-Wall`

Option	Purpose
`unused-function`	Warn of undefined but declared static function.
`unused-label`	Warn of declared but unused label.
`unused-parameter`	Warn of unused function argument.
`unused-variable`	Warn of unused locally declared variable
`unused-value`	Warn of computed but unused value.
`format`	Verify that the format strings of `printf` and so on have valid arguments based upon the types specified in the format string.
`implicit-int`	Warn when a declaration doesn't specify a type.
`implicit-function-declaration`	Warn of a function being used prior to its declaration.
`char-subscripts`	Warn if an array is subscripted by a `char` (common error considering that the type is signed).
`missing-braces`	Warn if an aggregate initializer is not fully bracketed.

→

Option	Purpose
parentheses	Warn of omissions of `()`s if they could be ambiguous.
return-type	Warn of function declarations that default to `int` or functions that lack a return, which note a return type.
sequence-point	Warn of code elements that are suspicious (such as `a[i] = c[i++];`).
switch	In `switch` statements that lack a default, warn of missing cases that would be present in the `switch` argument.
strict-aliasing	Use strictest rules for aliasing of variables (such as trying to alias a `void*` to a `double`).
unknown-pragmas	Warn of `#pragma` directives that are not recognized.
uninitialized	Warn of variables that are used but not initialized (enabled only with `-O2` optimization level).

Note that most options also have a negative form, so that they can be disabled (if on by default or covered by an aggregate option such as `-Wall`). For example, if we wanted to enable `-Wall` but disable the `unused` warning set, we could specify this as follows:

```
$ gcc -Wall -Wno-unused test.c -o test
```

Numerous other warnings can be enabled outside of `-Wall`. Table 4.3 provides a list of some of the more useful options and their descriptions.

TABLE 4.3 Other Useful Warning Options Not Enabled in `-Wall`

Option	Purpose
cast-align	Warn whenever a pointer is cast and the required alignment is increased.
sign-compare	Warn if a signed vs. unsigned compare could yield an incorrect result.
missing-prototypes	Warn if a global function is used without a previous prototype definition.
packed	Warn if a structure is provided with the `packed` attribute and no packing occurs.

→

Option	Purpose
padded	Warn if a structure is padded to align it (resulting in a larger structure).
unreachable-code	Warn if code is found that can never be executed.
inline	Warn if a function marked as inline could not be inlined.
disabled-optimization	Warn that the optimizer was not able to perform a given optimization (required too much time or resources to perform).

One final warning option that can be very useful is `-Werror`. This option specifies that instead of simply issuing a warning if one is detected, the compiler will instead treat all warnings as errors and abort the compilation process. This can be very useful to ensure the highest quality code and is therefore recommended.

GCC OPTIMIZER

The job of the optimizer is essentially to do one of three potentially orthogonal tasks. It can optimize the code to make it faster and smaller, it can optimize the code to make it faster but potentially larger, or it can simply reduce the size of the code but potentially make it slower. Luckily, we have control over the optimizer to instruct it on what we really want.

While the GCC optimizer does a good job of code optimization, it can sometimes result in larger or slower images (the opposite of what you may be after). It's important to test your image to ensure that you're getting what you expect. When you don't get what you expect, changing the options you provide to the optimizer can usually remedy the situation.

In this section, we'll look at the various mechanisms to optimize code using GCC.

In its simplest form, GCC provides a number of levels of optimization that can be enabled. The `-O` (oh) option permits the specification of five different optimization levels, listed in Table 4.4.

TABLE 4.4 Optimization Settings and Descriptions

Optimization Level	Description
-O0	No optimization (the default level).
-O, -O1	Tries to reduce both compilation time and image size.
-O2	More optimizations than -O1, but only those that don't increase size over speed (or vice-versa).
-Os	Optimize for resulting image size (all -O2, except for those that increase size).
-O3	Even more optimizations (-O2, plus a couple more).

Enabling the optimizer simply entails specifying the given optimization level on the GCC command line. For example, in the following command line, we instruct the optimizer to focus on reducing the size of the resulting image:

```
$ gcc -Os test.c -o test
```

Note that it is possible to specify different optimization levels for each file that is to make up an image. There are certain optimizations (not contained within the optimization levels) that require all files to be compiled with the option if one is compiled with it. We'll not address any of those here.

Let's now dig into the optimization levels and see what each does and also identify the individual optimizations that are provided.

-O0 Optimization

With -O0 optimization (or no optimizer spec specified at all), the compiler will simply generate code that provides the expected results and is easily debuggable within a source code debugger (such as the GNU Debugger, gdb). The compiler is also much faster when not optimizing, as the optimizer is not invoked at all.

-O1 Optimization (-O)

In the first level of optimization, the optimizer's goal is to compile as quickly as possible and also to reduce the resulting code size and execution time. Compilation may take more time with -O1 (over -O0), but depending upon the source being compiled, this is usually not noticeable.

The individual optimizations in -O1 are shown in Table 4.5.

TABLE 4.5 Optimizations Available in `-O1`

Optimization Level	Description
`defer-pop`	Defer popping function `args` from stack until necessary.
`thread-jumps`	Perform jump threading optimizations (to avoid jumps to jumps).
`branch-probabilities`	Use branch profiling to optimize branches.
`cprop-registers`	Perform a register copy-propagation optimization pass.
`guess-branch-probability`	Enable guessing of branch probabilities.
`omit-frame-pointer`	Do not generate stack-frames (if possible).

The `-O1` optimization is usually a safe level if you still desire to safely debug the resulting image.

When specifying optimizations explicitly, the `-f` *option is used to identify them. For example, to enable the* `defer-pop` *optimization, we would simply define this as* `-fdefer-pop`. *If the option is enabled via an optimization level, and you want it turned off, simply use the negative form* `-fno-defer-pop`.

-O2 Optimization

The second optimization level provides even more optimizations (while including those in `-O1`) but does not include any optimizations that will trade speed for space (or vice-versa). The optimizations that are present in `-O2` are listed in Table 4.6.

Note that Table 4.6 lists only those optimizations that are unique to `-O2`, but it doesn't list the `-O1` optimizations. It should be assumed that `-O2` is the collection of optimizations shown in Tables 4.5 and 4.6.

-Os Optimization

The `-Os` optimization level simply disables some `-O2` optimizations that would otherwise increase the size of the resulting image. Those optimizations that are disabled for `-Os` (that do appear in `-O2`) are `-falign-labels`, `-falign-jumps`, `-falign-labels`, and `-falign-functions`. Each of these has the potential to increase the size of the resulting image, and therefore they are disabled to help build a smaller executable.

TABLE 4.6 Optimizations Available in -02

Optimization	Description
`align-loops`	Align the start of loops.
`align-jumps`	Align the labels that are only reachable by jumps.
`align-labels`	Align all labels.
`align-functions`	Align the beginning of functions.
`optimize-sibling-calls`	Optimize sibling and tail recursive calls.
`cse-follow-jumps`	When performing CSE, follow jumps to their targets.
`cse-skip-blocks`	When performing CSE, follow conditional jumps.
`gcse`	Perform global common subexpression elimination.
`expensive-optimizations`	Perform a set of expensive optimizations.
`strength-reduce`	Perform strength reduction optimizations.
`rerun-cse-after-loop`	Rerun CSE after loop optimizations.
`rerun-loop-opt`	Rerun the loop optimizer twice.
`caller-saves`	Enable register saving around function calls.
`force-mem`	Copy memory operands into registers before using.
`peephole2`	Enable an `rtl` peephole pass before `sched2`.
`regmove`	Enable register move optimizations.
`strict-aliasing`	Assume that strict aliasing rules apply.
`delete-null-pointer-checks`	Delete useless null pointer checks.
`reorder-blocks`	Reorder basic blocks to improve code placement.
`schedule-insns`	Reschedule instructions before register allocation.
`schedule-insns2`	Reschedule instructions after register allocation.

-03 Optimization

The `-03` optimization level is the highest level of optimization provided by GCC. In addition to those optimizations provided in `-02`, this level also includes those shown in Table 4.7.

Architectural Optimizations

While standard optimization levels can provide meaningful improvements on software performance and code size, specifying the target architecture can also be very useful. The `-mcpu` option tells the compiler to generate instructions for the CPU type as specified. For the standard ×86 target, Table 4.8 lists some of the options.

TABLE 4.7 Optimizations Enabled in `-O3` (Above `-O2`)

Optimization	Description
`-finline-functions`	Inline simple functions into the calling function.
`-frename-registers`	Optimize register allocation for architectures with large numbers of registers (makes debugging difficult).

TABLE 4.8 Architectures (CPUs) Supported for x86

Target CPU	`-mcpu=`
i386 DX/SX/CX/EX/SO	`i386`
i486 DX/SX/DX2/SL/SX2/DX4	`i486`
487	`i486`
Pentium	`pentium`
Pentium MMX	`pentium-mmx`
Pentium Pro	`pentiumpro`
Pentium II	`pentium2`
Celeron	`pentium2`
Pentium III	`pentium3`
Pentium IV	`pentium4`
Via C3	`c3`
Winchip 2	`winchip2`
Winchip C6-2	`winchip-c6`
AMD K5	`i586`
AMD K6	`k6`
AMD K6 II	`k6-2`
AMD K6 III	`k6-3`
AMD Athlon	`athlon`
AMD Athlon 4	`athlon`
AMD Athlon XP/MP	`athlon`
AMD Duron	`athlon`
AMD Tbird	`athlon-tbird`

So if we were compiling specifically for the Intel Celeron architecture, we'd use the following command line:

```
$ gcc -mcpu=pentium2 test.c -o test
```

Of course, combining the `-mcpu` option with an optimization level can lead to additional performance benefits. One very important point to note is that once we compile for a given CPU, it may not run on another. Therefore, if we're more interested in an image running on a variety of CPUs, allowing the compiler to pick the default (i386) will support any of the ×86 architectures.

DEBUGGING OPTIONS

If we want to debug our code with a symbolic debugger, we can specify the `-g` flag to produce debugging information in the image for GDB. The `-g` option can specify an argument to specify which format to produce. To request debugging information to be produced using the `dwarf-2` format, we would provide the option as follows:

```
$ gcc -gdwarf-2 test.c -o test
```

OTHER TOOLS

In this final section, we'll look at some of the other GNU tools (usually called *binutils*) that help us in the development process.

First, how can we identify how large our executable image or intermediate object is? The `size` utility emits the text size (instruction count) and also the `data` and `bss` segments. Consider this example:

```
$ size test.o
text    data      bss     dec       hex filename
789     256       4       1049      419 test.o
$
```

Here we request the size of our intermediate object file, `test.o`. We find that the text size (instructions and constants) is 789 bytes, the `data` segment is 256 bytes, and the `bss` segment (which will be automatically initialized to zero) is 4 bytes. If we want more detailed information on the image, we can use the `objdump` utility. We can explore the symbol table of the image or object using the `-syms` argument, as:

```
$ objdump -syms test.o
```

This results in a list of symbols available in the object, their type (`text`, `bss`, `data`), lengths, offset, and so on. We can also disassemble the image using the `–disassemble` argument, as:

```
# objdump –disassemble test.o
```

This provides a list of the functions found in the object, along with the instructions that were generated for each by GCC.

Finally, the `nm` utility can also be used to understand the symbols that are present in an object file. This utility lists not only each symbol but also detailed information on the type of the symbol. Numerous other options are available, which can be found in the `nm` main page.

SUMMARY

In this chapter, we explored the GCC compiler and some of the related tools. We investigated some of the commonly used patterns with GCC and detailed the use of the warning options. We also reviewed the various optimization levels provided by GCC in addition to the architectural specifier that provides even greater optimization. Finally, we reviewed a few tools that relate to the GCC products, such as `size`, `objdump`, and `nm`.

5 Building Software with GNU `make`

by Curtis Nottberg

In This Chapter

- Compiling C
- Basic Makefile
- Makefile Constructs
- Dependency Tracking

INTRODUCTION

Creating a binary in a compiled language often involves lots of steps to compile all of the source files into object code and then invoking the linker to put the object code modules together into an executable. The necessary steps can all be performed by hand, but this becomes tedious very quickly. Another solution would be to write a shell script to perform the commands each time. This is a better solution but has a number of drawbacks for larger projects and tends to be hard to maintain over the life of the project. Building software has a number of unique requirements that justify the development of a tool that is specifically targeted at automating the software build process. The developers of UNIX recognized this requirement early on and developed a utility named `make` to solve the problem. This chapter is an introduction to GNU `make`, the open source implementation of the `make` utility commonly used in Linux software development.

An Example Project

The approach used in this chapter will be to introduce a simple example project and then show how to build the project starting with a command line solution and progressing to a fairly complete GNU `make` implementation. The examples in this chapter will show various ways to build a project consisting of four source files. The diagram shown in Figure 5.1 illustrates the directory layout of the project.

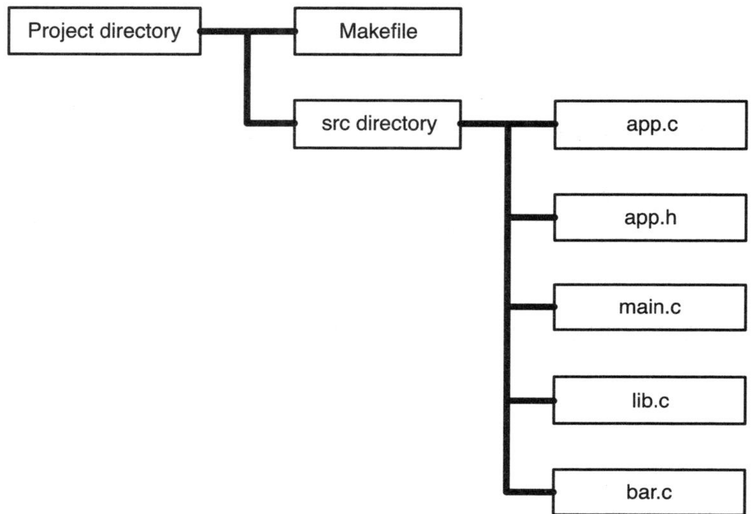

FIGURE 5.1 Directory structure of example project.

Compiling by Hand

The simplicity of the example project makes it very easy to compile by hand. Executing the following command in the top-level project directory will generate the application named appexp with a single command.

```
gcc -o appexp src/main.c src/app.c src/bar.c src/lib.c
```

This command runs the GCC wrapper program that invokes the preprocessor, compiler, and linker to turn these four c-files into an executable. The single command approach is acceptable for such a simple project, but for a larger project this would be impractical. The following set of commands breaks the compilation into the incremental steps commonly seen in a more complicated build process.

```
gcc -c -o main.o src/main.c
gcc -c -o app.o src/app.c
gcc -c -o bar.o src/bar.c
gcc -c -o lib.o src/lib.c
gcc -o appexp main.o app.o bar.o lib.o
```

The first four commands in this series turn the c-files in the `src` directory into object files in the top-level directory. The last command invokes the linker to combine the four generated object files into an executable.

A Build Script

Typing any of the commands described in the previous section would get pretty tedious if it had to be done every time the application needed to be rebuilt. The next obvious step would be to put these commands into a script so that a single command could perform all of the steps needed to build the application. The following listing shows the contents of the `buildit` script that might be written to automate the build process.

Listing 5.1 The `buildit` Script (on the CD-ROM at `./source/ch5/buildit`)

```
#!/bin/sh
# Build the chapter 5 example project

gcc -c -o main.o src/main.c
gcc -c -o app.o src/app.c
gcc -c -o bar.o src/bar.c
gcc -c -o lib.o src/lib.c
gcc -o appexp main.o app.o bar.o lib.o
```

The script collects the commands outlined in the previous section into one place. It allows the developer and user of the source code to build the application with the simple `./buildit` command line. Also, it can be revision controlled and distributed with the source code to ease the understanding needed by those trying to build an application. One of the disadvantages of the build script is that it rebuilds the entire project every time it is invoked. For the small example in this chapter, this does not cause a significant time increase in the development cycle, but as the number of files increases, rerunning the entire build process turns into a significant burden. One of the major enhancements of the `make` utility over a shell script solution is its capability to understand the dependencies of a project. Understanding the dependencies allows the `make` utility to rebuild only the parts of the project that need updating due to source file changes.

A Simple Makefile

The make utility uses a developer-created input file to describe the project to be built. GNU make uses the name Makefile as the default name for its input file. Thus, when the make utility is invoked by typing the make command, it will look in the current directory for a file named Makefile to tell it how to build the project. The following listing illustrates a basic Makefile used to build the example project.

Listing 5.2 Simple Makefile (on the CD-ROM at ./source/ch5/Makefile.simple)

```
 1:  appexp: main.o app.o bar.o lib.o
 2:  gcc -o appexp main.o app.o bar.o lib.o
 3:
 4:  main.o : src/main.c src/lib.h src/app.h
 5:  gcc -c -o main.o src/main.c
 6:
 7:  app.o : src/app.c src/lib.h src/app.h
 8:  gcc -c -o app.o src/app.c
 9:
10:  bar.o : src/bar.c src/lib.h
11:  gcc -c -o bar.o src/bar.c
12:
13:  lib.o : src/lib.c src/lib.h
14:  gcc -c -o lib.o src/lib.c
```

Line 1 illustrates the basic construct in Makefile syntax, the *rule*. The portion of the rule before the colon is called the *target*, while the portion after the colon is the rule *dependencies*. Rules generally have commands associated with them that turn the prerequisites into the target. In the specific case of line 1, the target of the rule is the appexp application that depends on the existence of main.o, app.o, bar.o, and lib.o before it can be built. Line 2 illustrates the command used to invoke the linker to turn the main.o, app.o, bar.o, and lib.o object files into the appexp executable. It is important to note that one of the idiosyncrasies of Makefile syntax is the need to put a hard tab in front of the commands in a Makefile; this is how the make utility differentiates commands from other Makefile syntax. The make utility uses rules to determine how to build a project. When the make utility is invoked on the command line, it parses the Makefile to find all of the target rules. It then attempts to build a target with the name all; if the all target has not been defined, then make will build the first target it encounters in the Makefile.

For the Makefile in Listing 5.2, the default target is the appexp application because its rule (line 1) occurs first in the Makefile. The neat part about the Makefile rule syntax is that the make utility can chain the rules together to create the whole build process. When the make utility is invoked, it will begin to build the project by trying to build the default rule: rule 1 in the example Makefile. The rule on line 1 tells the make utility that to build appexp, it must have the files main.o, app.o, bar.o, and lib.o. make will check for the existence of those files to determine if it has everything needed to build the application. If one of the prerequisite files is missing or is newer than the target, then make will start searching the target rules to determine if it has a rule to create the prerequisite.

Once an appropriate rule is identified, then the process will start again by ensuring that the new rule's prerequisites exist. Thus the make utility chains rules together into a tree of dependencies that must be satisfied to build the original target. make will then start executing the commands associated with the rules at the leaves of the tree to build the prerequisites needed to move back toward the root of the tree. In the example Makefile, the process would start with the default rule on line 1. If the object files didn't exist yet, then the make utility would find the rules to make them. First it would find a rule to create main.o, which occurs on line 4. Next, make would examine the rule on line 4 and realize that the main.c, lib.h, and app.h prerequisites all exist, so the commands to create main.o (line 5) would be executed. The next prerequisite needing to be created would be app.o, so make would move to the rule on line 7 and execute the command on line 8. This process would then continue to create bar.o and lib.o, and finally the command on line 2 would be executed to create the application. If the make command is executed in a clean directory, then it would be expected that make would automatically execute the following series of commands:

```
gcc -c -o main.o src/main.c
gcc -c -o app.o src/app.c
gcc -c -o bar.o src/bar.c
gcc -c -o lib.o src/lib.c
gcc -o appexp main.o app.o bar.o lib.o
```

Comparing this to the script in Listing 5.1, we can see that we have reimplemented the simple build script using GNU make. So why use GNU make instead of a build script? After all, it seems like a more complicated way to implement the same steps the build script took care of for us. The answer is the capability of GNU make to build things in an incremental fashion based upon the dependencies that are set up in the Makefile. For example, suppose after building the appexp program we discover an error in the app.c source file. To correct this, we would edit the app.c source file and then rebuild the executable. If we were using the Listing 5.1 script, then all of the source files would get rebuilt into objects, and those objects would get linked into an executable. On the other hand, with the make utility, the rebuild would be accomplished with the following two commands.

```
gcc -c -o app.o src/app.c
gcc -o appexp main.o app.o bar.o lib.o
```

How did GNU `make` eliminate the need for the other commands? Since GNU `make` understands each step that goes into the creation of the executable, it can examine the dates on the files to determine that nothing in the dependency tree for `main.o`, `bar.o`, and `lib.o` changed, and thus these object files don't need to be recreated. Conversely, it examines `app.o` and realizes one of its dependencies, namely `app.c`, did change, so it needs to be rebuilt. If you understand the dependency tree that is represented in the Makefile syntax, then you understand the power of `make` over a simple build script. The Makefile syntax also provides other capabilities beyond those that have been discussed in this section; the rest of this chapter will consider some of the more useful capabilities of GNU `make`.

MAKEFILE VARIABLES

GNU `make` provides support for a form of variable that closely resembles the variables provided in the standard UNIX shells. The variables allow a name to be associated with an arbitrarily long text string. The basic syntax to assign a value to a variable is:

```
MY_VAR = A text string
```

This results in the creation of a variable named `MY_VAR` with a value of `A text string`. The value of a variable can be retrieved in subsequent portions of the Makefile by using the following syntax: `${var-name}`, where `var-name` is replaced with the name of the variable that is being retrieved. Listing 5.3 provides an example Makefile to illustrate the basic use of variables in a Makefile.

ON THE CD

Listing 5.3 Simple Variable Makefile (on the CD-ROM at `./source/ch5/Makefile.simpvar`)

```
1:
2:   MY_VAR=file1.c file2.c
3:
4:   all:
5:   echo ${MY_VAR}
6:
```

Line 2 assigns the value `file1.c file2.c` to the `MY_VAR` variable. When the `make` utility is invoked, it will attempt to build the `all` target defined on line 4. The command on line 5 will be executed, and the output is illustrated in Listing 5.4.

Listing 5.4 Simple Variable Makefile Output

```
1:
2:  $ make
3:  echo file1.c file2.c
4:  file1.c file2.c
5:
```

Line 2 is the invocation of the make utility. Line 3 is output by the make utility when it runs the command on line 5 of Listing 5.3. Notice that the variable replacement has already occurred by this point in the execution. Line 4 is the actual output from the echo command that make invokes.

GNU make allows the value in a variable to be constructed incrementally by providing the ability to concatenate strings to the existing value of a variable. GNU make syntax uses the addition symbol to indicate a concatenation operation. Listing 5.5 illustrates this by modifying the simple Makefile in Listing 5.3 to use the concatenation operation when creating the value in the MY_VAR variable.

ON THE CD

Listing 5.5 Variable Concatenation Makefile (on the CD-ROM at ./source/ch5/Makefile.varconcat)

```

MY_VAR=file1.c
MY_VAR+=file2.c

all:
echo ${MY_VAR}

```

Running make against the Makefile in Listing 5.5 should generate the exact same output, illustrated in Listing 5.4, as the Makefile in Listing 5.3.

GNU make provides a wide range of built-in functionality to operate on defined variables. Functions are provided for general string handling operations such as substitution, stripping, and tokenizing. A common use of variables in Makefiles is to store and manipulate filenames and paths that are involved in the make process. To facilitate this use of variables, GNU make provides specialized functions that operate on the contents of a variable as a path or filename. The contrived Makefile in Listing 5.6 illustrates the use of some of the functions provided by GNU make. Listing 5.7 illustrates the expected output from running the make utility on the Makefile in Listing 5.6.

ON THE CD

Listing 5.6 Variable Manipulation Makefile (on the CD-ROM at `./source/ch5/Makefile.varmanip`)

```
 1:
 2:  SRC_VAR=My test string for variable manipulation.
 3:
 4:  TEST1_VAR=$(subst for,foo,${SRC_VAR})
 5:  TEST2_VAR=$(patsubst t%t,T%T, ${SRC_VAR})
 6:  TEST3_VAR=$(filter %ing %able, ${SRC_VAR})
 7:  TEST4_VAR=$(sort ${SRC_VAR})
 8:  TEST5_VAR=$(words ${SRC_VAR})
 9:  TEST6_VAR=$(word 2,${SRC_VAR})
10:  TEST7_VAR=$(wordlist 2, 3, ${SRC_VAR})
11:
12:  all:
13:  @echo original str: ${SRC_VAR}
14:  @echo substitution: ${TEST1_VAR}
15:  @echo pattern sub : ${TEST2_VAR}
16:  @echo filter   : ${TEST3_VAR}
17:  @echo sort    : ${TEST4_VAR}
18:  @echo word count : ${TEST5_VAR}
19:  @echo word 2   : ${TEST6_VAR}
20:  @echo word 2 thru 4: ${TEST7_VAR}
```

Line 2 sets up the source string that will be used to exercise the GNU `make` string manipulation functions. Lines 4–10 illustrate the use of the Makefile functions: `subst`, `patsubst`, `filter`, `sort`, `words`, `word`, and `wordlist`. Notice that the `make` file syntax for a function call takes the general form of `${func arg1,arg2,...)`. Lines 13–20 output the results when the Makefile is evaluated. Notice the use of the `@` before the `echo` commands in lines 13–20, which tells the `make` utility to suppress the printing of the command line before it executes.

Listing 5.7 Output from the Makefile in Listing 5.6

```

original str: My test string for variable manipulation.
substitution: My test string foo variable manipulation.
pattern sub : My TesT string for variable manipulation.
filter : string variable
sort : My for manipulation. string test variable
word count : 6
word 2 : test
word 2 thru 4: test string
```

Line 2 outputs the value of the original string that will be manipulated with the Makefile functions. Line 3 illustrates the output of the `subst` function, substituting the word `for` with the word `foo`. Line 4 illustrates the output of the `patsubst` function; the word `test` has been modified to `TesT`. Notice how the `patsubst` function uses the wildcard `%` to match one or more characters in a pattern. The result of the wildcard match can then be substituted back into the replacement string by using the `%` character. Line 5 illustrates the use of the `filter` function, which again makes use of the `%` wildcard character to match words ending in `ing` or `able`. Line 6 shows the output of the `sort` function that rearranges the input variables into lexical order. Line 7 illustrates the output of the `words` function that performs a word count operation on the input string. Line 8 illustrates the `word` function that allows the subsetting of a string by numerical index. In this case we have selected the second word, `test`. Notice that the indexes start at one rather than zero. Line 5 illustrates the `wordlist` function that allows a substring to be extracted based on word indexes. In the example, words 2 through 3 have been extracted, resulting in the substring `test string`.

The example in Listing 5.7 provides a taste of some of the string manipulation functions provided by GNU `make`. GNU `make` provides many others that can be found by consulting the GNU `make` documentation.

One of the primary uses of variables in a Makefile is to contain and manipulate the filename and path information associated with the build process. Take note of the use of variables in the next couple of sections, where they are employed in more realistic Makefile examples.

Pattern Matching Rules

The conversion of one type of file into another in a Makefile often follows a very specific pattern that varies only with the conversion and not the specific file. If this is the case, then GNU `make` provides a special type of rule that allows the target and dependencies to both be specified as patterns. Listing 5.8 contains a Makefile for the example project introduced at the beginning of this chapter. This example uses a pattern-matching rule in combination with the GNU `make` `VPATH` capability to simplify the Makefile. Still problematic is the header file dependencies that have been sequestered to the bottom of the new Makefile. The next section will introduce one mechanism for automating the processing of header dependencies.

ON THE CD

Listing 5.8 More Realistic Makefile (on the CD-ROM at `./source/ch5/Makefile.realistic`)

```

SRC_FILES=main.c app.c bar.c lib.c
OBJ_FILES=$(patsubst %.c, %.o, ${SRC_FILES})
```

```
 4:
 5:  VPATH = src
 6:
 7:  CFLAGS = -c -g
 8:  LDFLAGS = -g
 9:
10:  appexp: ${OBJ_FILES}
11:  gcc ${LDFLAGS} -o appexp ${OBJ_FILES}
12:
13:  %.o:%.c
14:  gcc ${CFLAGS} -o $@ $15:
16:  clean:
17:  rm *.o appexp
18:
19:  MAIN_HDRS=lib.h app.h
20:  LIB_HDRS=lib.h
21:
22:  main.o : $(addprefix src/, ${MAIN_HDRS})
23:  app.o : $(addprefix src/, ${MAIN_HDRS})
24:  bar.o : $(addprefix src/, ${LIB_HDRS})
25:  lib.o : $(addprefix src/, ${LIB_HDRS})
```

This Makefile is quite different than the first one introduced in Listing 5.2, but it accomplishes the same basic task as the original Makefile. First you should notice the introduction of variables into the Makefile to allow the control parameters (compile flags, file lists, and so on) to be enumerated and set at the top of the Makefile. Lines 2 through 8 set up the variables that will control the build process. Line 3 illustrates the use of the pattern substitution function to generate a list of object files that the final application is dependent on. The pattern substitution will result in the `OBJ_FILES` variable being assigned the value `main.o app.o bar.o lib.o`. Therefore the `OBJ_FILES` variable now contains a list of the object files that will be needed to link the `appexp` executable. Line 10 is the rule that explicitly states this relationship; it uses the value of the `OBJ_FILES` variable as its dependency list. Lines 16 and 17 illustrate a target that is often found in standard Makefiles. The name of the target is `clean`, and its responsibility is to remove files that are generated by the `make` process. This provides a nice mechanism for a developer to clean up generated files so that a complete rebuild of the entire project can occur. Lines 15 through 25 represent the header file dependencies. It will be left to the reader to understand how this works, because in the next section a different mechanism will be introduced to handle include file dependencies automatically.

The most interesting part of Listing 5.8 is the addition on lines 5, 13, and 14. Line 13 introduces a pattern rule to indicate to the `make` utility how to transform an arbitrary file ending with a .c extension into a corresponding file ending in a .o extension. The transformation is accomplished by executing the commands associated with the pattern rule, in this case the command on line 14. Notice that the command on line 14 uses some special variable references to indicate the files that GCC should operate on. GNU make provides a large number of these special variables to allow the commands in a pattern rule to gain access to information about the matched files. In the specific case of line 14, the $@ variable contains the filename matched for the left side of the rule, and the $ variable contains the filename matched for the right side of the variable. GNU make provides a large number of these special variables for pattern rules that are documented in the GNU make manual. This one pattern rule replaces the four specific rules used in Listing 5.2 to build the object files.

The previous paragraph breezed over an important detail concerning the use of pattern matching rules. The pattern matching rules don't really take into account filename and path when performing comparisons. They assume that the left and right sides of the pattern rule both occur in the same directory. In the example provided in this chapter that is not the case, because the source files are one level removed from the location of the Makefile. The source files are all contained in the src directory. To resolve the situation, the VPATH feature of GNU make is used to provide the pattern matching rules with a list of search directories to use when the right side of a rule isn't found in the current directory. In line 5, the special VPATH variable is set to src so that the pattern rule on line 13 can find the source files it needs when trying to generate the object files.

So what have we gained between Listing 5.2 and Listing 5.8? The Makefile in Listing 5.8 scales much better. It attempts to make a distinction between the operations used to build an application and the files those operations are performed on. All of the variable aspects of the build process are controlled in the variables at the top of the Makefile, while the actions are contained in the rules lower in the file. Thus, to add or delete files from the build, one needs only to modify the variables at the top of the file. Listing 5.8 still has a problem with the include file tracking, but the next section illustrates how to resolve this situation.

Automatic Dependency Tracking

There are problems with including header file dependencies in a Makefile. For instance, keeping the Makefile current with the #include directives in the source files becomes problematic as the project grows. Since the preprocessor in the tool chain already has to resolve the #includes, most modern compilers provide a mechanism to output these rules automatically. The generated rules can then be used in the Makefile to track changes to the #include structure and rebuild the project appropriately. Listing 5.9 illustrates the Listing 5.8 Makefile, modified to use the automatic dependency tracking mechanism proposed in the GNU make manual.

Listing 5.9 Makefile with Dependency Tracking (on the CD-ROM at `./source/ch5/Makefile.deptrack`)

```
 1:
 2:  SRC_FILES=main.c app.c bar.c lib.c
 3:  OBJ_FILES=$(patsubst %.c, %.o, ${SRC_FILES})
 4:  DEP_FILES=$(patsubst %.c, %.dep, ${SRC_FILES})
 5:
 6:  VPATH = src
 7:
 8:  CFLAGS = -c -g
 9:  LDFLAGS = -g
10:
11:  appexp: ${OBJ_FILES}
12:  gcc ${LDFLAGS} -o appexp ${OBJ_FILES}
13:
14:  %.o:%.c
15:  gcc ${CFLAGS} -o $@ $
16:
17:  clean:
18:  rm *.o appexp
19:
20:  include ${DEP_FILES}
21:
22:  %.dep: %.c
23:  @set -e; rm -f $@; \
24:  gcc -MM $(CFLAGS) $< > $@.$$$$; \
25:  sed ‘s,\($*\)\.o[ :]*,\1.o $@ : ,g’ < $@.$$$$ > $@; \
26:  rm -f $@.$$$$
```

Listing 5.9 is very similar to Listing 5.8 except for line 4 and lines 20 through 26. Line 4 is creating a variable based on the source file list by replacing the `.c` extension with a `.dep` extension. The `.dep` files will contain the generated dependency rules for each source file. Line 20 is the most important line in the new Makefile because it establishes a dependency between the main Makefile and all the generated `.dep` files. When `make` goes to evaluate Listing 5.9, it will first realize that to evaluate the Makefile it needs to find all of the `*.dep` files that are included by line 20. `make` must also ensure that all of the `*.dep` files are up to date before it can proceed. So how do the `.dep` files get generated? When `make` is trying to `include` the `*.dep` files, it will also evaluate the rules in the current Makefile to see if it knows how to create the `*.dep` files from other files it knows about. Lines 22 through 26 are a pattern-matching rule that tell `make` how to create `.dep` files given a `.c` file. Line 24 invokes the c compiler to generate the basic `#include` dependency rule. The result is dumped into a temporary file. Line 25 then uses sed to massage the output so that the .dep file itself is dependent on the same dependencies; when any of the dependent files changes, the .dep file gets rebuilt as well.

The following is the content of the main.dep file generated by the Makefile in Listing 5.9.

```
main.o main.dep : src/main.c src/lib.h src/app.h
```

This generated rule indicates that main.o and main.dep are dependent on the source files main.c, lib.h, and app.h. Comparing the generated rule against the same hand-written rule in previous listings illustrates that we have automated this process in Listing 5.9.

The method chosen for automatic dependency tracking in this section is the one proposed in the GNU make manual. There are numerous other mechanisms that have been employed to accomplish the same thing; please consult the resources to find the one that works best for your application.

SUMMARY

A good understanding of make and how it operates are an essential skill in any modern software development environment. This chapter provided a brief introduction to some of the capabilities of the GNU make utility. GNU make has a rich set of capabilities beyond what was discussed in this chapter; a consultation of the resources should help the user who will be using make extensively.

Most of the large projects in Linux software development do not employ make directly but instead employ the GNU automake/autoconf utilities that are layered on top of GNU make. Chapter 9 will introduce GNU automake/autoconf, which should be used when starting a new Linux development project. Don't worry; the understanding of GNU make introduced in this chapter will be invaluable when understanding and debugging problems in the GNU automake/autoconf environment.

6 Building and Using Libraries

In This Chapter

- Introduction to Libraries
- Building and Using Static Libraries
- Building and Using Shared Libraries
- Building and Using Dynamic Libraries
- GNU/Linux Library Commands

INTRODUCTION

In this chapter, we'll explore the topic of program libraries. First we'll investigate static libraries and their creation using the `ar` command. Then we'll look at shared libraries (also called *dynamically linked libraries*) and some of the advantages (and complexities) they provide. Finally, we'll look at some of the utilities that manipulate libraries in GNU/Linux.

WHAT IS A LIBRARY?

A library is really nothing more than a collection of object files. When the collection of object files provides related behavior to solve a given problem, the objects can be integrated into a library to simplify their access for application developers.

Static libraries are created using the ar, or archive, utility. Once the application developer compiles and then links with the library, the needed elements of the library are integrated into the resulting executable image. From the perspective of the application, the external library is no longer relevant because it's been combined with the application image.

Shared, or dynamic, libraries are also linked with an application image, but in two separate stages. In the first stage (at the application's build time), the linker verifies that all of the symbols necessary to build the application (functions or variables) are available within either the application or libraries. Rather than pull in the elements from the shared library into the application image (as was done with the static library), at the second stage (at runtime) a dynamic loader pulls the necessary shared libraries into memory and then dynamically links the application image with them. These steps result in a smaller image, as the shared library is separate from the application image (see Figure 6.1).

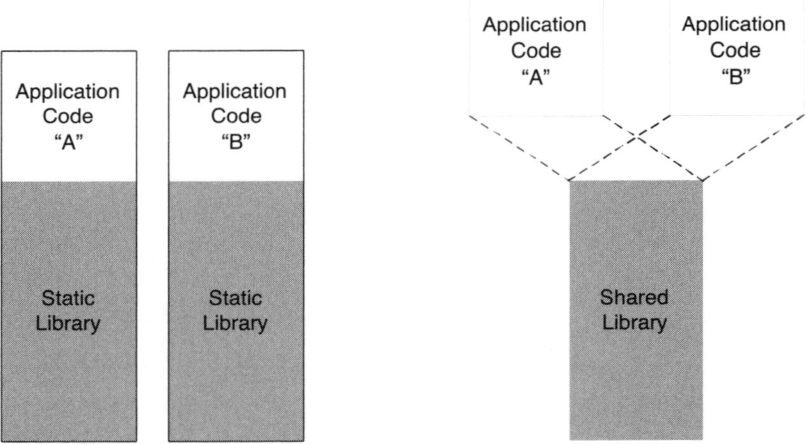

FIGURE 6.1 Memory savings of static versus shared libraries.

The tradeoff to this memory saving for shared libraries is that the libraries must be resolved at runtime. This requires a small amount of time to figure out which libraries are necessary, find them, and then bring them in to memory.

In the next sections, we'll build a couple of libraries using both the static and shared methods to see how they're built and how the program changes to support them.

BUILDING STATIC LIBRARIES

Let's first look at the simplest type of library development in GNU/Linux. The static library is linked statically with the application image. This means that once the image is built, the external library is not necessary to be present for the image to execute properly as its part of the resulting image.

To demonstrate the construction of libraries, let's look at a sample set of source files. We'll build a simple random number generator wrapper library using the GNU/Linux random functions. Let's first look at the API for our library. The header file, `randapi.h`, is shown in Listing 6.1.

Listing 6.1 Random Number Wrapper API Header (on the CD-ROM at `./source/ch6/statshrd/randapi.h`)

```
/*
 * randapi.h
 *
 * Random Functions API File
 *
 */


#ifndef __RAND_API_H
#define __RAND_API_H

extern void initRand( void );

extern float getSRand( void );

extern int getRand( int max );

#endif /* __RAND_API_H */
```

Our API consists of three functions. The first function, `initrand`, is an initialization function that prepares our wrapper libraries for use. It must be called once prior to calling any of the random functions. Function `getSRand()` returns a random floating-point value between 0.0 and 1.0. Finally, function `getRand(x)` returns a random integer between 0 and (x–1).

While this functionality could be implemented in a single file, we'll split it between two files for the purposes of demonstration. The next file, `initrand.c`, provides the initialization function for the wrapper API (see Listing 6.2). The single function, `initRand()`, simply initializes the random number generator using the current time as a seed.

ON THE CD

Listing 6.2 Random Number Initialization API (on the CD-ROM at `./source/ch6/statshrd/initapi.c`)

```
/*
 * Random Init Function API File
 *
 */

#include <stdlib.h>
#include <time.h>


/*
 *  initRand() initializes the random number generator.
 *
 */

void initRand()
{
  time_t seed;

  seed = time(NULL);

  srand( seed );

  return;
}
```

Our final API file, `randapi.c`, provides the random number functions (see Listing 6.3). The integer and floating-point random number wrapper functions are provided here.

ON THE CD

Listing 6.3 Random Number Wrapper Functions (on the CD-ROM at `./source/ch6/statshrd/randapi.c`)

```
/*
 * randapi.c
 *
 * Random Functions API File
 *
 */

#include <stdlib.h>
```

```
/*
 *  getSRand() returns a number between 0.0 and 1.0.
 *
 */

float getSRand()
{
  float randvalue;

  randvalue = ((float)rand() / (float)RAND_MAX);

  return randvalue;
}


/*
 *  getRand() returns a number between 0 and max-1.
 *
 */

int getRand( int max )
{
  int randvalue;

  randvalue = (int)((float)max * rand() / (RAND_MAX+1.0));

  return randvalue;
}
```

That's it for our API. Note that both `initapi.c` and `randapi.c` use the single header file `randapi.h` to provide their function prototypes. Let's now take a quick look at the test program that utilizes the API and then get back to the task at hand—libraries!

Listing 6.4 provides the test application that uses the wrapper function API. This application provides a quick test of the API by identifying the average value provided, which should represent the average around the middle of the random number range.

ON THE CD

Listing 6.4 Test Application for the Wrapper Function API (on the CD-ROM at `./source/ch6/statshrd/test.c`)

```
#include "randapi.h"

#define ITERATIONS      1000000L
```

```
int main()
{
  long  i;
  long  isum;
  float fsum;

  /* Initialize the random number API */
  initRand();

  /* Find the average of getRand(10) returns (0..9) */
  isum = 0L;
  for (i = 0 ; i < ITERATIONS ; i++) {

    isum += getRand(10);

  }

  printf( "getRand() Average %d\n", (int)(isum / ITERATIONS) );


  /* Find the average of getSRand() returns */
  fsum = 0.0;
  for (i = 0 ; i < ITERATIONS ; i++) {

    fsum += getSRand();

  }

  printf( "getSRand() Average %f\n", (fsum / (float)ITERATIONS) );

  return;
}
```

If we wanted to build all source files discussed here and integrate them into a single image, we could do the following:

```
$ gcc initapi.c randapi.c test.c -o test
```

This would compile all three files and then link them together into a single image called `test`. This use of `gcc` provides not only compilation of the source files, but also linking to a single image. Upon executing the image, we'd see the averages for each of the random number functions:

```
$ ./test
getRand() Average 4
getSRand() Average 0.500001
$
```

As expected, the random number generated generates an average value that's in the middle of the random number range.

Let's now get back to the subject of libraries, and rather than build the entire source together, we'll build a library for our random number functions. This is achieved using the `ar` utility (archive). Below, we'll demonstrate the building of our static library along with the construction of the final image.

```
$ gcc -c -Wall initapi.c
$ gcc -c -Wall randapi.c
$ ar -cru libmyrand.a initapi.o randapi.o
$
```

In this example, we first compile our two source files (`initapi.c` and `randapi.c`) using `gcc`. We specify the `-c` option to tell gcc to compile only (don't link) and also to turn on all warnings. Next, we use the `ar` command to build our library (`libmyrand.a`). The `cru` options are a standard set of options for creating or adding to an archive. The `c` option specifies to create the static library (unless it already exists, in which case the option is ignored). The `r` option tells `ar` to replace existing objects in the static library (if they already exist). Finally, the `u` option is a safety option to specify that objects are replaced in the archive only if the objects to be inserted are newer than existing objects in the archive (of the same name).

We now have a new file called `libmyrand.a`, which is our static library containing two objects: `initapi.o` and `randapi.o`. Let's now look at how we can build our application using this static library. Consider the following:

```
$ gcc test.c -L. -lmyrand -o test
$ ./test
getRand() Average 4
getSRand() Average 0.499892
$
```

Here we use `gcc` to first compile the file `test.c` and then link the `test.o` object with `libmyrand.a` to produce the test image file. The `-L.` option tells `gcc` that libraries can be found in the current subdirectory (`.` represents the directory). Note that we could also provide a specific subdirectory for the library, such as `-L/usr/mylibs`. The `-L` option identifies the library to use. Note that `myrand` isn't the name of our library, but instead `libmyrand.a`. When the `-L` option is used, it automatically surrounds the name specified with `lib` and `.a`. Therefore, if the user had specified `-ltest`, `gcc` would look for a library called `libtest.a`.

Now that we see how to create a library and use it to build a simple application, let's return to the `ar` utility to see what other uses it has. We can inspect a static library to see what's contained within it by using the `-t option`:

```
$ ar -t libmyrand.a
initapi.o
randapi.o
$
```

If desired, we can also remove objects from a static library. This is done using the `-d` option, such as:

```
$ ar -d libmyrand.a initapi.o
$ ar -t libmyrand.a
randapi.o
$
```

It's important to note that `ar` won't actually show a failure for a delete. To see an error message if the delete fails, add a `-v` option as shown below:

```
$ ar -d libmyrand.a initapi.o
$ ar -dv libmyrand.a initapi.o
No member named 'initapi.o'
$
```

In the first case, we try to delete the object `initapi.o`, but no error message is generated (even though it doesn't exist in the static library). In the second case, we add the verbose option and the corresponding error message results.

Rather than remove the object from the static library, we can extract it using the `-x` option.

```
$ ar -xv libmyrand.a initapi.o
x - initapi.o
$ ls
initapi.o  libmyrand.a
$ ar -t libmyrand.a
randapi.o
initapi.o
    $
```

The extract option is coupled with verbose (`-v`) so that we can see what `ar` is doing. The `ar` utility responds with the file being extracted (`x - initapi.o`), which we can see after doing a `ls` in the subdirectory. Note here that we also list the contents of the static library after extraction, and our `initapi.o` object is still present. The extract option doesn't actually remove the object, it only copies it externally to the archive. The delete (`-d`) option must be used to remove it outright from the static library.

The `ar` utility option list is shown in Table 6.1.

TABLE 6.1 Important Options for the `ar` Utility

Option	Name	Example
-d	Delete	ar -d <archive> <objects>
-r	Replace	ar -r <archive> <objects>
-t	Table list	ar -t <archive>
-x	Extract	ar -x <archive> <objects>
-v	Verbose	ar -v
-c	Create	ar -c <archive>
-ru	Update object	ar -ru <archive> <objects>

BUILDING SHARED LIBRARIES

Now let's try our test application again, this time using a shared library instead of a static library. The process is essentially just as simple. Let's first build a shared library using our `initapi.o` and `randapi.o` objects. One change is necessary when building source for a shared library. Since the library and application are not tied together as they are in a static library, the resulting library can't assume anything about the binding application. For example, in addressing, references must be relative (through the use of a GOT, or Global Offset Table). The loader automatically resolves all GOT addresses as the shared library is loaded. To build our source files for position-independence, we use the `PIC` option of `gcc`:

```
$ gcc -fPIC -c initapi.c
$ gcc -fPIC -c randapi.c
```

This results in two new object files containing position-independent code. We can create a shared library of these using `gcc` and the `-shared` flag. This flag tells `gcc` that a shared library is to be created:

```
$ gcc -shared initapi.o randapi.o -o libmyrand.so
```

We specify our two object modules with an output file (`-o`) as our shared library. Note that we use the `.so` suffix to identify that the file is a shared library (shared object).

To build our application using the new shared object, we link the elements back together as we did with the static library:

```
$ gcc test.c -L. -lmyrand -o test
```

We can tell that our new image is dependent upon our shared library by using the `ldd` command. The `ldd` command prints shared library dependencies for the given application. For example:

```
$ ldd test
    libmyrand.so => not found
    libc.so.6 => /lib/tls/libc.so.6 (0x42000000)
    /lib/ld-linux.so.2 => /lib/ld-linux.so.2 (0x40000000)
$
```

The `ldd` command identifies the shared libraries that will be used by `test`. The standard C library is shown (`libc.so.6`) as is the dynamic linker/loader (`ld-linux.so.2`). Note that our `libmyrand.so` file is shown as not found. It's present in the current subdirectory with our application, but it must be explicitly specified to GNU/Linux. We do this through the `LD_LIBRARY_PATH` environment variable. After exporting the location of our shared library, we try our `ldd` command again:

```
$ export LD_LIBRARY_PATH=./
$ ldd test
        libmyrand.so => ./libmyrand.so (0x40017000)
        libc.so.6 => /lib/tls/libc.so.6 (0x42000000)
        /lib/ld-linux.so.2 => /lib/ld-linux.so.2 (0x40000000)
$
```

We specify that our shared libraries are found in the current directory (`./`), and then after performing another `ldd`, our shared library is successfully found.

If we had tried to execute our application without having done this, a reasonable error message would have resulted, telling us that the shared library could not be found:

```
$ ./test
./test: error while loading shared libraries: libmyrand.so:
cannot find shared object file: No such file or directory.
$
```

DYNAMICALLY LOADED LIBRARIES

The final type of library that we'll explore is the dynamically loaded (and linked) library. This library can be loaded at any time during the execution of an application, unlike a shared object that is loaded immediately upon application start-up. We'll build our shared object file as we did before, as:

```
$ gcc -fPIC -c initapi.c
$ gcc -fPIC -c randapi.c
$ gcc -shared initapi.o randapi.o -o libmyrand.so
$ su -
<provide your root password>
$ cp libmyrand.so /usr/local/lib
$ exit
```

In this example, we'll move our shared library to a common location (`/usr/local/lib`). This is a standard directory for libraries, rather than relying on the image and shared library always being in the same location (as was assumed in the previous example). Note that this library is identical to our original shared library. What is different is how our application deals with the library.

In order to copy our library to `/usr/local/lib`, *we must first gain root privileges. To do so, we use the* `su` *command to create a login shell for the root user.*

Now that we have our shared library re-created, how do we access this in a dynamic fashion from our test application? The answer is that we must modify our test application to change the way that we access the API. Let's first look at our updated test app (modified from Listing 6.4). Then we'll investigate how this is built for dynamic loading. Our updated test application is shown in Listing 6.5. We'll walk through this, identifying what changed from our original application, and then look at the dynamically loaded (DL) API in more detail.

Listing 6.5 Updated Test Application for Dynamic Linkage (on the CD-ROM at `./source/ch6/dynamic/test.c`)

```
/*
 * Dynamic rand function test.
 *
 */

#include <stdlib.h>
#include <stdio.h>
#include <dlfcn.h>
```

```
#include "randapi.h"

#define ITERATIONS      1000000L


int main()
{
  long  i;
  long  isum;
  float fsum;
  void  *handle;
  char  *err;

  void (*initRand_d)(void);
  float (*getSRand_d)(void);
  int (*getRand_d)(int);

  /* Open a handle to the dynamic library */
  handle = dlopen( "/usr/local/lib/libmyrand.so", RTLD_LAZY );
  if (handle == (void *)0) {
    fputs( dlerror(), stderr );
    exit(-1);
  }

  /* Check access to the initRand() function */
  initRand_d = dlsym( handle, "initRand" );
  err = dlerror();
  if (err != NULL) {
    fputs( err, stderr );
    exit(-1);
  }

  /* Check access to the getSRand() function */
  getSRand_d = dlsym( handle, "getSRand" );
  err = dlerror();
  if (err != NULL) {
    fputs( err, stderr );
    exit(-1);
  }

  /* Check access to the getRand() function */
  getRand_d = dlsym( handle, "getRand" );
  err = dlerror();
  if (err != NULL) {
```

```
53:         fputs( err, stderr );
54:         exit(-1);
55:       }
56:
57:
58:       /* Initialize the random number API */
59:       (*initRand_d)();
60:
61:       /* Find the average of getRand(10) returns (0..9) */
62:       isum = 0L;
63:       for (i = 0 ; i < ITERATIONS ; i++) {
64:
65:         isum += (*getRand_d)(10);
66:
67:       }
68:
69:       printf( "getRand() Average %d\n", (int)(isum / ITERATIONS) );
70:
71:
72:       /* Find the average of getSRand() returns */
73:       fsum = 0.0;
74:       for (i = 0 ; i < ITERATIONS ; i++) {
75:
76:         fsum += (*getSRand_d)();
77:
78:       }
79:
80:       printf( "getSRand() Average %f\n", (fsum / (float)ITERATIONS) );
81:
82:       /* Close our handle to the dynamic library */
83:       dlclose( handle );
84:
85:       return;
86:     }
```

This code may appear a little convoluted given the earlier implementation, but it's actually quite simple once you understand how the DL API works. All that's really going on is that we're opening the shared object file using **`dlopen`**, and then assigning a local function pointer to the function within the shared object (using **`dlsym`**). This allows us then to call it from our application. When we're done, we close the shared library using **`dlclose`**, and the references are removed (freeing any used memory for the interface).

We make the DL API visible to us by including the `dlfcn.h` (DL function) header file. The first step in using a dynamic library is opening it with **`dlopen`** (line

27). We specify the library we need to use (`/usr/local/lib/libmyrand.so`) and also a single flag. Of the two flags that are possible (`RTLD_LAZY` and `RTLD_NOW`), we specify `RTLD_LAZY` to resolve references as we go, rather than immediately upon loading the library, which would be the case with `RTLD_NOW`. The function `dlopen` returns an opaque handle representing the opened library. Note that if an error occurs, we can use the `dlerror` function to provide an error string suitable for emitting to `stdout` or `stderr`.

While not necessary in this example, if we desired to have an initialization function called when our shared library was opened via `dlopen`*, we could create a function called* `_init` *in our shared library. The* **`dlopen`** *function will ensure that this* `_init` *function is called before* `dlopen` *returns to the caller.*

Getting the references for the functions that we need to address is the next step. Let's look at one below (taken from lines 34–39 of Listing 6.5).

```
34:     initRand_d = dlsym( handle, "initRand" );
35:     err = dlerror();
36:     if (err != NULL) {
37:       fputs( err, stderr );
38:       exit(-1);
39:     }
```

The process, as can be seen from this code snippet, is very simple. The API function `dlsym` searches our shared library for the function defined (in this case, our initialization function `"initRand"`). Upon locating it, a `(void *)` pointer is returned and stored in a local function pointer. This may then be called (as shown at line 59) to perform the actual initialization. We automatically check our error status (at line 35), and if an error string was returned, we emit it and exit the application.

That's really it for identifying the functions that we desire to call in the shared library. After we grab the `initRand` function pointer, we grab `getSRand` (lines 42–47) and then `getRand` (lines 49–55).

Our test application is now fundamentally the same, except that instead of calling functions directly, we call them indirectly using the pointer-to-function interface. That's a small price to pay for the flexibility that the dynamically loaded interface provides.

Our last step in the new test application is to close out the library. This is done with the `dlclose` API function (line 83). If the API finds that there are no other users of the shared library, then it is unloaded.

As was provided with **`dlopen`**, **`dlclose`** *provides a mechanism by which the shared object can export a completion routine that is called when the* **`dlclose`** *API function is called. The developer must simply add a function called* `_fini` *to the shared library, and* `dlclose` *will ensure that* `_fini` *is called prior to* `dlclose return`.

And that's it! For the small amount of pain involved in creating an application that utilizes dynamically loaded shared libraries, it provides a very flexible environment that ultimately can save on memory use. Note also that it's not always necessary to make all dynamic functions visible when your application starts. You can instead make only those that are necessary for normal operation and load other dynamic libraries as they become necessary.

The dynamically loaded library API is very simple and is shown here for completeness:

```
void *dlopen( const char *filename, int flag );
const char *dlerror( void );
void *dlsym( void *handle, char *symbol );
int dlclose( void *handle );
```

How a library is made up depends upon what it's representing. The library should contain all functions that are necessary for the given problem domain. Functions that are not specifically associated with the domain should be excluded and potentially included in another library.

UTILITIES

Now let's look at some of the other utilities that are useful when creating static, shared, or dynamic libraries.

file

The `file` utility tests the file argument for the purposes of identifying what it is. This utility is very useful in a number of different scenarios, but in this case it will provide us with a small amount of information about the shared object. Let's look at an interactive example:

```
$ file /usr/local/lib/libmyrand.so
/usr/local/lib/libmyrand.so: ELF 32-bit LSB shared object,
Intel 80386, version 1 (SYSV), not stripped
$
```

So, using `file`, we see that our shared library is a 32-bit ELF object for the Intel 80386 processor family. It has been defined as "not stripped," which simply means that debugging information is present.

size

The `size` command provides us with a very simple way to understand the `text`, `data`, and `bss` section sizes for an object. An example of the `size` command on our shared library is shown here:

```
$ size /usr/local/lib/libmyrand.so
   text    data     bss     dec     hex filename
   2013     264       4    2281     8e9 /usr/local/lib/libmyrand.so
$
```

nm

To dig into the object, we use the `nm` command. This commands permits us to look at the symbols that are available within a given object file. Let's look at a simple example using `grep` to filter our results:

```
$ nm -n /usr/local/lib/libmyrand.so | grep " T "
00000608 T _init
0000074c T initRand
00000784 T getSRand
000007be T getRand
00000844 T _fini
$
```

In this example, we use `nm` to print the symbols within the shared library, but then only emit those with the tag `" T "` to `stdout` (those symbols that are part of the `.text` section, or code segments). We also use the `-n` option to sort the output numerically by address, rather than the default, which is alphabetically by symbol name. This gives us relative address information within the library; if we wanted to know the specific sizes of these `.text` sections, we could use the `-S` option, as:

```
$ nm -n -S /usr/local/lib/libmyrand.so | grep " T "
00000608 T _init
0000074c 00000036 T initRand
00000784 0000003a T getSRand
000007be 00000050 T getRand
00000844 T _fini
$
```

From this example, we can see that the initRand is located at relative offset 0×74c in the library and its size is 0×36 (decimal 54) bytes. Many other options are available; the nm mainpage provides more detail on this.

objdump

The objdump utility is similar to nm in that it provides the ability to dig in and inspect the contents of an object. Let's now look at some of the specialized functions of objdump.

One of the most interesting features of objdump is its ability to disassemble the object into the native instruction set. Here's an excerpt of objdump performing this capability:

```
$ objdump -disassemble -S /usr/local/lib/libmyrand.so
...
0000074c <initRand>:
  74c:   55                      push   %ebp
  74d:   89 e5                   mov    %esp,%ebp
  74f:   53                      push   %ebx
  750:   83 ec 04                sub    $0x4,%esp
  753:   e8 00 00 00 00          call   758 <initRand+0xc>
  758:   5b                      pop    %ebx
  759:   81 c3 f8 11 00 00       add    $0x11f8,%ebx
  75f:   83 ec 0c                sub    $0xc,%esp
  762:   6a 00                   push   $0x0
  764:   e8 c7 fe ff ff          call   630 <_init+0x28>
  769:   83 c4 10                add    $0x10,%esp
  76c:   89 45 f8                mov    %eax,0xfffffff8(%ebp)
  76f:   83 ec 0c                sub    $0xc,%esp
  772:   ff 75 f8                pushl  0xfffffff8(%ebp)
  775:   e8 d6 fe ff ff          call   650 <_init+0x48>
  77a:   83 c4 10                add    $0x10,%esp
  77d:   8b 5d fc                mov    0xfffffffc(%ebp),%ebx
  780:   c9                      leave
  781:   c3                      ret
  782:   90                      nop
  783:   90                      nop
...
$
```

In addition to -disassemble (to disassemble to the native instruction set), we also specified -S to output interspersed source code. The problem is that we compiled our object to exclude this information. We can easily fix this as follows, by adding -g to the compilation process.

```
    $ gcc -c -g -fPIC initapi.c
    $ gcc -c -g -fPIC randapi.c
    $ gcc -shared initapi.o randapi.o -o libmyrand.so
    $ objdump -disassemble -S libmyrand.so
    ...
00000790 <initRand>:
  *
  */
void initRand()
{
  790:   55                       push   %ebp
  791:   89 e5                    mov    %esp,%ebp
  793:   53                       push   %ebx
  794:   83 ec 04                 sub    $0x4,%esp
  797:   e8 00 00 00 00           call   79c <initRand+0xc>
  79c:   5b                       pop    %ebx
  79d:   81 c3 fc 11 00 00        add    $0x11fc,%ebx
  time_t seed;
  seed = time(NULL);
  7a3:   83 ec 0c                 sub    $0xc,%esp
  7a6:   6a 00                    push   $0x0
  7a8:   e8 c7 fe ff ff           call   674 <_init+0x28>
  7ad:   83 c4 10                 add    $0x10,%esp
  7b0:   89 45 f8                 mov    %eax,0xfffffff8(%ebp)
  srand( seed );
  7b3:   83 ec 0c                 sub    $0xc,%esp
  7b6:   ff 75 f8                 pushl  0xfffffff8(%ebp)
  7b9:   e8 d6 fe ff ff           call   694 <_init+0x48>
  7be:   83 c4 10                 add    $0x10,%esp
  return;
}
  7c1:   8b 5d fc                 mov    0xfffffffc(%ebp),%ebx
  7c4:   c9                       leave
  7c5:   c3                       ret
  7c6:   90                       nop
  7c7:   90                       nop
...
$
```

Having compiled our source code with `-g`, we now have the ability to understand the C source to machine code mapping.

Numerous other capabilities are provided with `objdump`. The GNU/Linux mainpage lists the plethora of other options.

ranlib

The ranlib utility is one of the most important utilities when creating static libraries. This utility creates an index of the contents of the library and stores it in the library file itself. When this index is present in the library, the linking stage of building an image can be sped up considerably. Therefore, the ranlib utility should be performed whenever a new static library is created. An example of using ranlib is shown here:

```
$ ranlib libmyrand.a
$
```

Note that the same thing can be performed using the ar command with the -s option, as:

```
$ ar -s libmyrand.a
$
```

SUMMARY

In this chapter, we explored the creation and use of program libraries. Traditional static libraries were discussed first, followed by shared libraries and finally dynamically loaded libraries. Source code was also investigated to demonstrate the methods for creating libraries using the ar command as well as using libraries with gcc. Finally, a number of library-based utilities were discussed, including ldd, objdump, nm, size, and ranlib.

DYNAMIC LIBRARY APIS

```
#include <dlfcn.h>
void *dlopen( const char *filename, int flag );
const char *dlerror( void );
void *dlsym( void *handle, char *symbol );
int dlclose( void *handle );
```

7 Coverage Testing with GNU gcov

In This Chapter

- Understanding GNU's gcov Tool
- Explore the Different Uses for gcov
- Build Software for gcov
- Understand gcov's Various Data Products
- Illustrate Problems with gcov and Optimization

INTRODUCTION

In this chapter, we'll explore the gcov utility and see how it can be used to both help test and support software profiling and optimization. We'll learn how to build software for use with gcov and then understand the various types of data that are provided. Finally, we'll investigate things to avoid when performing coverage testing.

WHAT IS gcov?

Let's begin with an overview of what gcov can do for us. The gcov utility is a coverage testing tool. When built with an application, the gcov utility monitors an application under execution and identifies which source lines have been executed and which have not. Further, gcov can identify the number of times a particular line has been executed, making it useful for performance profiling (where an application is spending most of its time). Because gcov can tell which lines have not been executed, it is useful as a coverage testing tool. In concert with a test suite, gcov can identify whether all source lines have been adequately covered [FSF 2002].

We'll discuss the use of gcov bundled with version 3.2.2 of the GNU compiler tool chain.

Preparing the Image

Let's first look at how an image is prepared for use with gcov. We'll provide more detail of gcov options in the coming sections, so this will serve as an introduction. We'll use the simple bubblesort source file shown in Listing 7.1.

Listing 7.1 Sample Source File to Illustrate the gcov Utility (on the CD-ROM at ./source/ch7/bubblesort.c)

```
#include <stdio.h>

void bubbleSort( int list[], int size )
{
  int i, j, temp, swap = 1;

  while (swap) {

    swap = 0;

    for ( i = (size-1) ; i >= 0 ; i– ) {

      for ( j = 1 ; j <= i ; j++ ) {

        if ( list[j-1] > list[j] ) {

          temp = list[j-1];
          list[j-1] = list[j];
          list[j] = temp;
          swap = 1;

        }

      }

    }

  }

```

```
}

int main()
{
  int theList[10]={10, 9, 8, 7, 6, 5, 4, 3, 2, 1};
  int i;

  /* Invoke the bubble sort algorithm */
  bubbleSort( theList, 10 );

  /* Print out the final list */
  for (i = 0 ; i < 10 ; i++) {
    printf("%d\n", theList[i]);
  }

}
```

The `gcov` utility is used in conjunction with the compiler tool chain. This means that the image that we're to do coverage testing on must be compiled with a special set of options. These are illustrated below for compiling the source file `bubblesort.c`:

```
gcc bubblesort.c -o bubblesort -ftest-coverage -fprofile-arcs
```

The resulting image, when executed, produces a number of files containing statistics about the application (along with statistics emitted to standard-out). These files are then used by the `gcov` utility to report statistics and coverage information to the developer. When the `-ftest-coverage` option is specified, two files are generated for each source file. These files use the extension `.bb` (basic-block) and `.bbg` (basic block graph) and are used to reconstruct the program flow graph of the executed application. For the option `-fprofile-arcs`, a `.da` file is generated that contains the execution count for each instrument branch. These files are used after execution, along with the original source file, to identify the execution behavior of the source.

Using the gcov Utility

Now that we have our image, let's continue to walk through the rest of the process. Executing our new application yields the set of statistics files discussed previously (`.bb`, `.bbg`, and `.da`). We then execute the `gcov` application with the source file that we wish to examine, as:

```
$ ./bubblesort
...
```

```
$ gcov bubblesort.c
100.00% of 17 source lines executed in file bubblesort.c
Creating bubblesort.c.gcov.
```

This tells us that all source lines within our sample application were executed at least once. We can see the actual counts for each source line by reviewing the generated file `bubblesort.c.gcov` (see Listing 7.2).

Listing 7.2 File `bubblesort.c.gcov` Resulting from Invocation of `gcov` Utility

```
               #include <stdio.h>

               void bubbleSort( int list[], int size )
        1      {
        1        int i, j, temp, swap = 1;

        3        while (swap) {

        2          swap = 0;

       22           for ( i = (size-1) ; i >= 0 ; i– ) {

      110              for ( j = 1 ; j <= i ; j++ ) {

       90                 if ( list[j-1] > list[j] ) {

       45                   temp = list[j-1];
       45                   list[j-1] = list[j];
       45                   list[j] = temp;
       45                   swap = 1;

                         }

                      }

                   }

                 }

               }

             int main()
        1      {
        1       int theList[10]={10, 9, 8, 7, 6, 5, 4, 3, 2, 1};
        1       int i;
```

```
36:
37:              /* Invoke the bubble sort algorithm */
38:     1        bubbleSort( theList, 10 );
39:
40:             /* Print out the final list */
41:     11        for (i = 0 ; i < 10 ; i++) {
42:     10          printf("%d\n", theList[i]);
43:               }
44:
45:             }
```

Let's now walk through some of the major points of Listing 7.2 to see what's provided. The first column shows the execution count for each line of source (line 4 shows a count of one execution, the call of the `bubbleSort` function). In some cases execution counts aren't provided. These are simply C source elements that don't result in code (for example, lines 22 through 30).

The counts can provide some information about the execution of the application. For example, the test at line 15 was executed 90 times, but the code executed within the test (lines 17–20) was executed only 45 times. This tells you that while the test was invoked 90 times, the test succeeded only 45. In other words, half of the tests resulted in a swap of two elements. This behavior is due to the ordering of the test data at line 34.

The `gcov` *files (*`.bb`*,* `.bbg`*, and* `.da`*) should be removed before running the application again. If the* `.da` *file isn't removed, the statistics will simply accumulate rather than start over. This can be useful but, if unexpected, problematic.*

The code segment executed most often, not surprisingly, is the inner loop of the sort algorithm. This is because line 13 is invoked one time more than line 15 due to the exit test (to complete the loop).

Looking at Branch Probabilities

We can also see the branch statistics for the application using the `-b` option. This option writes branch frequencies and summaries for each branch in the instrumented application. For example, when we invoke `gcov` with the `-b` option, we now get the following:

```
$ gcov -b bubblesort.c
100.00% of 17 source lines executed in file bubblesort.c
100.00% of 12 branches executed in file bubblesort.c
100.00% of 12 branches taken at least once in file bubblesort.c
100.00% of 2 calls executed in file bubblesort.c
Creating bubblesort.c.gcov.
$
```

The resulting `bubblesort.c.gcov` file is shown in Listing 7.3. Here we see a similar listing to 7.2, but this time the branch points have been labeled with their frequencies.

Listing 7.3 File `bubblesort.c.gcov` Resulting from Invocation of `gcov` Utility with `-b`

```
                #include <stdio.h>

                void bubbleSort( int list[], int size )
          1     {
          1       int i, j, temp, swap = 1;

          3       while (swap) {
      branch 0 taken = 67%
      branch 1 taken = 100%

          2         swap = 0;

         22         for ( i = (size-1) ; i >= 0 ; i– ) {
      branch 0 taken = 91%
      branch 1 taken = 100%
      branch 2 taken = 100%

        110           for ( j = 1 ; j <= i ; j++ ) {
      branch 0 taken = 82%
      branch 1 taken = 100%
      branch 2 taken = 100%

         90             if ( list[j-1] > list[j] ) {
      branch 0 taken = 50%

         45               temp = list[j-1];
         45               list[j-1] = list[j];
         45               list[j] = temp;
         45               swap = 1;

                        }

                      }

                    }

                  }

                }

```

```
41:            int main()
42:        1   {
43:        1     int theList[10]={10, 9, 8, 7, 6, 5, 4, 3, 2, 1};
44:        1     int i;
45:
46:              /* Invoke the bubble sort algorithm */
47:        1     bubbleSort( theList, 10 );
48:   call 0 returns = 100%
49:
50:              /* Print out the final list */
51:       11     for (i = 0 ; i < 10 ; i++) {
52:   branch 0 taken = 91%
53:   branch 1 taken = 100%
54:   branch 2 taken = 100%
55:       10       printf("%d\n", theList[i]);
56:   call 0 returns = 100%
57:              }
58:
59:            }
```

The branch points are very dependent upon the target architecture's instruction set. Line 23 is a simple `if` statement and therefore has one branch point represented. Note that this is 50%, which cross-checks with our observation of line execution counts previously. Other branch points are a little more difficult to parse. For example, line 7 represents a `while` statement and has two branch points. In ×86 assembly, this line compiles to what you see in Listing 7.4.

Listing 7.4 ×86 Assembly for the First Branch Point of `bubblesort.c.gcov`

```
1:   cmpl     $0, -20(%ebp)
2:   jne     .L4
3:   jmp     .L1
```

The `swap` variable is compared at line 1 to the value 0 in Listing 7.4. If it's not equal to zero, the jump at line 2 is taken (jump-nonzero) to `.L4` (line 11 from Listing 7.3). Otherwise, the jump at line 3 is taken to `.L1`. The branch probabilities show that line 2 (branch 0) was taken 67% of the time. This is because the line was executed three times, but the `jne` (line 2 of Listing 7.3) was taken only twice (2/3 or 67%). When the `jne` at line 2 is not taken, we do the absolute jump (`jmp`) at line 3. This is executed once, and once executed the application ends. Therefore, branch 1 (line 9 of Listing 7.3) is taken 100% of the time.

So the branch probabilities are useful in understanding program flow, but consulting the assembly can be required to understand what the branch points represent.

Incomplete Execution Coverage

When gcov encounters an application whose test coverage is not 100%, the lines that are not executed are labeled with ###### rather than an execution count. Listing 7.5 shows a source file created by gcov that illustrates less than 100% coverage.

Listing 7.5 A Sample Program with Incomplete Test Coverage (on the CD-ROM at ./source/ch7/incomptest.c)

```
                #include <stdio.h>

                int main()
          1     {
          1       int a=1, b=2;

          1       if (a == 1) {
          1         printf("a = 1\n");
                  } else {
     ######         printf("a != 1\n");
                  }

          1       if (b == 1) {
     ######         printf("b = 1\n");
                  } else {
          1         printf("b != 1\n");
                  }

          1       return 0;
                }
```

The gcov utility also reports this information to standard-out when it is run. It emits the number of source lines possible to execute (in this case 9) and the percentage that were actually executed (here, 78%):

```
$ gcov incomptest.c
 77.78% of 9 source lines executed in file incomptest.c
Creating incomptest.c.gcov.
$
```

If our sample application had multiple functions, we could see the breakdown per function through the use of the -f option (or -function-summaries). This is illustrated using our previous bubblesort application as:

```
$ gcov -f bubblesort.c
100.00% of 11 source lines executed in function bubbleSort
```

```
100.00% of 6 source lines executed in function main
100.00% of 17 source lines executed in file bubblesort.c
Creating bubblesort.c.gcov.
$
```

Options Available for gcov

Now that we've seen gcov in action in a few scenarios, let's look at gcov's full list of options (see Table 7.1). The gcov utility is invoked with the source file to be annotated, as:

```
gcov [options] sourcefile
```

TABLE 7.1 gcov Utility Options

Option	Purpose
-v, –version	Emit version information (no further processing).
-h, –help	Emit help information (no further processing).
-b, –branch-probabilities	Emit branch frequencies to the output file (with summary).
-c, –branch-counts	Emit branch counts rather than frequencies.
-n, –no-output	Do not create the gcov output file.
-l, –long-file-names	Create long filenames.
-f, –function-summaries	Emit summaries for each function.
-o, –object-directory	Directory where .bb, .bbg, and .da files are stored.

From Table 7.1, we can see a short single letter option, and a longer option. The short option is useful when using gcov from the command line, but when gcov is part of a Makefile, the longer options should be used as they're more descriptive.

To retrieve version information about the gcov utility, the -v option is used. Since gcov is tied to a given compiler tool chain (it's actually built from the gcc tool chain source), the versions for gcc and gcov will be identical.

An introduction to gcov and option help for gcov can be displayed using the -h option.

The branch probabilities can be emitted to the annotated source file using the -b option (see the section "Looking at Branch Probabilities," earlier in this chapter). Rather than producing branch percentages, branch counts can be emitted using the -c option.

If the annotated source file is not important, the -n option can be used. This can be useful if all that's important is to understand the test coverage of the source. This information is emitted to standard-out.

When including source in header files, it can be useful to use the `-l` option to produce long filenames. This helps make filenames unambiguous if multiple source files include headers containing source (each getting its own `gcov` annotated header file).

Coverage information can be emitted to standard-out for each function rather than the entire application using the `-f` option. This is discussed in the section "Incomplete Execution Coverage," earlier in this chapter.

The final option, `-o`, tells `gcov` where the `gcov` object files are stored. By default, `gcov` will look for the files in the current directory. If they're stored elsewhere, this option specifies where `gcov` can find them.

Considerations

Certain capabilities should be avoided when using `gcov` for test coverage. Optimization should be disabled when using `gcov`. Since optimization can result in source lines being moved or removed, coverage is less meaningful. Coverage testing is also less meaningful when using source macro expansion in the source after the preprocessor stage. These aren't shown in `gcov` and therefore miss identification of full test coverage.

For GNU/Linux kernel developers, `gcov` can be used for certain architectures within the kernel. A patch is available from IBM to allow `gcov` use in the kernel. Its availability is provided in the Resources section.

SUMMARY

In this chapter, we introduced GNU's `gcov` test coverage tool. We explored the capabilities for `gcov`, including coverage testing, identifying branch probabilities, and emitting summaries for each function under review. We investigated building software for use with `gcov` and some considerations for options to avoid, such as optimization and source macro expansion.

REFERENCES

[FSF 2002] "Using the GNU Compiler Collection (GCC)," Free Software Foundation at *http://gcc.gnu.org/onlinedocs/gcc-3.2.3/gcc/gcov.html#gcov*

RESOURCES

The LTP GCOV-kernel extension (GCOV-kernel) at *http://ltp.sourceforge.net/coverage/gcov-kernel.php*

8 Profiling with GNU gprof

In This Chapter

- An Introduction to Performance Profiling
- An Introduction to the GNU gprof Profiler Utility
- Preparing an Image for Use with gprof
- Discussing the Data Products Provided by gprof
- Exploring Some of the Most Important gprof Utility Options

INTRODUCTION

In this chapter, we'll investigate the gprof utility and explore how it can be used to help build efficient programs. We'll learn how software must be built for use with gprof and then understand the data products that are provided. Finally, we'll investigate the variety of options that gprof provides and how they can be used.

WHAT IS PROFILING?

Profiling is the art of analyzing the performance of an application. By identifying where a program spends the majority of its time, we can better isolate where our modifications can yield the biggest performance gains. The most common result of profiling is a better understanding of where a given program spends its time. By looking at where the program spends the majority of its time, we can yield significant gains by improving that portion of code, rather than fine-tuning code that doesn't affect the bottom line.

WHAT IS gprof?

The `gprof` utility is the GNU profiler, a tool that identifies how much time is spent in a function of an operating program. The GNU profiler also identifies which functions were called by a given function. Similar to the `gcov` utility (the topic of Chapter 7, "Coverage Testing with GNU `gcov`"), the compiler introduces profiling code into the target image, which generates a statistics file upon execution. This file (`gmon.out`) contains histogram records, call-graph arc records, and basic-block execution records that illustrate the execution profile of an application. When read by the `gprof` utility, the performance behavior of the application can be readily understood.

We'll discuss use of the `gprof` utility bundled with version 3.2.2 of the GNU compiler tool chain.

PREPARING THE IMAGE

Let's now look at how an image is prepared for profiling with `gprof`. We'll first look at some basic uses of profiling with `gprof`, and in later sections we'll discuss some of the other options available. For our profiling example, we'll use the following sorting demo shown in Listing 8.1. This example source (`sort.c`) illustrates two sorting algorithms, the insert-sort (function `insertSort`, lines 5–21) and the bubble-sort (function `bubbleSort`, lines 23–50). Each is run with identical data to unambiguously understand their profiling properties for a given data set (as provided by function `init_list`, lines 53–62).

Listing 8.1 Sample Source to Explore the `gprof` Utility (on the CD-ROM at `./source/ch8/sort.c`)

```
1:     #include <stdio.h>
2:
3:     #define MAX_ELEMENTS 10000
4:
5:     void insertSort( int list[], int size )
6:     {
7:       int i, j, temp;
8:
9:       for ( i = 1 ; i <= size-1 ; i++ ) {
```

```

    temp = list[i];

    for ( j = i-1 ; j >= 0 && (list[j] > temp) ; j– ) {
      list[j+1] = list[j];
    }

    list[j+1] = temp;

  }

}

void bubbleSort( int list[], int size )
{
  int i, j, temp, swap = 1;

  while (swap) {

    swap = 0;

    for ( i = (size-1) ; i >= 0 ; i– ) {

      for ( j = 1 ; j <= i ; j++ ) {

        if ( list[j-1] > list[j] ) {

          temp = list[j-1];
          list[j-1] = list[j];
          list[j] = temp;
          swap = 1;

        }

      }

    }

  }

}


void init_list( int list[], int size )
{
```

```
    int i;

    for ( i = 0 ; i < size ; i++ ) {
      list[i] = (size-i);
    }

    return;
  }


  int main()
  {
    int list[MAX_ELEMENTS];  int i;

    /* Invoke the bubble sort algorithm */
    init_list( list, MAX_ELEMENTS );
    bubbleSort( list, MAX_ELEMENTS );
    init_list( list, MAX_ELEMENTS );
    insertSort( list, MAX_ELEMENTS );

  }
```

The `gprof` utility uses information from the executable image and the profiler output file, `gmon.out`, to generate its profiling data. In order to collect the profiling data, the image must be compiled and linked with a special set of compiler flags. These are illustrated below for compiling our sample source file, `sort.c`:

```
gcc sort.c -o sort -pg
```

The result is an image, `sort`, which is instrumented to collect profiling information. When the image is executed and completes normally, a file called `gmon.out` results, containing the profiling data.

The `gmon.out` *file is written upon normal exit of the application. If the program exits abnormally or the user forces an exit with a Ctrl+C, the* `gmon.out` *file will not be written.*

USING THE gprof UTILITY

Upon execution of the profiler-instrumented image, the `gmon.out` file is generated. This file is used in conjunction with the original image for `gprof` to generate human-readable statistics information. Let's look at a simple example and the data products that result. First, we invoke the image and then generate the `gprof` summary:

```
$ ./sort
$ gprof sort gmon.out > sort.gprof
```

The `gprof` utility writes its human-readable output to standard-out, so the user must redirect this to a file to save it. The first element of the `gprof` output is what's called the "flat profile." This provides the basic timing summary of the executable, as shown in Listing 8.2. Note that this is not the complete output of `gprof`. We'll look at other data products shortly.

Listing 8.2 Sample Flat Profile Output from `gprof`

```
$ gprof sort gmon.out | more
Flat profile:
Each sample counts as 0.01 seconds.
  %    cumulative    self              self     total
 time    seconds   seconds     calls   s/call   s/call  name
 71.66      3.11      3.11         1     3.11     3.11  bubbleSort
 28.11      4.33      1.22         1     1.22     1.22  insertSort
  0.23      4.34      0.01         2     0.01     0.01  init_list
...
```

As we saw in Listing 8.1, our application is made up of three functions. Each function is represented here with a variety of timing data. The first column represents the percentage of time spent in each function in relation to the whole. What's interesting to note from this column is that the bubble sort algorithm requires 2.5 times as much execution time to sort the identical list as the insert sort. The next column, `cumulative seconds`, is a running sum of the number of seconds, and the next, `self seconds`, is the number of seconds taken by this function alone. Note that the table itself is sorted by this column in descending order. The column entitled `calls` represents the total number of times that this function was called (if the function itself was profiled, otherwise the element will be blank. The `self s/call` represents the average number of seconds spent in this function (including functions that it calls), while the `total s/call` represents the total number of seconds spent in the function (including functions that it calls). Finally, the name of the function is provided.

The next element provided by `gprof` is the call graph. This summary (shown in Listing 8.3) shows each function and the calls that are made by it, including the time spent within each function. It illustrates the timing as a hierarchy, which is useful in understanding timing for individual functions down the chain and their effect on higher layers of the call chain.

Listing 8.3 Sample Call Graph Output from `gprof`

```
index % time    self  children    called     name
                                                 <spontaneous>
[1]    100.0    0.00    2.11                 main [1]
                1.44    0.00       1/1           bubbleSort [2]
                0.67    0.00       1/1           insertSort [3]
                0.00    0.00       2/2           init_list [4]
-----------------------------------------------
                1.44    0.00       1/1           main [1]
[2]     68.2    1.44    0.00       1         bubbleSort [2]
-----------------------------------------------
                0.67    0.00       1/1           main [1]
[3]     31.8    0.67    0.00       1         insertSort [3]
-----------------------------------------------
                0.00    0.00       2/2           main [1]
[4]      0.0    0.00    0.00       2         init_list [4]
-----------------------------------------------
```

The `index` column is a unique number given to each element. The column marked `% time` represents the percentage of the total amount of time that was spent in the given function and its children calls. The `self` column is the amount of time spent in the function, with `children` as the amount of time spent in the children's functions. Note that in the first row, `children` is 2.11 (a sum of the two `sort` functions 1.44 + 0.67). This illustrates that the children functions took all of the time, and no meaningful time was spent in the `main` function itself. The `called` field identifies how many times the function was called by the parent. The first number is the number of times the particular parent called the function, and the second number is the total number of times that the child function was called altogether. Finally, the `name` column represents the names of the functions. Note that in the first row (after the row headings), the name `<spontaneous>` is used. This simply means that the parent could not be determined (very likely the C-start initialization).

Now that we have a performance baseline of our application, let's rebuild our application to see how we can improve it. In this example, we'll build using `-O2` optimization to see how well it improves this very simple application:

```
$ gcc -o sort sort.c -pg -O2
$ ./sort
$ gprof sort gmon.out | more
```

From Listing 8.4, we see that the performance of our application improved significantly (a five times improvement for the `insert-sort`) from Listing 8.2 (pre-optimization). The function `bubbleSort` saw only a modest four times improvement.

Listing 8.4 Sample Flat Profile Output from gprof for the Optimized Application

```
Flat profile:
Each sample counts as 0.01 seconds.
%    cumulative   self              self     total
time   seconds   seconds    calls  ms/call  ms/call  name
76.47      0.78      0.78        1   780.00   780.00  bubbleSort
23.53      1.02      0.24        1   240.00   240.00  insertSort
0.00       1.02      0.00        2     0.00     0.00  init_list
```

This clearly illustrates the utility of the gprof utility (and the gcc optimizer). We can see how the -O2 optimization level improves the source.

Options Available for gprof

Now that we've covered the basic uses of gprof, we'll look at the variety of options that are provided. While gprof provides a large number of options, we'll discuss some of the more useful ones here. For a complete list of options, see the gprof help.

Source Annotation

The gprof utility can be used to annotate the source with frequency of execution. The image must be built for this purpose. The -g and -a options must be specified along with -pg as:

```
gcc -o sort sort.c -g -pg
```

This results in an image with not only the profile information (via the -pg option), but also debugging information (through the -g option). Upon executing the image and checking the resulting output, as

```
gprof -A -l sort gmon.out
```

we get Listing 8.5. The -A option tells gprof to emit an annotated source listing. The -l option specifies to emit a function execution profile, as shown below:

Listing 8.5 Sample Source Annotation from gprof for the Sort Application (Incomplete)

```
           #include <stdio.h>
           #define MAX_ELEMENTS 10000
           void insertSort( int list[], int size )
      1 -> {
             int i, j, temp;
             for ( i = 1 ; i <= size-1 ; i++ ) {
               temp = list[i];
```

```
            for ( j = i-1 ; j >= 0 && (list[j] > temp) ; j-- ) {
              list[j+1] = list[j];
            }
            list[j+1] = temp;
          }
        }
```

The -x option can also be used with gprof to extend execution counts to all source lines.

Ignore Private Functions

For private functions that are statically declared, we can ignore the statistics for these through the use of the -a option (or -no-static). This attributes time spent in the private function to the caller, with the private function never appearing in the flat profile or the call graph.

Recommending Function Ordering

The gprof utility can recommend a function ordering that can improve cache and translation lookaside buffer (TLB) performance on systems that require functions to be contiguous in memory. For systems that include multiway caches, this may not provide much improvement.

To recommend a function ordering (via –function-ordering), the following command sequences can be used:

```
$ gcc -o sort sort.c -pg -g
$ ./sort
$ gprof sort gmon.out –function-ordering
```

A list of functions is then suggested with specific ordering (using the source from Listing 8.1). The result is shown in Listing 8.6.

Listing 8.6 Sample Function Ordering from gprof

```
    insertSort
    bubbleSort
    init_list
    _init
    _start
    __gmon_start__
. . .
```

Note the grouping of the sorting and `init` functions, which are all called in proximity of one another. In some cases, reordering functions can be done as simply as coexisting contiguous functions within a single source file. The linker can also provide this capability.

The file ordering option (`–file-ordering`) provides a similar capability by recommending the order in which objects should be linked to the target image.

Minimizing gprof Summary

The `-b` option is useful to minimize the amount of superfluous description data that is emitted. Using `-b` removes the field discussion in the output, as shown in Listing 8.7.

Listing 8.7 Sample Brief Output from `gprof`

```
Flat profile:
Each sample counts as 0.01 seconds.
  %   cumulative   self              self     total
 time   seconds   seconds    calls  s/call   s/call  name
 71.59      3.10     3.10        1     3.10     3.10  bubbleSort
 28.41      4.33     1.23        1     1.23     1.23  insertSort
  0.00      4.33     0.00        2     0.00     0.00  init_list
            Call graph
granularity: each sample hit covers 4 byte(s) for 0.23% of 4.33
seconds
index % time    self  children    called     name
                                                 <spontaneous>
[1]    100.0    0.00    4.33                 main [1]
                3.10    0.00       1/1           bubbleSort [2]
                1.23    0.00       1/1           insertSort [3]
                0.00    0.00       2/2           init_list [4]
-----------------------------------------------
                3.10    0.00       1/1           main [1]
[2]     71.6    3.10    0.00       1         bubbleSort [2]
-----------------------------------------------
                1.23    0.00       1/1           main [1]
[3]     28.4    1.23    0.00       1         insertSort [3]
-----------------------------------------------
                0.00    0.00       2/2           main [1]
[4]      0.0    0.00    0.00       2         init_list [4]
-----------------------------------------------
Index by function name
 [2] bubbleSort (sort.c)  [4] init_list (sort.c)  [3] insertSort
(sort.c)
```

To further minimize the output, the `–no-flat-profile` can be used if the flat profile is not needed (the default is `–flat-profile`). The call graph can be disabled using the `–no-graph` option (default is `–graph`).

Finding Unused Functions

The `gprof` utility can identify functions that are not called in a given run. The `–display-unused-functions` is used in conjunction with the `–static-call-graph` option to list those functions, as:

```
gprof sort gmon.out –display-unused-functions –static-call-graph
```

Increasing gprof Accuracy

In some cases, a program may differ in timing or may represent such a small timing sample that its accuracy is left in question. To increase the accuracy of applications profiling, the application can be performed numerous times and then averaged. A sample bash script that provides this capability is shown in Listing 8.8 [GNU `gprof`]:

Listing 8.8 Analyzing Multiple Invocations of an Application (on the CD-ROM at `./source/ch8/script`)

```
#!/bin/bash
for i in 'seq 1 5'; do
  ./sort
  mv gmon.out gmon.out.$i
done
gprof –sum sort gmon.out.*
gprof -b –no-graph sort gmon.sum
```

Running this script results in a single flat profile over five invocations of the sort application. This uses the `-sum` option of `gprof` to summarize the collection of input `gmon.out` files into a single summary file `gmon.sum`. The per/call measurements of the flat profile can then be used as higher accuracy function timings.

Considerations

Profiling with `gprof` is a sampling process that is subject to statistical inaccuracies. Recall from Listing 8.7 that each sample counts as 0.01 seconds. This is the sampling period. The closer the sampling time to this period, the larger the error will be for the profile. For this reason, increasing the accuracy of the profile is recommended (see the previous section, "Increasing `gprof` Accuracy"). When `gprof` is enabled, it does introduce extra code into the image which can also affect its behavior and performance. Therefore, simply by measuring the performance of the code, we can affect it. This should always be kept in mind when using performance and coverage tools.

SUMMARY

In this chapter, profiling with GNU's `gprof` was discussed, identifying some of the most useful options that are provided. Building an application for use with `gprof` and then gathering a profile from it were explored, including options for the various `gprof` data products and recommendations for improving the performance of an application from a caching perspective.

REFERENCES

[GNU gprof] The GNU Profiler, Jay Fenlason and Richard Stallman at *http://www.gnu.org/manual/gprof-2.9.1/html_mono/gprof.html*

9 Building Packages with automake/autoconf

by Curtis Nottberg

In This Chapter

- make Review
- Introduction to the GNU Autotools
- Quick Introduction to autoconf
- Quick Introduction to automake
- Converting a Project to Use Autotools

INTRODUCTION

The standard GNU make utility eases many of the burdens associated with building an executable from multiple source files. It enables incremental building of the source and allows the commands and processes needed to maintain a source package to be collected in a single location. GNU make is excellent at implementing the steps needed to build a moderately complex project. GNU make starts to become cumbersome as projects grow in complexity. Examples of factors that cause Makefile maintenance to become cumbersome are these:

- Projects with a large number of files that have varied build requirements.
- Dependencies on external libraries.
- A desire to build in a multiplatform environment.
- Installing built software in multiple environments.
- Distributing a source package.

The solution to these complexities is to move up one level and automatically generate the appropriate Makefiles to build the project. This allows GNU `mmake` to focus on the things it is good at while still allowing the capability to configure for the current build environment. The GNU Autotools are an attempt to provide this next level of build functionality on top of the GNU `mmake` utility.

An Example Project

The examples in this chapter will show various ways to build a project consisting of four source files. Two of the source files will be used to build a library, and the other files will build an application that uses the library (see Figure 9.1).

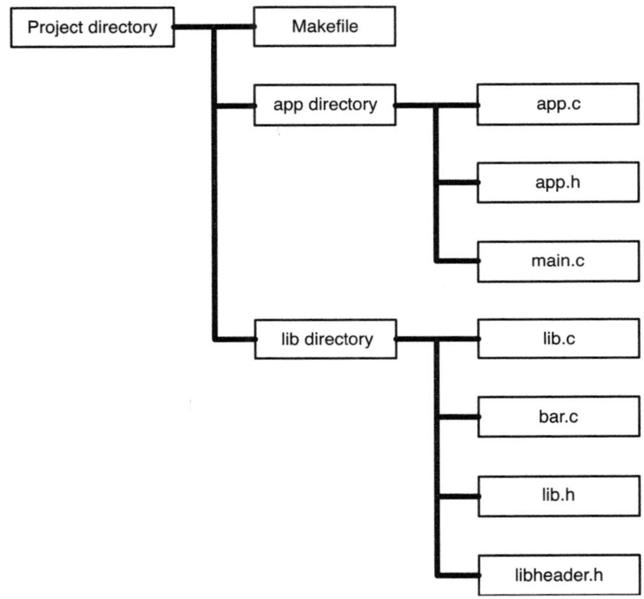

FIGURE 9.1 Directory structure of example project.

A Makefile Solution

The following listing shows a simple Makefile that builds the library and an application that uses the library. This Makefile will be used to provide a basis for comparing how Autotools would build the same project. Keep in mind that the example considered here is very simple. The application of `automake/autoconf` to this project will seem like more work than payoff, but as the project grows, the payback from using the Autotools will increase.

Listing 9.1 Simple Makefile to Build Example (on the CD-ROM at `./source/ch9/Makefile.simple`)

```
1:   VPATH= lib app
2:
3:   LIBSRC= lib.c bar.c
4:   LIBOBJ= $(LIBSRC:.c=.o)
5:
6:   APPSRC= main.c app.c
7:   APPOBJ= $(APPSRC:.c=.o)
8:
9:   CFLAGS=
10:  INCLUDES= -I ./lib
11:
12:  all: libexp.a appex
13:
14:  %.o:%.c
15:          $(CC) -c $(CFLAGS) $(INCLUDES) -o $@ $
16:
17:  libexp.a: $(LIBOBJ)
18:          $(AR) cru libexp.a $(LIBOBJ)
19:
20:  appex: $(APPOBJ) libexp.a
21:          $(CC) -o appex $(APPOBJ) -lexp -L .
```

Line 1 of the listing sets the `VPATH` variable. The `VPATH` variable specifies search paths that will be used by the `make` utility to find the source files for the build rules. The `VPATH` capability of `make` allows the use of a single Makefile to reference the source files in the `lib` and `app` subdirectories. Lines 3 through 7 create source and object file lists for use by the build rules. Notice that the file list doesn't include paths because this is taken care of by the `VPATH` search variable. Lines 9 and 10 set up the necessary compiler flags to perform the build. Line 12 sets up the default build target to build both the library and application. Lines 14 and 15 define the rules used to turn c-source files into object files. Lines 17 and 18 describe how to build the library. Finally, lines 20 and 21 describe how to build the application.

The simplified Makefile is missing a few features that would need to be included in a real build system to make it useable: a clean target, dependency tracking on included header files, a method for installing the generated binaries, and so forth. The Autotools implementation will provide many of these missing features with only a little bit of additional work when compared to the effort needed to create the simplified Makefile.

A Simple Implementation Using Autotools

The initial implementation using the Autotools will require the creation of five files to replace the simple Makefile described in Listing 9.1. Although this seems like a lot of files to replace a single file, each of the replacement files is generally simpler. Both the simple Makefile and the Autotool files will contain roughly the same information, but Autotools chooses to distribute the information differently with a project's directory structure. Figure 9.2 illustrates the directory structure from Figure 9.1 with the additional of the Autotools files.

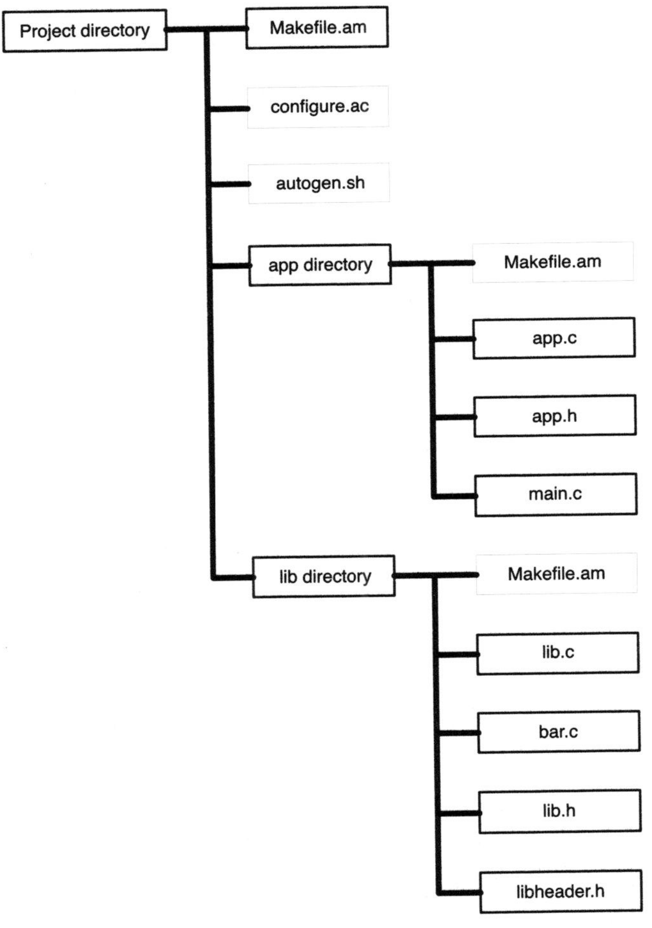

FIGURE 9.2 Directory structure of example project with Autotool files.

The additional files added to support a simple Autotools project are these:

autogen.sh: A shell script to run Autotools to generate the build environment.

configure.ac: The input file for the autoconf tool.

Makefile.am: The top-level Makefile template.

app/Makefile.am: The Makefile template for appexp executable.

lib/Makefile.am: The Makefile template for building the libexp.a library.

These files describe the intended build products and environment to Autotools. Autotools will take this input and generate a build environment template that will be further configured on the build system to generate the final set of Makefiles. Assuming that we are developing and building on the same machine, the following commands should configure and build our example project:

```
# ./autogen.sh
# ./configure
# make
```

Running the autogen.sh script will execute the Autotool utilities to convert the input files into a build environment template that can be configured on the host system. Executing the configure script causes the build environment template to be customized for the build machine. The output of the configure script is a set of GNU Makefiles that can be used to build the system. Executing the make command in the root directory will cause both the library and application to be built.

An examination of autogen.sh should be the starting point for understanding how this process works. Listing 9.2 shows a very simple autogen.sh script that just executes the Autotools utilities. Generally the autogen.sh script in a real project will be much more complicated to first check that the Autotools exist and are of the appropriate version. To find an example of a more complex autogen.sh script, you should examine this file in the source repositories of your favorite open source project.

ON THE CD

Listing 9.2 Simple autogen.sh Script (on the CD-ROM at ./source/ch9/autogen.sh)

```
#!/bin/sh
# Run this to generate all the initial makefiles, etc.

aclocal
libtoolize —automake
automake -a
autoconf
```

Line 1 indicates the shell to use when running this script. Line 4 runs the `aclocal` utility. The `aclocal` utility creates the local environment needed by the `automake` and `autoconf` tools to work. Specifically `aclocal` makes sure the `m4` macro environment that `automake` and `autoconf` use to implement their functionality is set up appropriately. Line 5 executes the `libtoolize` utility, which enables the `libtool` functionality in `automake`. The `libtool` functionality will be discussed in a subsequent section. Line 6 executes the `automake` utility, which turns the `Makefile.am` files into `Makefile.in` files. This operation is discussed more in the next section. Line 7 executes the `autoconf` utility that takes the `configure.ac` input file and turns it into a pure shell script named `configure`.

automake

The input to the `automake` utility is a series of `Makefile.am` files that describe the targets to be built and the parameters used to build them. The `automake` utility transforms the `Makefile.am` files into `makefile.in` files. The `Makefile.in` file is a GNU `make` format file that acts as a template that the configure script will transform into the final Makefile. `automake` has built-in support for building binaries and libraries, and with the support of `libtool` can also be used to build shared libraries. The example project required three separate `automake` files: one in the root directory and one for each subdirectory. Let's examine the root `Makefile.am` to see how `automake` handles subdirectories.

ON THE CD

Listing 9.3 Listing of the Root `Makefile.am` (on the CD-ROM at `./source/ch9/Makefile.am`)

```
1:   SUBDIRS = lib app
```

The contents of the root `Makefile.am` simply indicate that all the work for this project will be done in the subdirectories. Line 1 tells the `automake` utility that it should descend into the subdirectories and process the `Makefile.am` files it finds there. The ordering of directories in the `SUBDIRS` variable is significant; the subdirectories will be built in the left to right order specified in the subdirectories list. The sample project uses this to ensure that the `lib` directory is built before the `app` directory; a requirement of the sample project because the application is dependent on the library being built first. Let's move on to the `lib/Makefile.am` file that shows `automake` how to build the `libexp.a` library.

ON THE CD

Listing 9.4 Listing of `lib/Makefile.am` (on the CD-ROM at `./source/ch9/lib/Makefile.am`)

```
lib_LIBRARIES = libexp.a
libexp_a_SOURCES = bar.c lib.c
```

Line 1 is a list of the static libraries to be built in this directory. In this case the only library being built is libexp.a. The syntax of line 1 is more complex than it first appears. The lib_LIBRARIES variable name indicates two pieces of information. The lib portion indicates that when the library is installed it will be put in the lib directory. The LIBRARIES portion indicates that the listed targets should be built as static libraries. Line 2 lists the source files that go into building the libexp.a static library. Again automake uses the format of the variable name to provide the association between both the library that the variable applies to and content of the variable. The libexp_a portion of the name indicates that this variable's value applies to building libexp.a. The SOURCES portion of the name implies that the value of this variable will be a space-separated list of source files. The app/Makefile.am file is very similar to the one in the lib directory but includes a few additional variables to take care of using the libexp.a library that was previously built in the lib directory.

ON THE CD

Listing 9.5 Listing of app/Makefile.am (on the CD-ROM at ./source/ch9/app/Makefile.am)

```
1:    bin_PROGRAMS = appexp
2:    appexp_SOURCES = app.c main.c
3:    appexp_LDADD = $(top_builddir)/lib/libexp.a
4:    appexp_CPPFLAGS = -I $(top_srcdir)/lib
```

Line 1 of Listing 9.5 should look similar to line 1 in Listing 9.4 in that it is a list of things to be built. In this case the bin_PROGRAMS variable name indicates to automake that the result will be installed in the bin directory and listed targets should be built as executables. The appexp_ prefix on the variable in lines 2 through 4 indicates that these variables apply to building the appexp executable. Line 2 has the SOURCES variable that lists the source files that will be compiled into the appexp executable. Line 3 specifies the LDADD variable, which are things that will be included during linking. In this example, the LDADD variable is used to add the library that was previously built in the lib directory. The $(top_builddir) is set by the configure script when it is run and provides a mechanism for Makefiles to reference the build directories in a relative manner. Line 4 specifies the CPPFLAGS variable that is passed to the preprocessor when it runs. This variable should contain the -I include paths and the -D defines that would normally be passed to the preprocessor in a Makefile. In this example it is being used to get access to the library header file contained in the lib directory. The $(top_srcdir) variable is set by the configure script; it provides a mechanism for Makefiles to reference source files in a relative manner.

autoconf

The autoconf utility converts the configure.ac input file into a shell script named configure. The configure script is responsible for collecting information about the current build system and using the information to transform the Makefile.in template files into the Makefiles used by the GNU make utility. The configure script performs the transformation by replacing occurrences of configuration variables in the Makefile.in template file with values for those variables as determined by the configure script. The configure.ac input file contains macros that describe the types of configuration checks the configure script should perform when it is run. The configure.ac for our example project illustrates the very simple series of checks needed to compile c-files and create static libraries.

Listing 9.6 Listing of configure.ac (on the CD-ROM at ./source/ch9/configure.ac)

```
1:    dnl Process this file with autoconf to produce a configure script
2:    AC_PREREQ(2.53)
3:    AC_INIT(app)
4:    AM_INIT_AUTOMAKE(appexp, 0.1.00)
5:    AC_PROG_CC
6:    AC_PROG_RANLIB
7:    AC_OUTPUT(app/Makefile lib/Makefile Makefile)
```

Line 1 illustrates the comment format used by autoconf. Line 2 is a macro defined by autoconf to ensure that the version of the autoconf utility being used to create the configure script is new enough. This macro results in a check to make sure that autoconf utility has a version number equal to or greater than 2.53. If the version isn't recent enough, an error will be generated, and the configure script will not be generated. Line 3 is a required autoconf macro that must be called before any other macros are invoked; it gives autoconf a chance to initialize itself and parse its command line parameters. The parameter to AC_INIT is a name for the project. Line 4 is the initialization macro for automake. autoconf can be used independently of automake, but if they are to be used together, then the AM_INIT_AUTOMAKE macro is required in the project's configure.ac file. Line 5 and 6 are the first macros that actually check for tools used during the make process. Line 5 indicates that the project will cause the configure script to find and prepare the make files to use the C compiler. Line 6 does the checks to find the tools needed to build static libraries. Line 7 is the other required macro that must exist in a configure.ac file. This macro indicates the output files that should be generated by the configure script. When the configure script is ready to generate its output files, it will iterate through the files in the AC_OUTPUT macro and look for a corresponding file with an .in suffix. It will then perform a substitution step on the .in file to generate the output file.

The configure Script

The output of the autoconf utility is a shell script named configure. The example configure.ac in Listing 9.7 generates a configure script with approximately 4,000 lines when run through the autoconf utility. Executing the configure script will collect the required configuration information from the executing system and generate the appropriate Makefiles by performing a substitution step on the Makefile.in files generated by automake. Let's examine the output of running the configure script generated by our example.

Listing 9.7 Output from the Example configure Script

```
1:  checking for a BSD-compatible install... /usr/bin/install -c
2:  checking whether build environment is sane... yes
3:  checking for gawk... gawk
4:  checking whether make sets $(MAKE)... yes
5:  checking for gcc... gcc
6:  checking for C compiler default output file name... a.exe
7:  checking whether the C compiler works... yes
8:  checking whether we are cross compiling... no
9:  checking for suffix of executables... .exe
10: checking for suffix of object files... o
11: checking whether we are using the GNU C compiler... yes
12: checking whether gcc accepts -g... yes
13: checking for gcc option to accept ANSI C... none needed
14: checking for style of include used by make... GNU
15: checking dependency style of gcc... gcc3
16: checking for ranlib... ranlib
17: configure: creating ./config.status
18: config.status: creating app/Makefile
19: config.status: creating lib/Makefile
20: config.status: creating Makefile
21: config.status: executing depfiles commands
```

Lines 1 through 4 are checks that occur to ensure that the build environment has the appropriate tools to support the make process. Lines 5 through 15 are checks generated by the AC_PROG_CC macro that locate and ready the compiler toolchain for processing C-source code. Line 16 is a check generated by the AC_PROG_RANLIB macro to ensure that the ranlib utility exists for generating static libraries. Lines 18 through 20 indicate that the substitution step to turn the Makefile.in templates into the actual Makefiles is occurring.

Once the configure script has completed successfully, then all of the Makefiles needed to build the project should have been successfully created. Typing **make** in the root directory at this point should build the project.

The Generated Makefiles

The generated Makefiles have a number of nice characteristics that were lacking in the simple Makefile of Listing 9.1, such as:

- Automatic dependency tracking. For example, when a header file is modified, only the source files that are affected will be rebuilt.
- A clean target that will clean up all the generated output.
- The automated ability to install the generated binaries into the appropriate system directories for execution.
- The automated ability to generate a distribution of the source code as a compressed tar file.

The generated Makefiles have numerous predefined targets that allow the user to invoke these capabilities. Let's examine the common targets used in the `automake`-generated Makefiles:

`make:` The default target. This will cause the project binaries to be built.

`make clean:` The clean target. This will remove all of the generated build files so that the next call to make will rebuild everything.

`make distclean:` This target removes all generated files, including those generated by the `configure` script. After using this target, the `configure` script will need to be run again before another build can take place.

`make install:` This target will move the generated binaries and supporting files into the system directory structure. The location of the installation can be controlled with the `-enable-prefix` parameter that can be passed to the `configure` script when it is run.

`make dist:` This target generates a `.tar.gz` file that can be used to distribute the source code and build setup. Included in the tarball will be all of the source files, the `makefile.in` files, and the `configure` script.

Just looking at the default targets provided by the standard `automake` Makefile should indicate some of the power that exists in using the Autotools to generate the Makefiles for your project. The initial setup to use the Autotools can be a bit cumbersome, but once the infrastructure is in place, then future updates to the files are very incremental, and the payback is large compared to implementing the same capabilities by hand in a developer-maintained Makefile.

SUMMARY

This chapter presented the GNU Autotools by illustrating how they can be used to build a simple project. The example makes little use of the advanced features of `automake` and `autoconf`, but it should provide a good illustration of the big-picture concepts needed to understand how the Autotools work. The GNU Autotools provide a wealth of features that are quite useful to larger software projects, and the effort of integrating them into your project should be expended early on. The downside of the tools is that they can be somewhat difficult to employ properly, and the documentation for them is a bit arcane. On balance, the uses of the Autotools are well worth the effort, but expect to put a little bit of time into getting things working correctly. One of the best ways to learn about the more advanced usage of `automake` and `autoconf` is to look at the existing implementations used in current open source projects. The simple example presented in this chapter should provided the basis needed to examine and learn from the more complex use of the GNU Autotools found in the larger open source projects.

Part III Application Development Topics

In this part of the book, we'll review a number of important topics that are important to application development. This includes using the most important elements of GNU/Linux including various IPC mechanisms, Sockets, and multiprocess and multithreaded programming.

Chapter 10, "File Handling in GNU/Linux"

The file handling APIs are important in GNU/Linux because there are patterns for many other types of I/O, such as sockets and pipes. This chapter demonstrates the proper use of the file handling APIs using binary, character, and string interfaces. Numerous examples illustrate the APIs in their different modes.

Chapter 11, "Programming with Pipes"

The pipe model of communication is an older aspect of UNIX, but it is still an important one, considering its wide use in shell programming. The pipe model is first reviewed, with discussion of anonymous and named pipes. The API to create pipes is discussed, along with examples of using pipes for multiprocess communication. Shell-level creation and use of pipes complete this chapter.

Chapter 12, "Introduction to Sockets Programming"

Network programming using the standard Sockets API is the focus of this chapter. Each of the API functions is detailed, illustrating their use in both client and server systems. After a discussion of the Sockets programming paradigm and each of the API functions, other elements of Sockets programming are discussed, including multilanguage aspects.

Chapter 13, "GNU/Linux Process Model"

The GNU/Linux process model refers to the standard multiprocessing environment. We discuss the `fork` function (to create child processes) and the other process-related API functions (such as `wait`). The topic of signals is also discussed, including the range of signals and their uses. Finally, the GNU/Linux process commands (such as `ps`) are detailed.

Chapter 14, "POSIX Threads (Pthreads) Programming"

Programming with threads using the `pthreads` library is the topic of this chapter. The functions in the pthreads library are discussed, including thread creation and destruction, synchronization (with mutexes and condition variables), communication, and other thread-related topics. Problems in multithreaded applications are also discussed, such as reentrancy.

Chapter 15, "IPC with Message Queues"

Message queues are a very important paradigm for communication in multiprocess applications. The model permits one-to-many and many-to-one communication and a very simple and intuitive API. This chapter details the message queue APIs for creating, configuring, and then sending and receiving messages. Advanced topics such as conditional message receipt are also discussed, along with user layer utilities for message queue inspection.

Chapter 16, "Synchronization with Semaphores"

Semaphores in GNU/Linux and the ability to create critical sections are the topics of this chapter. After a discussion of the problems that semaphores solve, the API for semaphores is detailed, including creation, acquisition, release, and removal. The advanced features provided by GNU/Linux such as semaphore arrays are discussed, including user-level commands to inspect and remove semaphores.

Chapter 17, "Shared Memory Programming"

One of the most important process communication mechanisms available in GNU/Linux is shared memory. The shared memory APIs allow segments of memory to be created and then shared between two or more processes. This chapter details the shared memory APIs for creating, attaching, detaching, locking, and unlocking shared memory segments.

Chapter 18, "Other Application Development Topics"

In this final chapter of Part III, we explore some of the important application development topics that were not covered in the preceding chapters. The topics explored here include command-line parsing with the `getopt` and `getopt_long` APIs, time conversion functions, `sysinfo`, memory mapping with `mmap`, and locking and unlocking memory pages for performance.

10 File Handling in GNU/Linux

In This Chapter

- Understand File Handling APIs in GNU/Linux
- Explore the Character Access Mechanisms
- Explore the String Access Mechanisms
- Investigate Both Sequential and Nonsequential (Random Access) Methods
- Review Alternate APIs and Methods for File Access

INTRODUCTION

In this chapter, we'll look at the file handling APIs of GNU/Linux and explore a number of applications to demonstrate the proper use of the file handling APIs. We'll look at a number of different file handling functions, including character interfaces, string interfaces, and ASCII-mode and binary interfaces. The emphasis on this chapter will be to discuss the APIs and then use them in applications to illustrate their use.

FILE HANDLING WITH GNU/LINUX

File handling within GNU/Linux is accomplished through the standard C Library. We can create and manipulate ASCII text or binary files with the same API. We can append to files or seek within them.

In this chapter, we'll look at the fopen call (to open or create a file), the **fwrite** and **fread** functions (to write to or read from a file), **fseek** (to position ourselves at a given position in an existing file), the **feof** call (to test whether we're at the end of a file while reading), and some other lower-level calls (such as **open**, **write**, and **read**).

FILE HANDLING API EXPLORATION

Let's now get our hands dirty by working through some examples of GNU/Linux stream file I/O programming.

Creating a File Handle

To write an application that performs file handling, the first step is to make visible the file I/O APIs. This is done by simply including the stdio.h header file, as:

```
#include <stdio.h>
```

Not doing so will result in compiler errors (undeclared symbols). The next step is to declare our handle to be used in file I/O operations. This is often called a *file pointer* and is a transparent structure that should not be accessed by the developer.

```
FILE *my_fp;
```

We'll build on this in the next sections to illustrate ASCII and binary applications.

Opening a File

Let's now open a file and illustrate the variety of modes that can be used. Recall that opening a file can also be the mechanism to create a file. We'll investigate this first.

The **fopen** function is very simple and provides the following API:

```
FILE *fopen( const char *filename, const char *mode );
```

We specify the filename that we wish to access (or create) through the first argument (filename) and then the mode we wish to use (mode). The result of the **fopen** operation is a FILE pointer, which could be NULL, indicating that the operation failed.

The key to the **fopen** call is the mode that is provided. Table 10.1 provides an initial list of access modes.

The mode is simply a string that the fopen call uses to determine how to open (or create) the file. If we wanted to create a new file, we could simply use the fopen call as follows:

```
my_fp = fopen( "myfile.txt", "w" );
```

TABLE 10.1 Simple File Access Modes

Mode	Description
r	Open an existing file for read
w	Open a file for write (create new if exists)
a	Open a file for append (create if file doesn't exist)
rw	Open for read and write (create if it doesn't exist)

The result would be the creation of a new file (or the destruction of the existing file) in preparation for **write** operations. If instead we wanted to read from an existing file, we'd open it as follows:

```
my_fp = fopen( "myfile.txt", "r" );
```

Note that we've simply used a different mode here. The `read` mode assumes that the file exists, and if not, a `NULL` is returned.

In both cases, it is assumed that our file `myfile.txt` will either exist or be created in the current working directory. The current directory is the directory from which we invoked our application.

It's very important that the results of all file I/O operations be checked for success. For the **fopen** call, we simply test the response for `NULL`. What happens upon error is ultimately application dependent (you decide). An example of one mechanism is provided in Listing 10.1.

ON THE CD

Listing 10.1 Catching an Error in an `fopen` Call (on the CD-ROM at `./source/ch10/test.c`)

```
#include <stdio.h>
#include <errno.h>
#include <string.h>

#define MYFILE    "missing.txt"

main()
{

  FILE *fin;

  /* Try to open the file for read */
  fin = fopen( MYFILE, "r" );
```

```
14:
15:      /* Check for failure to open */
16:      if (fin == (FILE *)NULL) {
17:
18:        /* Emit an error message and exit */
19:        printf("%s: %s\n", MYFILE, strerror( errno ) );
20:        exit(-1);
21:
22:      }
23:
24:      /* All was well, close the file */
25:      fclose( fin );
26:
27:    }
```

In Listing 10.1, we use a couple of new calls not yet discussed. After trying to open the file at line 13, we check to see if our new file handle is `NULL` (zero). If it is, then we know that either the file is not present or we're not able to access (we don't have proper access to the file). In this case, we emit an error message that consists of the file that we attempted to open for `read` and then the error message that resulted. We capture the error number (integer) with the `errno` variable. This is a special variable that is set by system calls to indicate the last error that occurred. We pass this value to the `strerror` function, which turns the integer error number into a string suitable for printing to standard-out. Executing our sample application results in the following:

```
$ ./app
missing.txt: No such file or directory
$
```

Let's now move on to writing and then reading data from a file.

Reading and Writing Data

A number of methods exist for both reading and writing data to a file. More options can be a blessing, but it's also important to know where to use which mechanism. For example, we could read or write on a character basis or on a string basis (for ASCII text only). We could also use a more general API that permits reading and writing records, which supports both ASCII and binary representations. We'll look at each here, but we'll focus primarily on the latter mechanism.

The standard I/O library presents a buffered interface. This has two very important properties. First, system reads and writes are in blocks (typically 8KB in size). Character I/O is simply written to the `FILE` buffer, where the buffer is written to the media automatically when it's full. Second, **`fflush`** is necessary, or nonbuffered I/O must be set if the data is being sent to an interactive device such as the console terminal.

Character Interfaces

The character interfaces are demonstrated in Listings 10.2 and 10.3. In Listing 10.2, we illustrate character output using **`fputc`** and in Listing 10.3, character input using **`fgetc`**. These functions have the following prototypes:

```
int fputc( int c, FILE *stream );
int fgetc( FILE *stream );
```

In this example, we'll generate an output file using **`fputc`** and then use this file as the input to **`fgetc`**. In Listing 10.2, we open our output file at line 11 and then work our way through our sample string. Our simple loop walks through the entire string until a `NULL` is detected, at which point we exit and close the file (line 21). At line 16, we use **`fputc`** to emit our character (as an `int`, per the **`fputc`** prototype) as well as specifying our output stream (`fout`).

ON THE CD

Listing 10.2 The `fputc` Character Interface Example (on the CD-ROM at `./source/ch10/charout.c`)

```
1:     #include <stdio.h>
2:
3:     int main()
4:     {
5:       int i;
6:       FILE *fout;
7:       const char string[]={"This\r\nis a test\r\nfile.\r\n\0"};
8:
9:       fout = fopen("inpfile.txt", "w");
10:
11:      if (fout == (FILE *)NULL) exit(-1);
12:
13:      i = 0;
14:      while (string[i] != NULL) {
15:
16:        fputc( (int)string[i], fout );
17:        i++;
18:
19:      }
20:
21:      fclose( fout );
22:
23:      return 0;
24:    }
```

The function to read this file using the character interface is shown in Listing 10.3. This function is very similar to our file creation example. We open the file for read at line 8 and follow with a test at line 10. We then enter a loop to get the characters from the file (lines 12–22). The loop simply reads characters from the file using **`fgetc`** and stops when the special EOF symbol is encountered. This is the indication that we've reached the end of the file. For all characters that are not EOF (line 16), we emit the character to standard-out using the `printf` function. Upon reaching the end of the file, we close it using **`fclose`** at line 24.

ON THE CD

Listing 10.3 The `fgetc` Character Interface Example (on the CD-ROM at `./source/ch10/charin.c`)

```
 1:     #include <stdio.h>
 2:
 3:     int main()
 4:     {
 5:       int c;
 6:       FILE *fin;
 7:
 8:       fin = fopen("inpfile.txt", "r");
 9:
10:       if (fin == (FILE *)0) exit(-1);
11:
12:       do {
13:
14:         c = fgetc( fin );
15:
16:         if (c != EOF) {
17:
18:           printf("%c", (char)c);
19:
20:         }
21:
22:       } while (c != EOF);
23:
24:       fclose( fin );
25:
26:       return 0;
27:     }
```

Executing our applications is illustrated as follows:

```
$ ./charout
$ ./charin
This
```

```
is a test
file.
$
```

The character interfaces are obviously simple, but they are also inefficient and should be used only if a string-based method cannot be used. We'll look at this interface next.

String Interfaces

In this section, we'll look at four library functions in particular that provide the means to read and write strings. The first two (**fputs** and **fgets**) are simple string interfaces, and the second two (**fprintf** and **fscanf**) are more complex and provide additional capabilities.

The **fputs** and **fgets** interfaces mirror our previously discussed **fputc** and **fgetc** functions. They provide the means to write and read variable-length strings to files in a very simple way. Prototypes for the `fputs` and `fgets` are defined as:

```
int fputs( int c, FILE *stream );
char *fgets( char *s, int size, FILE *stream );
```

Let's first look at a sample application that accepts strings from the user (via standard-input) and then writes them to a file (see Listing 10.4). We'll halt the input process once a blank line has been received.

Listing 10.4 Writing Variable Length Strings to a File (on the CD-ROM at `./source/ch10/strout.c`)

```
#include <stdio.h>

#define LEN    80

int main()
{
  char line[LEN+1];
  FILE *fout, *fin;

  fout = fopen( "testfile.txt", "w" );
  if ( fout == (FILE *)0 ) exit(-1);

  fin = fdopen( 0, "r" );

  while ( (fgets( line, LEN, fin )) != NULL ) {

    fputs( line, fout );
```

```
18:
19:      }
20:
21:      fclose( fout );
22:      fclose( fin );
23:
24:      return 0;
25:    }
```

The application shown in Listing 10.4 gets a little trickier. Let's walk through this one line by line to cover all of the points. We declare our line string (used to read user input) at line 7, called oddly enough, `line`. Next, we declare two `FILE` pointers, one for input (called `fin`) and one for output (called `fout`).

At line 10, we open our output file using `fopen` to a new file called `testfile.txt`. We check the error status of this line at line 11, exiting if a failure occurred. At line 13, we use a special function **`fdopen`** to associate an existing file descriptor with a stream. In this case, we associate in the standard-input descriptor with a new stream called `fin` (returned by **`fdopen`**). Whatever we now type in (standard-in) will be routed to this file stream. Next, we enter a loop that attempts to read from the `fin` stream (standard-in) and write this out to the output stream (`fout`). At line 15, we read using **`fgets`** and check the return with `NULL`. The `NULL` will appear when we close the descriptor (which is achieved through pressing Ctrl+D at the keyboard). The `line` read is then emitted to the output stream using **`fputs`**. Finally, when the input stream has closed, we exit our loop and close the two streams at lines 21 and 22.

Let's now look at another example of the read side, **`fgets`**. In this example (Listing 10.5), we read the contents of our test file using `fgets` and then `printf` it to standard-out.

ON THE CD

Listing 10.5 Reading Variable Length Strings from a File (on the CD-ROM at `./source/ch10/strin.c`)

```
#include <stdio.h>

#define LEN    80

int main()
{
  char line[LEN+1];
  FILE *fin;

  fin = fopen( "testfile.txt", "r" );
  if ( fin == (FILE *)0 ) exit(-1);

  while ( (fgets( line, LEN, fin )) != NULL ) {
```

```
14:
15:         printf( "%s", line );
16:
17:       }
18:
19:       fclose( fin );
20:
21:       return 0;
22:     }
```

In this example, we open our input file and create a new input stream handle called `fin`. We use this at line 13 to read variable-length strings from the file, and when one is read, we emit it to standard-out via `printf` at line 15.

This demonstrates writing and reading strings to and from a file, but what if our data is more structured than simply strings? If our strings are actually made up of lower-level structures (such as integers, floating-point values, or other types), we can use another method to more easily deal with them. This is the next topic of discussion.

Consider the problem of reading and writing data that takes a regular form but consists of various data types. Let's say that we want to store an integer item (an `id`), two floating-point values (2d coordinates), and a string (an object name). Let's look first at the application that creates this file (see Listing 10.6). Note that in this example we ultimately deal with strings, but using the API functions, the ability to translate to the native data types is provided.

ON THE CD

Listing 10.6 Writing Structured Data in ASCII Format (on the CD-ROM at `./source/ch10/strucout.c`)

```
#include <stdio.h>

#define MAX_LINE    40

#define FILENAME "myfile.txt"

typedef struct {
  int id;
  float x_coord;
  float y_coord;
  char name[MAX_LINE+1];
} MY_TYPE_T;

#define MAX_OBJECTS    3

/* Initialize an array of three objects */
```

```
17:    MY_TYPE_T objects[MAX_OBJECTS]={
18:      { 0, 1.5, 8.4, "First-object" },
19:      { 1, 9.2, 7.4, "Second-object" },
20:      { 2, 4.1, 5.6, "Final-object" }
21:    };
22:
23:    int main()
24:    {
25:      int i;
26:      FILE *fout;
27:
28:      /* Open the output file */
29:      fout = fopen( FILENAME, "w" );
30:      if (fout == (FILE *)0) exit(-1);
31:
32:      /* Emit each of the objects, one per line */
33:      for ( i = 0 ; i < MAX_OBJECTS ; i++ ) {
34:
35:        fprintf( fout, "%d %f %f %s\n",
36:                 objects[i].id,
37:                 objects[i].x_coord, objects[i].y_coord,
38:                 objects[i].name );
39:
40:      }
41:
42:      fclose( fout );
43:
44:      return 0;
45:    }
```

In Listing 10.6, we illustrate another string method for creating data files. We create a test structure (lines 7–12) to represent our data that we're going to write and then read. We initialize this structure at lines 17–21 with three rows of data. Now let's get to the application. This one turns out to be very simple. At lines 29–30, we open and then check the `fout` file handle and then perform a `for` loop to emit our data to the file. We use the **`fprintf`** API function to emit this data. The format of the **`fprintf`** call is to first specify the output file pointer, followed by a format string, and then zero or more variables to be emitted. Our format string mirrors our data structure. We're emitting an `int` (`%d`), two floating-point values (`%f`), and then finally a string (`%s`). This converts all data to string format and writes it to the output file. Finally, we close the output file at line 42 with the **`fclose`** call.

We could have achieved this with a `sprintf` *call (to create our output string) and then written this out as follows:*

```
char line[81];
...
snprintf( line, 80, "%d %f %f %s\n",
              objects[i].id
              objects[i].x_coord, objects[i].y_coord,
              objects[i].name );
fputs( line, fout );
```

The disadvantage is that local space must be declared for the string being emitted. This would not be required with a call to **`fprintf`** *directly.*

The prototypes for both the `fprintf` and `sprintf` are shown here:

```
int fprintf( FILE* stream, const char *format, ... );
int sprintf( char *str, const char *format, ... );
```

From the file created in Listing 10.6, we read this file in Listing 10.7. This function utilizes the **`fscanf`** function to both read and interpret the data. After opening the input file (lines 21–22), we loop and read the data while the end of file has not been found. We detect the end of file marker using the `feof` function at line 25. The **`fscanf`** function utilizes the input stream (`fin`) and the format to be used to interpret the data. This string is identical to that used to write the data out (see Listing 10.6, line 35).

Once a line of data has been read, it's immediately printed to standard-out using the `printf` function at lines 32–35. Finally, the input file is closed using the **`fclose`** call at line 39.

ON THE CD

Listing 10.7 Reading Structured Data in ASCII Format (on the CD-ROM at `./source/ch10/strucin.c`)

```
 1:   #include <stdio.h>
 2:
 3:   #define MAX_LINE    40
 4:
 5:   #define FILENAME "myfile.txt"
 6:
 7:   typedef struct {
 8:     int id;
 9:     float x_coord;
10:     float y_coord;
11:     char name[MAX_LINE+1];
12:   } MY_TYPE_T;
13:
14:   int main()
15:   {
16:     int i;
```

```
    FILE *fin;
    MY_TYPE_T object;

    /* Open the input file */
    fin = fopen( FILENAME, "r" );
    if (fin == (FILE *)0) exit(-1);

    /* Read the records from the file and emit */
    while ( !feof(fin) ) {

      fscanf( fin, "%d %f %f %s\n",
                &object.id,
                &object.x_coord, &object.y_coord,
                object.name );

      printf("%d %f %f %s\n",
                object.id,
                object.x_coord, object.y_coord,
                object.name );

    }

    fclose( fin );

    return 0;
}
```

We could have achieved this functionality with an **sscanf** *call (to parse our input string).*

```
char line[81];
...
fgets( fin, 80, line );
sscanf( line, 80, "%d %f %f %s\n",
            objects[i].id
            objects[i].x_coord, objects[i].y_coord,
            objects[i].name );
```

The disadvantage is that local space must be declared for the parse to be performed on the input string. This would not be required with a call to `fscanf` *directly.*

The `fscanf` and `sscanf` function prototypes are both shown here:

```
int fscanf( FILE *stream, const char *format, ... );
int sscanf( const char *str, const char *format, ... );
```

All of the methods discussed thus far require that we're dealing with ASCII text data. In the next section, we'll look at API functions that permit dealing with binary data.

For survivability, it's important to not leave files open over long durations of time. When I/O is complete, the file should be closed with `fclose` (or at a minimum, flushed with `fflush`). This has the effect of writing any buffered data to the actual file.

Reading and Writing Binary Data

In this section, we'll look at a set of library functions that provide the ability to deal with both binary and ASCII text data. The **`fwrite`** and **`fread`** functions provide the ability to deal not only with the I/O of objects, but also with arrays of objects. The prototypes of the **`fwrite`** and **`fread`** functions are provided here:

```
size_t fread( void *ptr, size_t size, size_t nmemb, FILE *stream );
size_t fwrite( const void *ptr, size_t size,
                       size_t nmemb, FILE *stream );
```

Let's look at a couple of simple examples of `fwrite` and `fread` to explore their use (see Listing 10.8). In this first example, we'll emit the `MY_TYPE_T` structure first encountered in Listing 10.6.

Listing 10.8 Using `fwrite` to Emit Structured Data (on the CD-ROM at `./source/ch10/binout.c`)

```
#include <stdio.h>

#define MAX_LINE    40

#define FILENAME "myfile.bin"

typedef struct {
  int id;
  float x_coord;
  float y_coord;
  char name[MAX_LINE+1];
} MY_TYPE_T;

#define MAX_OBJECTS    3

MY_TYPE_T objects[MAX_OBJECTS]={
  { 0, 1.5, 8.4, "First-object" },
  { 1, 9.2, 7.4, "Second-object" },
  { 2, 4.1, 5.6, "Final-object" }
```

```
};

int main()
{
  int i;
  FILE *fout;

  /* Open the output file */
  fout = fopen( FILENAME, "w" );
  if (fout == (FILE *)0) exit(-1);

  /* Write out the entire object's structure */
  fwrite( (void *)objects, sizeof(MY_TYPE_T), 3, fout );

  fclose( fout );

  return 0;
}
```

What's interesting to note about Listing 10.8 is that a single `fwrite` emits the entire structure. We specify the object that we're emitting (variable object, passed as a void pointer) and then the size of a row in this structure (the type `MY_TYPE_T`) using the `sizeof` operator. We then specify the number of elements in our array of types (3) and finally the output stream to which we want this object to be written.

Let's look at the invocation of this application (called `binout`) and a method for inspecting the contents of the binary file (see Listing 10.9). After executing the `binout` executable, the file `myfile.bin` is generated. Attempting to use the `more` utility to inspect the file results in a blank line. This is because the first character in the file is a `NULL` character, which is interpreted by `more` as the end. Next, we use the `od` utility (octal dump) to emit the file without interpreting it. We specify `-x` as the option to emit the file in hexadecimal format. (For navigation purposes, the integer `id` field has been underlined.)

Listing 10.9 Inspecting the Contents of the Generated Binary File

```
$ ./binout
$ more myfile.bin
$ od -x myfile.bin
0000000 0000 0000 0000 3fc0 6666 4106 6946 7372
0000020 2d74 626f 656a 7463 0000 0000 0000 0000
0000040 0000 0000 0000 0000 0000 0000 0000 0000
0000060 0000 0000 0000 0000 0001 0000 3333 4113
0000100 cccd 40ec 6553 6f63 646e 6f2d 6a62 6365
0000120 0074 0000 0000 0000 0000 0000 0000 0000
```

```
0000140 0000 0000 0000 0000 0000 0000 0000 0000
0000160 0002 0000 3333 4083 3333 40b3 6946 616e
0000200 2d6c 626f 656a 7463 0000 0000 0000 0000
0000220 0000 0000 0000 0000 0000 0000 0000 0000
0000240 0000 0000 0000 0000
0000250
$
```

One important item to note about reading and writing binary data is the issue of portability and endianness. Consider that we create our binary data on a Pentium system, but the binary file is moved to a PowerPC system to read. The data will be in the incorrect byte order and therefore essentially corrupt. The Pentium uses little endian byte order (least significant byte first in memory), whereas the PowerPC uses big endian (most significant byte first in memory). For portability, endianness should always be considered when dealing with binary data. Also consider the use of host and network byte swapping functions, as discussed in Chapter 11, Programming with Pipes."

Now let's look at reading this file using **fread**, but rather than reading it sequentially, let's read it in a nonsequential way (otherwise known as random access). In this example, we'll read the records of the file in reverse order. This requires the use of two new functions that will permit us to seek into a file (**fseek**) and also rewind back to the start (**rewind**):

```
void rewind( FILE *stream );
int fseek( FILE *stream, long offset, int whence );
```

The **rewind** function simply resets the file read pointer back to the start of the file, while the `fseek` function allows us to the new position given an index. The `whence` argument defines whether the position is relative to the start of the file (`SEEK_SET`), the current position (`SEEK_CUR`), or the end of the file (`SEEK_END`). See Table 10.2. The **lseek** function operates like **fseek**, but instead on a file descriptor:

```
int lseek( FILE *stream, long offset, int whence );
```

TABLE 10.2 Function `fseek/lseek` whence Arguments

Name	Description
SEEK_SET	Moves the file position to the position defined by `offset`.
SEEK_CUR	Moves the file position the number of bytes defined by `offset` from the current file position.
SEEK_END	Moves the file position to the number of bytes defined by `offset` from the end of the file.

In this example (Listing 10.10), we open the file using **`fopen`**, which automatically sets the `read` index to the start of the file. Since we want to read the last record first, we seek into the file using **`fseek`** (line 26). The index that we specify is twice the size of the record size (`MY_TYPE_T`). This puts us at the first byte of the third record, which is then read with the **`fread`** function at line 28. Our `read` index is now at the end of the file, so we reset our `read` position to the top of the file using the **`rewind`** function.

We repeat this process, setting the file `read` position to the second element at line 38, and then read again with **`fread`**. The final step is reading the first element in the file. This requires no **`fseek`** because after the rewind (at line 48), we're at the top of the file. We can then **`fread`** the first record at line 50.

Listing 10.10 Using `fread` and `fseek/rewind` to Read Structured Data (on the CD-ROM at `./source/ch10/nonseq.c`)

```
1:     #include <stdio.h>
2:
3:     #define MAX_LINE    40
4:
5:     #define FILENAME "myfile.txt"
6:
7:     typedef struct {
8:       int id;
9:       float x_coord;
10:      float y_coord;
11:      char name[MAX_LINE+1];
12:    } MY_TYPE_T;
13:
14:    MY_TYPE_T object;
15:
16:    int main()
17:    {
18:      int i;
19:      FILE *fin;
20:
21:      /* Open the input file */
22:      fin = fopen( FILENAME, "r" );
23:      if (fin == (FILE *)0) exit(-1);
24:
25:      /* Get the last entry */
26:      fseek( fin, (2 * sizeof(MY_TYPE_T)), SEEK_SET );
27:
28:      fread( &object, sizeof(MY_TYPE_T), 1, fin );
29:
30:      printf("%d %f %f %s\n",
```

```
                object.id,
                object.x_coord, object.y_coord,
                object.name );

    /* Get the second to last entry */
    rewind( fin );

    fseek( fin, (1 * sizeof(MY_TYPE_T)), SEEK_SET );

    fread( &object, sizeof(MY_TYPE_T), 1, fin );

    printf("%d %f %f %s\n",
                object.id,
                object.x_coord, object.y_coord,
                object.name );

    /* Get the first entry */
    rewind( fin );

    fread( &object, sizeof(MY_TYPE_T), 1, fin );

    printf("%d %f %f %s\n",
                object.id,
                object.x_coord, object.y_coord,
                object.name );

    fclose( fin );

    return 0;
}
```

The process of reading the third record is illustrated graphically in Figure 10.1. We illustrate the **fopen**, **fseek**, **fread**, and finally the **rewind**.

The function `ftell` provides the means to identify the current position. This function returns the current position as a long type and can be used to pass as the offset to **fseek** (with SEEK_SET) to reset to that position. The **ftell** prototype is provided here:

```
long ftell( FILE *stream );
```

An alternate API exists to **ftell** and **fseek**. The **fgetpos** and **fsetpos** provide the same functionality, but in a different form. Rather than an absolute position, an

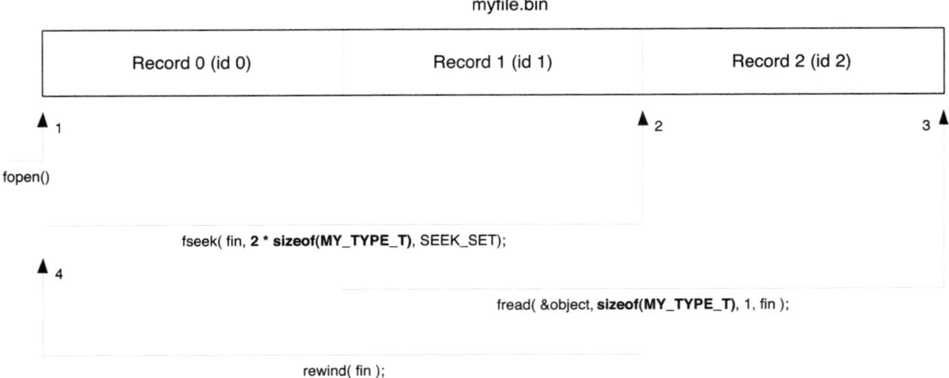

FIGURE 10.1 Nonsequential reads in a binary file.

opaque type is used to represent the position (returned by **fgetpos**, passed into **fsetpos**). The prototypes for these functions are provided here:

```
int fgetpos( FILE *stream, fpos_t *pos );
int fsetpos( FILE *stream, fops_t *pos );
```

An example code snippet of these functions is shown here:

```
fpos_t file_pos;
...
/* Get desired position */
fgetpos( fin, &file_pos );
...
rewind( fin );
/* Return to desired position */
fsetpos( fin, &file_pos );
```

It's recommended to use the **fgetpos** and **fsetpos** APIs over the **ftell** and **fseek** methods. Since the **ftell** and **fseek** methods don't abstract the details of the mechanism, the **fgetpos** and **fsetpos** functions are less likely to be deprecated in the future.

BASE API

The **open**, **read**, and **write** functions can also be used for file I/O. The API differs somewhat, but we'll also look here at how to switch between file and stream mode with **fdopen**.

These functions are referred to as the base API because they are the platform from which the standard I/O library is built.

The **open** function allows us to open or create a new file. Two variations are provided, with their APIs listed here:

```
int open( const char *pathname, int flags );
int open( const char *pathname, int flags, mode_t mode );
```

The `pathname` argument defines the file (with path) to be opened or created (such as `temp.txt` or `/tmp/myfile.txt`). The `flags` argument is one of `O_RDONLY`, `O_WRONLY`, or `O_RDWR`. One or more of the flags shown in Table 10.3 may also be OR'd in, depending on the needs of the **open** call.

TABLE 10.3 Additional Flags for the `open` Function

Flag	Description
O_CREAT	Create the file if it doesn't exist.
O_EXCL	If used with O_CREAT, will return an error if the file already exists, otherwise the file is created.
O_NOCTTY	If the file descriptor refers to a TTY device, this process will not become the controlling terminal.
O_TRUNC	The file will be truncated (if it exists) and the length reset to zero if write privileges are permitted.
O_APPEND	The file pointer is repositioned to the end of the file prior to each write.
O_NONBLOCK	Opens the file in nonblocking mode. Operations on the file will not block (such as read, write, and so on).
O_SYNC	write functions are blocked until the data is written to the physical device.
O_NOFOLLOW	Fail following symbolic links.
O_DIRECTORY	Fail the open if the file being opened is not a directory.
O_DIRECT	Attempts to minimize cache effects by doing I/O directly to/from user space buffers (synchronously as with O_SYNC).
O_ASYNC	Requests a signal when data is available on input or output of this file descriptor.
O_LARGEFILE	Request a large filesystem file to be opened on a 32-bit system whose size cannot be represented in 32 bits.

The third argument for the second open instance is a mode. This mode defines the permissions to be used with the file is created (used only with the flag O_CREAT). Table 10.4 lists the possible symbolic constants that can be OR'd together.

TABLE 10.4 Mode Arguments for the open System Call

Constant	Use
S_IRWXU	User has read/write/execute permissions.
S_IREAD	User has read permission.
S_IWRITE	User has write permission.
S_IEXEC	User has execute permission.
S_IRWXG	Group has read/write/execute permissions.
S_IRGRP	Group has read permission.
S_IWGRP	Group has write permission.
S_IXGRP	Group has execute permission.
S_IRWXO	Others have read/write/execute permissions.
S_IROTH	Others have read permission.
S_IWOTH	Others have write permission.
S_IXOTH	Others have execute permission.

To open a new file in the tmp directory, we could do this simply as:

```
int fd;
fd = open( "/tmp/newfile.txt", O_CREAT | O_WRONLY );
```

To instead open an existing file for read, we could open as follows:

```
int fd;
fd = open( "/tmp/newfile.txt", O_RDONLY );
```

Reading and writing to these files is done very simply with the **read** and **write** API functions.

```
ssize_t read( int fd, void *buf, size_t count );
ssize_t write( int fd, const void *buf, size_t count );
```

These are used simply with a buffer and a size to represent the number of bytes to read or write, such as:

```
unsigned char buffer[MAX_BUF+1];
int fd, ret;
...
ret = read( fd, (void *)buffer, MAX_BUF );
...
ret = write( fd, (void *)buffer, MAX_BUF );
```

We'll see more examples of these in Chapter 11. What's interesting here is that the same set of API functions to read and write data to a file can also be used for pipes and sockets. This represents a unique aspect of the UNIX-like operating systems, where many types of devices can be represented as files.

Finally, a file descriptor can be attached to a stream by using the **fdopen** system call. This call has the prototype:

```
FILE *fdopen( int filedes, const char *mode );
```

Therefore, if we've opened a device using the **open** function call, we can associate a stream with it using **fdopen** and then use stream system calls on the device (such as **fscanf** or **fprintf**). Consider the following example:

```
FILE *fp;
int fd;
fd = open( "/tmp/myfile.txt", O_RDWR );
fp = fdopen( fd, "rw" );
```

Once this is done, we can use `read/write` with the `fd` descriptor or `fscanf/fprintf` with the `fp` descriptor.

One other useful API to consider is the **pread/pwrite** API. These functions require an offset into the file to read or write, but they do not affect the file pointer. These functions have the prototype:

```
ssize_t pread( int filedes, void *buf, size_t nbyte, off_t offset );
ssize_t pwrite( int filedes, void *buf, size_t nbyte, off_t offset );
```

These functions require that the target be seekable (in other words, regular files) and are used regularly for record I/O in databases.

SUMMARY

In this chapter, the file handling APIs were discussed with examples provided for each. The character interfaces were first explored (**fputc**, **fgetc**), followed by the string interfaces (**fputs**, **fgets**). Some of the more structured methods for generating and parsing files were then investigated (such as the fprintf and fscanf functions), in addition to some of the other possibilities (sprintf and sscanf). Finally, the topics of binary files and random (nonsequential) access were discussed, including methods for saving and restoring file positions.

FILE HANDLING APIS

```
FILE *fopen( const char *filename, const char *mode );
FILE *fdopen( int filedes, const char *type );
int fputc( int c, FILE *stream );
int fgetc( FILE *stream );
int fputs( int c, FILE *stream );
char *fgets( char *s, int size, FILE *stream );
int fprintf( FILE* stream, const char *format, ... );
int sprintf( char *str, const char *format, ... );
int fscanf( FILE *stream, const char *format, ... );
int sscanf( const char *str, const char *format, ... );
void rewind( FILE *stream );
int fseek( FILE *stream, long offset, int whence );
int lseek( in filedes, long offset, int whence );
long ftell( FILE *stream );
int fgetpos( FILE *stream, fpos_t *pos );
int fsetpos( FILE *stream, fops_t *pos );
int fclose( FILE *stream );
int open( const char *pathname, int flags );
int open( const char *pathname, int flags, mode_t mode );
ssize_t read( int fd, void *buf, size_t count );
ssize_t write( int fd, const void *buf, size_t count );
ssize_t pread( int filedes, void *buf, size_t count, off_t offset );
ssize_t pwrite( int filedes, const void *buf,
size_t count, off_t offset );
```

11 Programming with Pipes

In This Chapter

- Review of the Pipe Model of IPC
- Differences Between (Anonymous) Pipes and Named Pipes
- Creating Anonymous and Named Pipes
- Communicating Through Pipes
- Command-line Creation and Use of Pipes

INTRODUCTION

In this chapter, we'll explore the GNU/Linux pipes. The pipe model is an older but still useful mechanism for inter-process communication. We'll look at what are known as half-duplex pipes and also named pipes. Each offers a FIFO queuing model to permit communication between processes.

THE PIPE MODEL

One way to visualize a pipe is a one-way connector between two entities. For example, consider the following GNU/Linux command:

```
ls -1 | wc -l
```

This command creates two processes, one for the `ls -1` and another for `wc -l`. It then connects the two together by setting the standard-input of the second process to the standard-output of the first process (see Figure 11.1). This has the effect of counting the number of files in the current subdirectory.

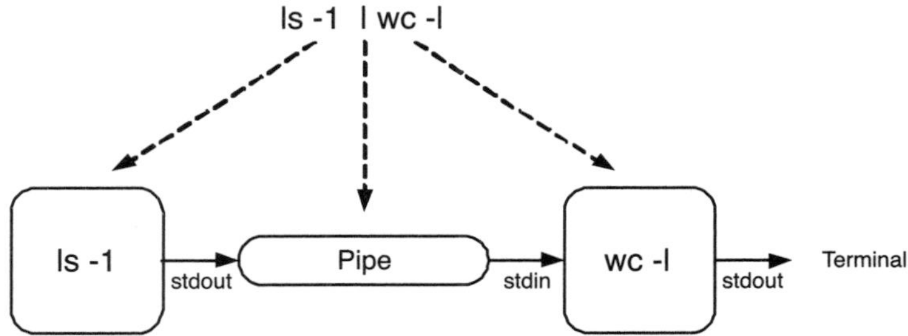

FIGURE 11.1 Simple pipe example.

Our command, as illustrated in Figure 11.1, sets up a pipeline between two GNU/Linux commands. The `ls` command is performed, which generates output that is used as the input to the second command, `wc` (word count). This is a half-duplex pipe as communication occurs in one direction. The linkage between the two commands is facilitated by the GNU/Linux kernel, which takes care of connecting the two together. We can achieve this as well in applications, which we'll demonstrate shortly.

Pipes and Named Pipes

A pipe, or half-duplex pipe, provides the means for a process to communicate with one of its ancestral subprocesses (of the anonymous variety). This is because there's no way in the operating system to locate the pipe (it's anonymous). It's most common use is to create a pipe at a parent process and then pass the pipe to the child so that they can communicate. Note that if full-duplex communication was required, the Sockets API should be considered instead.

Another type of pipe is called a *named pipe*. A named pipe works like a regular pipe but exists in the filesystem so that any process can find it. This means that processes not of the same ancestry are able to communicate with one another.

We'll look at both pipes and named pipes in the following sections. We'll first take a quick tour of pipes and then follow up with a more detailed look at the pipe API and GNU/Linux system-level commands that support pipes programming.

Whirlwind Tour

Let's begin with a simple example of the pipe programming model. In this simple example, we'll create a pipe within a process, write a message to it, read the message back from the pipe, and then emit it (see Listing 11.1).

Listing 11.1 Simple Pipe Example (on the CD-ROM at `./source/ch11/pipe1.c`)

```
 1:    #include <unistd.h>
 2:    #include <stdio.h>
 3:    #include <string.h>
 4:
 5:    #define MAX_LINE        80
 6:    #define PIPE_STDIN      0
 7:    #define PIPE_STDOUT     1
 8:
 9:    int main()
10:    {
11:      const char *string={"A sample message."};
12:      int ret, myPipe[2];
13:      char buffer[MAX_LINE+1];
14:
15:      /* Create the pipe */
16:      ret = pipe( myPipe );
17:
18:      if (ret == 0) {
19:
20:        /* Write the message into the pipe */
21:        write( myPipe[PIPE_STDOUT], string, strlen(string) );
22:
23:        /* Read the message from the pipe */
24:        ret = read( myPipe[PIPE_STDIN], buffer, MAX_LINE );
25:
26:        /* Null terminate the string */
27:        buffer[ ret ] = 0;
28:
29:        printf("%s\n", buffer);
30:
31:      }
32:
33:      return 0;
34:    }
```

In Listing 11.1, we create our pipe using the `pipe` call at line 16. We pass in a two-element `int` array that represents our pipe. The pipe is defined as a pair of separate file descriptors, an input and an output. We can write to one end of the pipe and read from the other. The **pipe** API function returns zero on success. Upon return, the `myPipe` array will contain two new file descriptors representing the input to the pipe (`myPipe[1]`) and the output from the pipe (`myPipe[0]`).

At line 21, we write our message to the pipe using the **write** function. We specify the `stdout` descriptor (from the perspective of the application, not the pipe). The pipe now contains our message and can be read at line 24 using the **read** function. Here again, from the perspective of the application, we use the `stdin` descriptor to read from the pipe. The **read** function stores what is read from the pipe in the `buffer` variable (argument three of the **read** function). We terminate it (add a `NULL` to the end) so that we can properly emit it at line 29 using `printf`. The pipe in this example is illustrated in Figure 11.2.

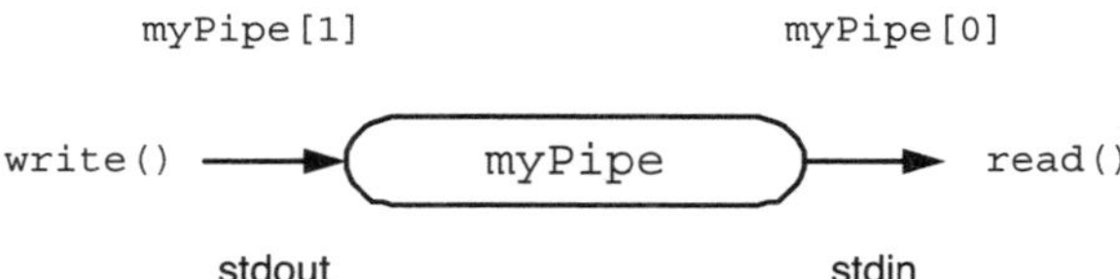

FIGURE 11.2 Half-duplex pipe example from Listing 11.1.

While this example was entertaining, communicating with ourselves could be performed using any number of mechanisms. In the detailed review, we'll look at more complicated examples that provide communication between processes (both related and unrelated).

DETAILED REVIEW

While the `pipe` function is the majority of the pipe model, there are a couple of other functions that we should discuss in their applicability toward pipe-based programming. Table 11.1 lists the functions that we'll detail in this chapter.

TABLE 11.1 API Functions for Pipe Programming

API Function	Use
`pipe`	Create a new anonymous pipe
`dup`	Create a copy of a file descriptor
`mkfifo`	Create a named pipe (fifo)

We'll also look at some of the other functions that are applicable to pipe communication, specifically those that can be used to communicate using a pipe.

Remember that a pipe is nothing more than a pair of file descriptors, and therefore any functions that operate on file descriptors can be used. This includes but is not restricted to select, read, write, fcntl, freopen, and such.

pipe

The `pipe` API function creates a new pipe, represented by an array of two file descriptors. The `pipe` function has the following prototype:

```
#include <unistd.h>
int pipe( int fds[2] );
```

The **`pipe`** function returns zero on success, or -1 on failure, with `errno` set appropriately. On successful return, the `fds` array (which was passed by reference) is filled with two active file descriptors. The first element in the array is a file descriptor that can be read by the application, and the second element is a file descriptor that can be written to.

Let's now look at a slightly more complicated example of pipe in a multiprocess application. In this application (see Listing 11.2), we'll create a pipe (line 14) and then fork our process into a parent and a child process (line 16). At the child, we attempt to read from the input file descriptor of our pipe (line 18), which suspends the process until something is available to read. When something is read, we terminate the string with a `NULL` and print out what was read. The parent simply writes a test string through the pipe using the write file descriptor (array offset 1 of the `pipe` structure) and then waits for the child to exit using the `wait` function.

Note that there isn't anything spectacular about this application except for the fact that our child process inherited the file descriptors that were created by the parent (using the **`pipe`** function) and then used them to communicate with one another. Recall that once the `fork` function is complete, our processes are independent (except that the child inherited features of the parent, such as the `pipe` file descriptors). Memory is separate, so the `pipe` method provides us with an interesting model to communication between processes.

Listing 11.2 Illustrating the Pipe Model with Two Processes (on the CD-ROM at `./source/ch11/fpipe.c`)

```
1:    #include <stdio.h>
2:    #include <unistd.h>
3:    #include <string.h>
4:    #include <wait.h>
```

```

#define MAX_LINE        80

int main()
{
  int thePipe[2], ret;
  char buf[MAX_LINE+1];
  const char *testbuf={"a test string."};

  if ( pipe( thePipe ) == 0 ) {

    if (fork() == 0) {

      ret = read( thePipe[0], buf, MAX_LINE );
      buf[ret] = 0;
      printf( "Child read %s\n", buf );

    } else {

      ret = write( thePipe[1], testbuf, strlen(testbuf) );
      ret = wait( NULL );

    }

  }

  return 0;
}
```

Note that in these simple programs, we've not discussed closing the pipe, because once the process finishes, the resources associated with the pipe will be automatically freed. It's good programming practice, nonetheless, to close the descriptors of the pipe using the `close` call, such as:

```
ret = pipe( myPipe );
...
close( myPipe[0] );
close( myPipe[1] );
```

If the write end of the pipe is closed and a process tries to read from the pipe, a zero is returned. This indicates that the pipe is no longer used and should be closed. If the read end of the pipe is closed and a process tries to write to it, a signal is generated. This signal (as discussed in Chapter 12, "Introduction to Sockets Programming") is called `SIGPIPE`. Applications that write to pipes commonly include a signal handler to catch just this situation.

dup and dup2

The **dup** and **dup2** calls are very useful functions that provide the ability to duplicate a file descriptor. They're most often used to redirect the stdin, stdout, or stderr of a process. The function prototypes for dup and dup2 are:

```
#include <unistd.h>
int dup( int oldfd );
int dup2( int oldfd, int targetfd );
```

The **dup** function allows us to duplicate a descriptor. We pass in an existing descriptor, and it returns a new descriptor that is identical to the first. This means that both descriptors share the same internal structure. For example, if we perform an **lseek** (seek into the file) for one file descriptor, the file position is the same in the second. Sample use of the **dup** function is illustrated in the following code snippet:

```
int fd1, fd2;
...
fd2 = dup( fd1 );
```

Creating a descriptor prior to the fork *call has the same effect as calling* **dup**. *The child process receives a duplicated descriptor, just like it would after calling* dup.

The **dup2** function is similar to dup but allows the caller to specify an active descriptor and the id of a target descriptor. Upon successful return of dup2, the new target descriptor is a duplicate of the first (targetfd = oldfd). Let's now look at a short code snippet that illustrates dup2:

```
int oldfd;
oldfd = open("app_log", (O_RDWR | O_CREATE), 0644 );
dup2( oldfd, 1 );
close( oldfd );
```

In this example, we open a new file called "app_log" and receive a file descriptor called fd1. We call dup2 with oldfd and 1, which has the effect of replacing the file descriptor identified as 1 (stdout) with oldfd (our newly opened file). Anything written to stdout now will go instead to the file named "app_log". Note that we close oldfd directly after duplicating it. This doesn't close our newly opened file, as file descriptor 1 now references it.

Let's now look at a more complex example. Recall that earlier in the chapter we investigated pipelining the output of ls -1 to the input of wc -l. We'll now explore this example in the context of a C application (see Listing 11.3).

We begin in Listing 11.3 by creating our pipe (line 9) and then forking the application into the child (lines 13–16) and parent (lines 20–23). In the child, we begin by closing the `stdout` descriptor (line 13). The child here will provide the `ls -1` functionality and will not write to `stdout` but instead to the input to our pipe (redirected using `dup`). At line 14, we use `dup2` to redirect the `stdout` to our pipe (`pfds[1]`). Once this is done, we close our input end of the pipe (as it will never be used). Finally, we use the `execlp` function to replace the child's image with that of the command `ls -1`. Once this command executes, any output that is generated is sent to the input.

Now let's look at the receiving end of the pipe. The parent plays this role and follows a very similar pattern. We first close our `stdin` descriptor at line 20 (since we'll accept nothing from it). Next, we use the `dup2` function again (line 21) to make the `stdin` the output end of the pipe. This is done by making file descriptor 0 (normal `stdin`) the same as `pfds[0]`. We close the `stdout` end of the pipe (`pfds[1]`) since we won't use it here (line 22). Finally, we `execlp` the command `wc -l`, which takes as its input the contents of the pipe (line 23).

Listing 11.3 Pipelining Commands in C (on the CD-ROM at `./source/ch11/dup.c`)

```
1:      #include <stdio.h>
2:      #include <stdlib.h>
3:      #include <unistd.h>
4:
5:      int main()
6:      {
7:        int pfds[2];
8:
9:        if ( pipe(pfds) == 0 ) {
10:
11:         if ( fork() == 0 ) {
12:
13:           close(1);
14:           dup2( pfds[1], 1 );
15:           close( pfds[0] );
16:           execlp( "ls", "ls", "-1", NULL );
17:
18:         } else {
19:
20:           close(0);
21:           dup2( pfds[0], 0 );
22:           close( pfds[1] );
23:           execlp( "wc", "wc", "-l", NULL );
24:
```

```
        }

    }

    return 0;
}
```

What's important to note in this application is that our child process redirects its output to the input of the pipe, and the parent redirects its input to the output of the pipe—a very useful technique that is worth remembering.

mkfifo

The **mkfifo** function is used to create a file in the filesystem that provides FIFO functionality (otherwise known as a *named pipe*). Pipes that we've discussed thus far are anonymous pipes. They're used exclusively between a process and its children. Named pipes are visible in the filesystem and therefore can be used by any (related or unrelated) process. The function prototype for **mkfifo** is defined as:

```
#include <sys/types.h>
#include <sys/stat.h>
int mkfifo( const char *pathname, mode_t mode );
```

The **mkfifo** command requires two arguments. The first (`pathname`) is the special file in the filesystem that is to be created. The second (`mode`) represents the read/write permissions for the FIFO. The **mkfifo** command returns zero on success or -1 on error (with `errno` filled appropriately). Let's look at an example of creating a `fifo` using the **mkfifo** function.

```
int ret;
...
ret = mkfifo( "/tmp/cmd_pipe", S_IFIFO | 0666 );
if (ret == 0) {
  // Named pipe successfully created
} else {
  // Failed to create named pipe
}
```

In this example, we create a `fifo` (named pipe) using the file `cmd_pipe` in the `/tmp` subdirectory. We can then open this file for read or write to communicate through it. Once we open a named pipe, we can read from it using the typical I/O commands. For example, here's a snippet reading from the pipe using **fgets**:

```
pfp = fopen( "/tmp/cmd_pipe", "r" );
...
ret = fgets( buffer, MAX_LINE, pfp );
```

We could write to the pipe for this snippet using:

```
pfp = fopen( "/tmp/cmd_pipe", "w+ );
...
ret = fprintf( pfp, "Here's a test string!\n" );
```

What's interesting about named pipes, which we'll explore in the discussion of the **mkfifo** system command, is that they work in what is known as a rendezvous model. A reader will be unable to open the named pipe unless a writer has actively opened the other end of the pipe. The reader is blocked on the `open` call until a writer is present. Despite this limitation, the named pipe can be a useful mechanism for interprocess communication.

System Commands

Let's now look at a system command that is related to the pipe model for IPC. The `mkfifo` command, just like the `mkfifo` API function, allows us to create a named pipe from the command line.

mkfifo

The `mkfifo` command is one of two methods for creating a named pipe (fifo special file) at the command line. The general use of the `mkfifo` command is:

```
mkfifo [options] name
```

where options are `-m` for mode (permissions) and `name` is the name of the named pipe to create (including path if needed). If permissions are not specified, the default is 0644. Here's a sample use, creating a named pipe in `/tmp` called `cmd_pipe`:

```
$ mkfifo /tmp/cmd_pipe
```

We can adjust the options simply by specifying them with the `-m` option. Here's an example setting the permissions to 0644 (but we delete the original first):

```
$ rm cmd_pipe
$ mkfifo -m 0644 /tmp/cmd_pipe
```

Once the permissions are created, we can communicate through this pipe via the command line. Consider the following scenario. In one terminal, we attempt to read from the pipe using the `cat` command:

```
$ cat cmd_pipe
```

Upon typing this command, we're suspended awaiting a writer opening the pipe. In another terminal, we write to the named pipe using the `echo` command, as:

```
$ echo Hi > cmd_pipe
```

When this command finishes, our reader wakes up and finishes (here's the complete reader command sequence again for clarity):

```
$ cat cmd_pipe
Hi
$
```

This illustrates that named pipes can be useful not only in C applications, but also in scripts (or combinations).

Named pipes can also be created with the `mknod` command (along with many other types of special files). We can create a named pipe (as `mkfifo` before) as

```
$ mknod cmd_pipe p
```

where the named pipe `cmd_pipe` is created in the current subdirectory (with type as `p` for named pipe).

SUMMARY

In this chapter, we took a very quick review of anonymous and named pipes. We reviewed application and command-line methods for creating pipes and also reviewed typical I/O mechanisms for communicating through them. We also reviewed the ability to redirect I/O using the **dup** and **dup2** commands. While useful for pipes, these commands are useful in many other scenarios as well (wherever a file descriptor is used, such as a socket or file).

PIPE PROGRAMMING APIS

```
#include <unistd.h>
int pipe( int filedes[2] );
int dup( int oldfd );
int dup2( int oldfd, int targetfd );
int mkfifo( const char *pathname, mode_t mode );
```

12 Introduction to Sockets Programming

In This Chapter

- Understand the Sockets Programming Paradigm
- Learn the BSD4.4 Sockets API
- See Sample Source for a TCP/IP Server and Client
- Explore the Various Capabilities of Sockets (Control, I/O, Notification)
- Investigate Socket Patterns that Illustrate Sockets API Use
- Examine Sockets Programming in Other Languages

INTRODUCTION

In this chapter, we'll take a quick tour of Sockets programming. We'll discuss the Sockets programming paradigm, elements of Sockets applications, and the Sockets API. The Sockets API allows us to develop applications that communicate over a network. The network can be a local private network or the public Internet. An important item to note about Sockets programming is that it's neither operating system specific nor language specific. Sockets applications can be written in the Ruby scripting language on a GNU/Linux host or in C on an embedded controller. This freedom and flexibility are the reasons that the BSD4.4 Sockets API is so popular.

LAYERED MODEL OF NETWORKING

Sockets programming uses the layered model of packet communication (see Figure 12.1). At the top is the Application layer, which is where applications exist (those that utilize Sockets for communication). Below the Application layer we define the Sockets layer. This isn't actually a layer, but it is shown here simply to illustrate where the API is located. The Sockets layer sits on top of the Transport layer. The Transport layer provides the transport protocols. Next is the Network layer, which provides among other things routing over the Internet. This layer is occupied by the Internet Protocol, or IP. Finally, the Physical layer driver is found, which provides the means to introduce packets onto the physical network.

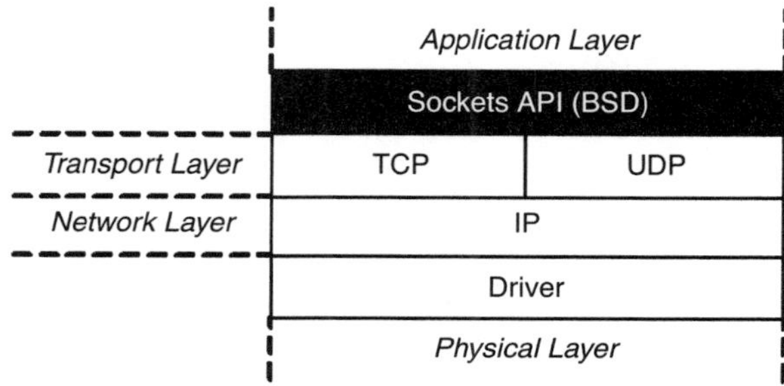

FIGURE 12.1 Layered model of communication.

SOCKETS PROGRAMMING PARADIGM

The Sockets paradigm involves a number of different elements that must be understood to use it properly. Let's look at the Sockets paradigm in a hierarchical fashion.

At the top of the hierarchy is the host. This is a source or destination node on a network to or from which packets are sent or received. (Technically, we would refer to interfaces as the source or destination, as a host may provide multiple interfaces, but we're going to keep it simple here.) The host implements a set of protocols. These protocols define the manner in which communication occurs. Within each protocol is a set of ports. Each port defines an endpoint (the final source of destination). See Table 12.1 for a list of these elements (and Figure 12.2 for a graphical view of these relationships).

TABLE 12.1 Sockets Programming Element Hierarchy

Element	Description
Host (Interface)	Network address (a reachable network node)
Protocol	Specific protocol (such as TCP or UDP)
Port	Client or server process endpoint

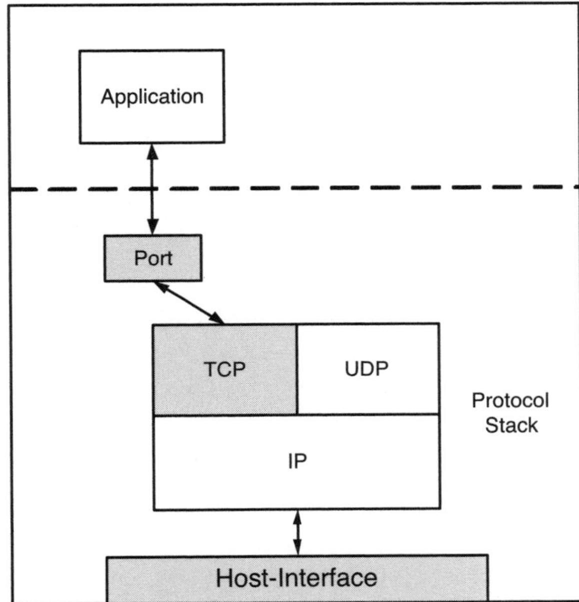

FIGURE 12.2 Graphical view of host/protocol/port relationship.

Hosts

Hosts are identified by addresses, and for IP (Internet Protocols), these are called IP addresses. An IPv4 address (of the version 4 class) is defined as a 32-bit address. This address is represented by four 8-bit values. A sample address can be illustrated as:

```
192.168.1.1    or    0xC0A80101
```

The first value shows the more popular form of IPv4 addresses, which is easily readable. The second notation is simply the first address in hexadecimal format (32 bits wide).

Protocol

The protocol specifies the details of communication over the socket. The two most common protocols used are the Transmission Control Protocol (TCP) and the User Datagram Protocol (UDP). TCP is a stream-based reliable protocol, and UDP is a datagram (message)-based protocol that can be unreliable. We'll provide additional details of these protocols in this chapter.

Port

The port is the endpoint for a given process (interface) for a protocol. This is the application's interface to the Socket interface. Ports are unique on a host (not interface) for a given protocol. Ports are commonly called "bound" when they are attached to a given socket.

Ports are numbers that are split basically into two ranges. Port numbers below 1024 are reserved for well-known services (called well-known addresses). Port numbers above 1024 are typically used by applications.

The original intent of service port numbers (such as FTP, HTTP, and DNS) was that they fall below port number 1024. Of course, the number of services exceeded that number long ago. Now, many system services occupy the port number space greater than 1024 (for example NFS at port number 2049 and X-11 at port number 6000).

Addressing

From this discussion, we see that a *tuple* uniquely identifies an endpoint from all other endpoints on a network. Consider the following tuple:

```
{ tcp, 192.168.1.1, 4097 }
```

This defines the endpoint on the host identified by the address `192.168.1.1` with the port `4097` using the TCP protocol.

The Socket

Simply put, a Socket can be defined as an endpoint of a communications channel between two applications. An example of this is defined as two tuples:

```
{ tcp, 192.168.1.1, 4097 }
{ tcp, 10.0.0.1, 5820 }
```

The first item to note is that a socket is defined as an association of two endpoints that share the same protocol. The IP addresses are different here, but they don't have to be. We could communicate via sockets in the same host. The port numbers are different here, but they could be the same unless they exist on the same host. Port numbers assigned by the TCP/IP stack are called *ephemeral ports*. This relationship is shown visually in Figure 12.3.

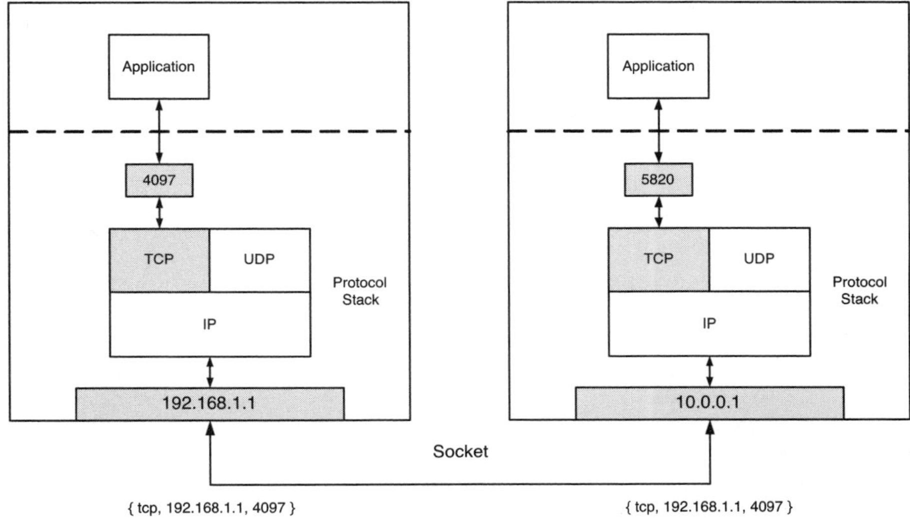

FIGURE 12.3 Visualization of a Socket between two hosts.

Client/Server Model

In most Sockets applications, there exists a server (responds to requests and provides responses) and a client (makes requests to the server). The Sockets API (which we'll explore in the next section) provides commands that are specific to clients and to servers. Figure 12.4 illustrates two simple applications that implement a client and a server.

The first step in a Sockets application is the creation of a socket. The socket is the communication endpoint that is created by the **`socket`** call. Note that in the sample flow (in Figure 12.4) both the server and client perform this step.

The server requires a bit more setup as part of registering a service to the host. The **`bind`** call binds an address and port to the server so that it's known. Letting the system choose the port can result in a service that can be difficult to find. If we choose the port, we know what it is. Once we've bound our port, we call the **`listen`** function for the server. This makes the server accessible (puts it in the `listen` mode).

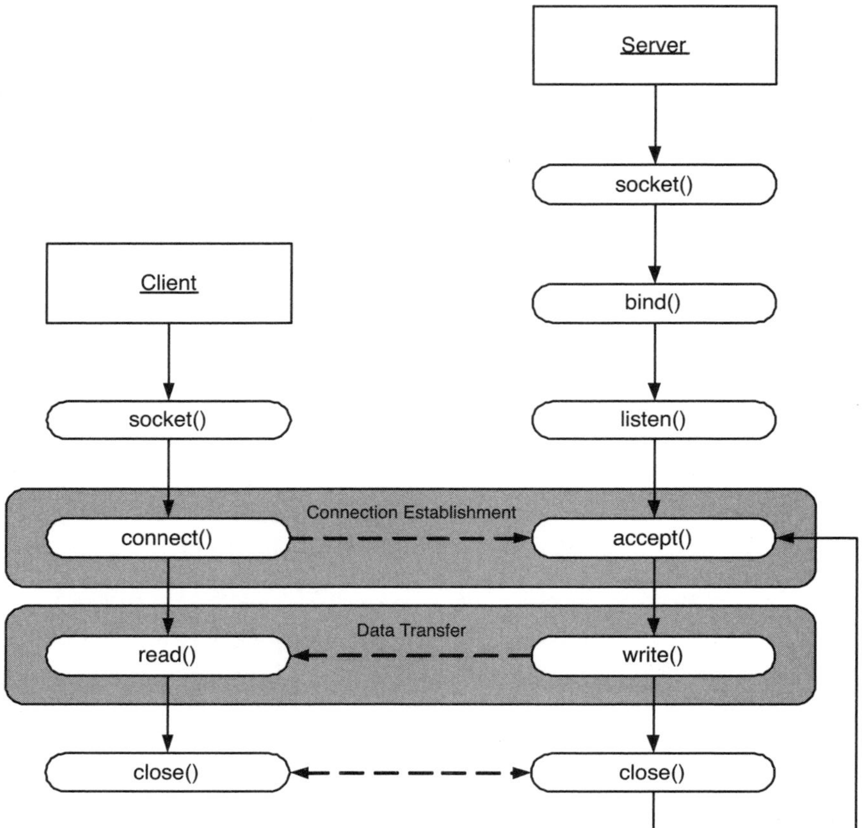

FIGURE 12.4 Client/server symmetry in Sockets applications.

We establish our socket next, using `connect` at the client and `accept` at the server. The `connect` call starts what's known as the three-way handshake, with the purpose of setting up a connection between the client and server. At the server, the `accept` call creates a new server-side client socket. Once `accept` finishes, a new socket connection exists between the client and server, and data flow can occur.

In the data transfer phase, we have an established socket for which communication can occur. Both the client and server can `send` and `recv` data asynchronously.

Finally, we can sever the connection between the client and server using the `close` call. This can occur asynchronously, but upon one endpoint closing the socket, the other side will automatically receive an indication of the closure.

SAMPLE APPLICATION

Now that we have a basic understanding of Sockets, let's look at a sample application that illustrates some of the functions available in the Sockets API. We'll look at the Sockets API from the perspective of two applications, a client and server that implement the Daytime protocol. This protocol server is ASCII based and simply emits the current date and time when requested by a client. The client connects to the server and emits what is read. This implements the basic flow shown previously in Figure 12.2.

Daytime Server

Let's now look at a C language server that implements the Daytime protocol. Recall that the Daytime server will simply emit the current date and time in ASCII string format through the socket to the client. Upon emitting the data, the socket is closed, and the server awaits a new client connection. Now that we understand the concept behind Daytime protocol server, let's look at the actual implementation (see Listing 12.1).

ON THE CD

Listing 12.1 Daytime Server Written in the C Language (on the CD-ROM at `./source/ch12/dayserv.c`)

```
#include <sys/socket.h>
#include <arpa/inet.h>
#include <stdio.h>
#include <time.h>
#include <string.h>
#include <unistd.h>

#define MAX_BUFFER        128
#define DAYTIME_SERVER_PORT    13

int main ( void )
{
 int serverFd, connectionFd;
 struct sockaddr_in servaddr;
 char timebuffer[MAX_BUFFER+1];
 time_t currentTime;

 serverFd = socket( AF_INET, SOCK_STREAM, 0 );

 memset( &servaddr, 0, sizeof(servaddr) );
 servaddr.sin_family = AF_INET;
```

```
22:      servaddr.sin_addr.s_addr = htonl(INADDR_ANY);
23:      servaddr.sin_port = htons(DAYTIME_SERVER_PORT);
24:
25:      bind( serverFd,
26:         (struct sockaddr *)&servaddr, sizeof(servaddr) );
27:
28:      listen( serverFd, 5 );
29:
30:      while ( 1 ) {
31:
32:       connectionFd = accept( serverFd,
33:                  (struct sockaddr *)NULL, NULL );
34:
35:       if (connectionFd >= 0) {
36:
37:        currentTime = time(NULL);
38:        snprintf( timebuffer, MAX_BUFFER,
39:            "%s\n", ctime(&currentTime) );
40:
41:        write( connectionFd, timebuffer, strlen(timebuffer) );
42:        close( connectionFd );
43:
44:       }
45:
46:      }
47:
48:    }
```

Lines 1–6 include the header files for necessary types, symbolic and function APIs. This includes not only the socket interfaces, but also `time.h`, which provides an interface to retrieve the current time. We specify the maximum size of the buffer that we'll operate upon using the symbolic constant `MAX_BUFFER` at line 8. The next symbolic constant at line 9, `DAYTIME_SERVER_PORT`, defines the port number to which we'll attach this socket server. This will allow us to define the well-known port for the Daytime protocol (`13`).

We declare our `main` function at line 11, and then a series of variables are created in lines 13–16. We create two Socket identifiers (line 13), a socket address structure (line 14), a buffer to hold our string time (line 15), and the GNU/Linux time representation structure (line 16).

The first step in any sockets program is to create our socket using the **`socket`** function (line 18). We specify that we're creating an IP socket (using the `AF_INET` domain) using reliable stream protocol type (`SOCK_STREAM`). The zero as the third argument specifies to use the default protocol of the stream type, which is TCP.

Now that we have our socket, we bind an address and a port to it (lines 20–26). At line 20, we initialize the address structure by setting all elements to zero. We specify our socket domain again with `AF_INET` (it's an IPv4 socket). The `s_addr` element specifies an address, which in this case is the address from which we'll accept incoming socket connections. The special symbol `INADDR_ANY` says that we'll accept incoming connections from any available interface on the host. We then define the port to use, our prior symbolic constant `DAYTIME_SERVER_PORT`. The `htonl` (host-to-network-long) and `htons` (host-to-network-short) take care of ensuring that the values provided are in the proper byte order for network packets. The final step is using the `bind` function to bind the address structure previously created with our socket. The socket is now bound with the address, which identifies it in the network stack namespace.

The Internet operates in big endian, otherwise known as network byte order. Hosts operate in host byte order, which, depending upon architecture, can be either big or little endian. For example, the PowerPC architecture is big endian, and the Intel x86 architecture is little endian. This is a small performance advantage to big endian architectures because they need not perform any byte-swapping to change from host byte order to network byte order (they're already the same).

Before a client can connect to our socket, we must call the **`listen`** function (line 28). This tells the protocol stack that we're ready to accept connections (a maximum of five pending connections, per the argument to **`listen`**).

We enter an infinite loop at lines 30–46 to accept client connections and provide them the current time data. At lines 32–33, we call the **`accept`** function with our socket (`serverFd`) to accept a new client connection. When a client connects to us, the network stack creates a new socket representing our end of the connection and returns this socket from the **`accept`** function. With this new client socket (`connectionFd`), we can communicate with the peer client.

At line 35, we check the return socket to see if it's valid (otherwise, an error has occurred and we ignore this client socket). If valid, we grab the current time at lines 37–39. We use the GNU/Linux **`time`** function to get the current time (the number of seconds that have elapsed from January 1, 1970). Passing this value to function `ctime` converts it into a human-readable format, which is used by `sprintf` to construct a response string. We send this to the peer client using the `connectionFd` socket using the `write` function. We pass our socket descriptor, the string to write (`timebuffer`), and its length. Finally, we close our client socket using the **`close`** function, which ends communication with that particular peer.

The loop then continues back to line 32, awaiting a new client connection with the `accept` function. When a new client connects, the process starts all over again.

From GNU/Linux, we could compile this application using GCC and execute it as follows (`filename server.c`):

```
[root@mtjones]$ gcc -o server server.c -Wall
[root@mtjones]$ ./server
```

When executing socket applications that bind to well-known ports (those under 1024), we must start from root. Otherwise, the application will fail with the inability to bind to the reserved port number.

We could now test this server very simply using the Telnet application available in GNU/Linux. As shown, we Telnet to the local host (identified by `localhost`) and port 13 (the port we registered previously in the server). The Telnet client connects to our server and then prints out what was sent to it (the time is shown in bold).

```
$ telnet localhost 13
Trying 127.0.0.1...
Connected to localhost.
Escape character is '^]'.
Sat Jan 17 13:33:57 2004
Connection closed by foreign host.
[root@mtjones]$
```

The final item to note is that once the time was received, we see the message reported to us from Telnet: "Connection closed by foreign host." Recall from the server source in Listing 12.1 that once the `write` has completed (sending the time to the client), the socket **`close`** is immediately performed. The Telnet application is reporting this event to us so that we know the socket is no longer active. We can reproduce Telnet's operation with a socket client, which we'll investigate next.

Daytime Client

The Daytime protocol client is shown in Listing 12.2. We'll avoid discussion of the source preliminaries and go directly to the sockets aspect of the application. As in the server, the first step in building a sockets application is the creation of a socket (of the TCP variety using `SOCK_STREAM`) using the **`socket`** function (at line 16).

Recall in the server application that we build an address structure (`servaddr`) that is then bound to the socket representing the service. In the client, we also build an address structure, but in this case it's to define to whom we're connecting (Listing 12.2, lines 18–21). Note the similarities here between the server address structure creation (shown in Listing 12.1). The only difference is that the interface address is specified here in the client as `localhost`, where in the server, we specify the wildcard to accept connections from any available interface.

Now that we have our socket and an address structure initialized with our destination, we can connect our socket to the server. This is done with the `connect` function shown in lines 23–24. In the `connect` function, we pass our socket descriptor (`connectionFd`), our address structure (`servaddr`), and its size. When this function returns, either we've connected to the server or an error has occurred. To minimize code size, we've omitted the error check here, but the error code should be checked upon return from `connect` to ensure that the socket is truly connected.

Now that we're connected to the server, we perform a socket **`read`** function. This allows us to read any data that has been sent to us. Given the Daytime protocol, we know as a client that the server will immediately send us the current date and time. Therefore, we immediately read from the socket and store the contents into `timebuffer`. This is then null-terminated, and the result printed to standard-out. If we read from the socket and no characters are received (or an error occurs, indicated by a `-1` return), then we know that the server has closed our connection, and we exit gracefully. The next step is closure of our half of the socket, shown at line 34 using the **`close`** function.

ON THE CD

Listing 12.2 Daytime Client Written in the C Language (on the CD-ROM at `./source/ch12/daycli.c`)

```
1:    #include <sys/socket.h>
2:    #include <arpa/inet.h>
3:    #include <stdio.h>
4:    #include <unistd.h>
5:    #include <time.h>
6:
7:    #define MAX_BUFFER       128
8:    #define DAYTIME_SERVER_PORT   13
9:
10:   int main ()
11:   {
12:    int connectionFd, in, index = 0.limit = MAX_BUFFER;
13:    struct sockaddr_in servaddr;
14:    char timebuffer[MAX_BUFFER+1];
15:
16:    connectionFd = socket(AF_INET, SOCK_STREAM, 0);
17:
18:    memset(&servaddr, 0, sizeof(servaddr));
19:    servaddr.sin_family = AF_INET;
20:    servaddr.sin_port = htons(DAYTIME_SERVER_PORT);
21:    servaddr.sin_addr.s_addr = inet_addr("127.0.0.1");
22:
23:    connect(connectionFd,
24:        (struct sockaddr *)&servaddr, sizeof(servaddr));
25:
```

```
    while ( (in = read(connectionFd, &timebuffer[index], limit)) >  0) {
     index += in;
     limit -= in;
    }

    timebuffer[index] = 0;
    printf("\n%s\n", timebuffer);

    close(connectionFd);

    return(0);
  }
```

SOCKETS API SUMMARY

The networking API for C provides a mixed set of functions for the development of client and server applications. Some functions are used by only server-side sockets, whereas others are used solely by client-side sockets (most are available to both).

Creating and Destroying Sockets

A Socket is necessary to be created as the first step of any socket-based application. The `socket` function provides the following prototype:

```
int socket( int domain, int type, int protocol );
```

The `socket` object is represented as a simple integer and is returned by the **`socket`** function. Three parameters must be passed to define the type of socket to be created. We're interested primarily in stream (TCP) and datagram (UDP) sockets, but many other types of sockets may be created. In addition to stream and datagram, a raw socket is also illustrated by the following code snippets:

```
myStreamSocket = socket( AF_INET, SOCK_STREAM, 0 );
myDgramSocket = socket( AF_INET, SOCK_DGRAM, 0 );
myRawSocket = socket( AF_INET, SOCK_RAW, IPPROTO_RAW );
```

The `AF_INET` symbolic constant defines that we are using the IPv4 Internet protocol. After this, the second parameter (`type`) defines the semantics of communication. For stream communication (using TCP), we use the `SOCK_STREAM` type, and for datagram communication (using UDP), we specify `SOCK_DGRAM`. The third parameter could define a particular protocol to use, but only the types exist for stream and datagram, so it's left as zero.

When we're finished with a socket, we must close it. The close prototype is defined as:

```
int close( sock );
```

After **close** is called, no further data may be received through the socket. Any data queued for transmission would be given some amount of time to be sent before the connection physically closes.

Note in these examples that the read *and* write *calls were used identically to the file I/O examples shown in Chapter 10, "File Handling in GNU/Linux." One of the interesting features of UNIX (and GNU/Linux) is that many types of devices are represented as files. After a socket is open, we can treat it just like a file or pipe (for* **read**, **write**, **accept**, *and so on).*

Socket Addresses

For socket communication over the Internet (domain AF_INET), the sockaddr_in structure is used for naming purposes.

```
struct sockaddr_in {
        int16_t sin_family;
        uint16_t sin_port;
        struct in_addr sin_addr;
        char sin_zero[8];
};
struct in_addr {
        uint32_t s_addr;
};
```

For Internet communication, we'll use AF_INET solely for sin_family. Field sin_port defines our specified port number in network byte order. Therefore, we must use **htons** to load the port and **ntohs** to read it from this structure. Field sin_addr is, through s_addr, a 32-bit field that represents an IPv4 Internet address. Recall that IPv4 addresses are four-byte addresses. We'll see quite often that the sin_addr is set to INADDR_ANY, which is the wildcard. When we're accepting connections (server socket), this wildcard says we accept connections from any available interface on the host. For client sockets, this is commonly left blank. For a client, if we set sin_addr to the IP address of a local interface, this restricts outgoing connections to that interface.

Let's now look at a quick example of addressing for both a client and a server. First, we'll create the socket address (later to be bound to our server socket) that permits incoming connections on any interface and port 48000.

```
int servsock;
struct sockaddr_in servaddr;
servsock = socket( AF_INET, SOCK_STREAM, 0);
memset( &servaddr, 0, sizeof(servaddr) );
servaddr.sin_family = AF_INET;
servaddr.sin_port = htons( 48000 );
servaddr.sin_addr.s_addr = inet_addr( INADDR_ANY );
```

Next, we'll create a socket address that permits a client socket to connect to our previously created server socket.

```
int clisock;
struct sockaddr_in servaddr;
clisock = socket( AF_INET, SOCK_STREAM, 0);
memset( &servaddr, 0, sizeof(servaddr) );
servaddr.sin_family = AF_INET;
servaddr.sin_port = htons( 48000 );
servaddr.sin_addr.s_addr = inet_addr( "192.168.1.1" );
```

Note the similarities between these two code segments. The difference, as we'll see later, is that the server uses the address to bind to itself as an advertisement. The client uses this information to define to whom it wants to connect.

Socket Primitives

In this section, we look at a number of other important server-side socket control primitives.

bind

The **bind** function provides a local naming capability to a socket. This can be used to name either client or server sockets, but it is used most often in the server case. The bind function is provided by the following prototype:

```
int bind( int sock, struct sockaddr *addr, int addrLen );
```

The socket to be named is provided by the `sock` argument, and the address structure previously defined is defined by `addr`. Note that the structure here differs from our address structure discussed previously. The `bind` function may be used with a variety of different protocols, but when using a socket created with `AF_INET`, the `sockaddr_in` structure must be used. Therefore, as shown in the following example, we cast our `sockaddr_in` structure as `sockaddr`.

```
err = bind( servsock, (struct sockaddr *)&servaddr,
sizeof(servaddr));
```

Using our address structure created in our server example in the previous address section, we bind the name defined by `servaddr` to our server socket `servsock`.

Recall that a client application can also call bind in order to name the client socket. This isn't used often, as the Sockets API will dynamically assign a port to us.

listen

Before a server socket can accept incoming client connections, it must call the **`listen`** function to declare this willingness. The **`listen`** function is provided by the following function prototype:

```
int listen( int sock, int backlog );
```

The `sock` argument represents the previously created server socket, and the `backlog` argument represents the number of outstanding client connections that may be queued. Within GNU/Linux, the `backlog` parameter (post 2.2 kernel version) represents the numbers of established connections waiting `accept` by the Application layer protocol. Other operating systems may treat this differently.

accept

The **`accept`** call is the final call made by servers to accept incoming client connections. Before **`accept`** can be called, the server socket must be created, a name must be bound to it, and `listen` must be called. The **`accept`** function returns a socket descriptor for a client connection and is provided by the following function prototype:

```
int accept( int sock, struct sockaddr *addr, int *addrLen );
```

In practice, two examples of `accept` are commonly seen. The first represents the case in which we need to know who connected to us. This requires the creation of an address structure that will not be initialized.

```
struct sockaddr_in cliaddr;
int cliLen;
cliLen = sizeof( struct sockaddr_in );
clisock = accept( servsock, (struct sockaddr *)cliaddr, &cliLen );
```

The call to **`accept`** will block until a client connection is available. Upon return, the `clisock` return value will contain the value of the new client socket, and `cliaddr` will represent the address for the client peer (host address and port number).

The alternate example is commonly found when the server application isn't interested in the client information. This one typically appears as:

```
cliSock = accept( servsock, (struct sockaddr *)NULL, NULL );
```

In this case, NULL is passed for the address structure and length. The accept function will then ignore these parameters.

connect

The **connect** function is used by client Sockets applications to connect to a server. Clients must have created a socket and then defined an address structure containing the host and port number to which they want to connect. The **connect** function is provided by the following function prototype:

```
int connect( int sock, (struct sockaddr *)servaddr, int addrLen );
```

The sock argument represents our client socket, created previously with the Sockets API function. The servaddr structure is the server peer to which we want to connect (as illustrated previously in the "Socket Addresses" section of this chapter). Finally, we must pass in the length of our servaddr structure so that connect knows we're passing in a sockaddr_in structure. The following code shows a complete example of connect:

```
int clisock;
struct sockaddr_in servaddr;
clisock = socket( AF_INET, SOCK_STREAM, 0);
memset( &servaddr, 0, sizeof(servaddr) );
servaddr.sin_family = AF_INET;
servaddr.sin_port = htons( 48000 );
servaddr.sin_addr.s_addr = inet_addr( "192.168.1.1" );
connect( clisock, (struct sockaddr_in *)&servaddr,
         sizeof(servaddr) );
```

The connect function blocks until either an error occurs or the three-way handshake with the server finishes. Any error is returned by the connect function.

Sockets I/O

A variety of API functions exist to read data from a socket or write data to a socket. Two of the API functions (**recv**, **send**) are used exclusively by sockets that are connected (such as stream sockets), whereas an alternative pair (**recvfrom**, **sendto**) is used exclusively by sockets that are unconnected (such as datagram sockets).

Connected Socket Functions

The **send** and **recv** functions are used to send a message to the peer socket endpoint and to receive a message from the peer socket endpoint. These functions have the following prototypes:

```
int send( int sock, const void *msg, int len, unsigned int flags );
int recv( int sock, void *buf, int len, unsigned int flags );
```

The **send** function takes as its first argument the socket descriptor from which to send the `msg`. The `msg` is defined as a `(const void *)` because the object referenced by `msg` will not be altered by the **send** function. The number of bytes to be sent in `msg` is contained by the `len` argument. Finally, a `flags` argument can be used to alter the behavior of the **send** call. An example of sending a string through a previously created stream socket is shown as:

```
strcpy( buf, "Hello\n");
send( sock, (void *)buf, strlen(buf), 0);
```

In this example, our character array is initialized by the `strcpy` function. This buffer is then sent through `sock` to the peer endpoint, with a length defined by the string length function, `strlen`. To illustrate flags use, let's look at one side effect of the **send** call. When **send** is called, it may block until all of the data contained within `buf` has been placed on the socket's send queue. If not enough space is available to do this, the **send** function blocks until space is available. If we want to avoid this blocking behavior and instead want the send call to simply return if sufficient space is available, we could set the `MSG_DONTWAIT` flag, such as:

```
send( sock, (void *)buf, strlen(buf), MSG_DONTWAIT);
```

The return value from **send** represents either an error (less than 0) or the number of bytes that were queued to be sent. Completion of the **send** function does not imply that the data was actually transmitted to the host, only queued on the socket's send queue waiting to be transferred.

The **recv** function mirrors the **send** function in terms of an argument list. Instead of sending the data pointed to be `msg`, the **recv** function fills the `buf` argument with the bytes read from the socket. We must define the size of the buffer so that the network protocol stack doesn't overwrite the buffer, which is defined by the `len` argument. Finally, we can alter the behavior of the read call using the flags argument. The value returned by the **recv** function is the number of bytes now contained in the `msg` buffer, or `-1` on error. An example of the **recv** function is:

```
#define MAX_BUFFER_SIZE        50
char buffer[MAX_BUFFER_SIZE+1];
...
numBytes = recv( sock, buffer, MAX_BUFFER_SIZE, 0);
```

At completion of this example, `numBytes` will contain the number of bytes that are contained within the `buffer` argument.

We could peek at the data that's available to read by using the MSG_PEEK flag. This performs a read, but it doesn't consume the data at the socket. This requires another **recv** to actually consume the available data. An example of this type of read is illustrated as:

```
numBytes = recv( sock, buffer, MAX_BUFFER_SIZE, MSG_PEEK);
```

This call requires an extra copy (the first to peek at the data, and the second to actually read and consume it). More often than not, this behavior is handled instead at the Application layer by actually reading the data and then determining what action to take.

Unconnected Socket Functions

The **sendto** and **recvfrom** functions are used to send a message to the peer socket endpoint and receive a message from the peer socket endpoint. These functions have the following prototypes:

```
int sendto( int sock, const void *msg, int len,
unsigned int flags,
const struct sockaddr *to, int tolen );
int recvfrom( int sock, void *buf, int len,
 unsigned int flags,
 struct sockaddr *from, int *fromlen );
```

The **sendto** function is used by an unconnected socket to send a datagram to a destination defined by an initialized address structure. The **sendto** function is similar to the previously discussed send function, except that the recipient is defined by the to structure. An example of the **sendto** function is shown in the following code example:

```
struct sockaddr_in destaddr;
int sock;
char *buf;
...
memset( &destaddr, 0, sizeof(destaddr) );
destaddr.sin_family = AF_INET;
destaddr.sin_port = htons(581);
destaddr.sin_addr.s_addr = inet_addr("192.168.1.1");
sendto( sock, buf, strlen(buf), 0,
 (struct sockaddr *)&destaddr, sizeof(destaddr) );
```

In this example, our datagram (contained with buf) is sent to an application on host 192.168.1.1, port number 581. The destaddr structure defines the intended recipient for our datagram.

Like the **send** function, the number of characters queued for transmission is returned, or -1 if an error occurs.

The **recvfrom** function provides the ability for an unconnected socket to receive datagrams. The **recvfrom** function is again similar to the **recv** function, but an address structure and length are provided. The address structure is used to return the sender of the datagram to the function caller. This information can be used with the **sendto** function to return a response datagram to the original sender.

An example of the **recvfrom** function is shown in the following code:

```
#define MAX_LEN    100
struct sockaddr_in fromaddr;
int sock, len, fromlen;
char buf[MAX_LEN+1];
...
fromlen = sizeof(fromaddr);
len = recvfrom( sock, buf, MAX_LEN, 0,
           (struct sockaddr *)&fromaddr, &fromlen );
```

This blocking call returns when either an error occurs (represented by a -1 return) or a datagram is received (return value of 0 or greater). The datagram will be contained within buf and have a length of len. The fromaddr will contain the datagram sender, specifically the host address and port number of the originating application.

Socket Options

Socket options permit an application to change some of the modifiable behaviors of sockets and the functions that manipulate them. For example, an application can modify the sizes of the send or receive socket buffers or the size of the maximum segment used by the TCP layer for a given socket.

The functions for setting or retrieving options for a given socket are provided by the following function prototypes:

```
int getsockopt( int sock, int level, int optname,
  void *optval, socklen_t *optlen );
int setsockopt( int sock, int level, int optname,
  const void *optval, socklen_t optlen );
```

First, we define the socket of interest using the sock argument. Next, we must define the level of the socket option that is being applied. The level argument can be SOL_SOCKET for socket-layer options, IPPROTO_IP for IP layer options, and IPPROTO_TCP for TCP layer options. The specific option within the level is applied using the optname argument. Arguments optval and optlen define the specifics of the value of the option. optval is used to get or set the option value, and optlen defines the length of the option. This slightly complicated structure is used because structures can be used to define options.

Let's now look at an example for both setting and retrieving an option. In the first example, we'll retrieve the size of the send buffer for a socket.

```
int sock, size, len;
...
getsockopt( sock, SOL_SOCKET, SO_SNDBUF,
            (void *)&size, (socklen_t *)&len );
printf( "Send buffer size is &d\n", size );
```

Now we'll look at a slightly more complicated example. In this case, we're going to set the linger option. Socket linger allows us to change the behavior of a stream socket when the socket is closed and data is remaining to be sent. After close is called, any data remaining will attempt to be sent for some amount of time. If after some duration the data cannot be sent, then the data to be sent is abandoned. The time after the close when the data is removed from the send queue is defined as the *linger time*. This can be set using a special structure called `linger`, as shown in the following example:

```
struct linger ling;
int sock;
...
ling.l_onoff = 1; /* Enable */
ling.l_linger = 10; /* 10 seconds */
setsockopt( sock, SOL_SOCKET, SO_LINGER,
    (void *)&ling, sizeof(struct linger) );
```

After this call is performed, the socket will wait 10 seconds after the socket `close` before aborting the `send`.

Other Miscellaneous Functions

Let's now look at a few miscellaneous functions from the Sockets API and the capabilities they provide. The three function prototypes we discuss in this section are shown in the following code:

```
struct hostent *gethostbyname( const char *name );
int getsockname( int sock, struct sockaddr *name,
                 socklen_t *namelen );
int getpeername( int sock, struct sockaddr *name,
                 socklen_t *namelen );
```

Function **gethostbyname** provides the means to resolve a host and domain name (otherwise known as a Fully Qualified Domain Name, or FQDN) to an IP address. For example, the FQDN of *www.microsoft.com* might resolve to the IP address 207.46.249.27. Converting an FQDN to an IP address is important because all of the Sockets API functions work with number IP addresses (32-bit addresses) rather than FQDNs. An example of the **gethostbyname** function is shown next:

```
struct hostent *hptr;
hptr = gethostbyname( "www.microsoft.com");
if (hptr == NULL) // can't resolve...
else {
   printf( "Binary address is %x\n", hptr-> h_addr_list[0] );
}
```

Function **gethostbyname** returns a pointer to a structure that represents the numeric IP address for the FQDN (`hptr->h_addr_list[0]`). Otherwise, **gethostbyname** returns a `NULL`, which means that the FQDN could not be resolved by the local resolver. This call blocks while the local resolver communicates with the configured DNS servers.

Function **getsockname** permits an application to retrieve information about the local socket endpoint. This function, for example, can identify the dynamically assigned ephemeral port number for the local socket. An example of its use is shown in the following code:

```
int sock;
struct sockaddr localaddr;
int laddrlen;
// Socket for sock created and connected.
...
getsockname( sock, (struct sockaddr_in *)&localaddr, &laddrlen );
printf( "local port is %d\n", ntohs(localaddr.sin_port) );
```

The reciprocal function of **getsockname** is **getpeername**. This permits us to gather addressing information about the connected peer socket. An example, similar to the **getsockname** example, is shown in the following code:

```
int sock;
struct sockaddr remaddr;
int raddrlen;
// Socket for sock created and connected.
...
getpeername( sock, (struct sockaddr_in *)&remaddr, &raddrlen );
printf( "remote port is %d\n", ntohs(remaddr.sin_port) );
```

In both examples, the address can also be extracted using the `sin_addr` field of the `sockaddr` structure.

MULTILANGUAGE PERSPECTIVES

In this chapter, we've focused on the Sockets API from the perspective of the C language, but the Sockets API is available for any worthwhile language.

Consider first the Ruby language. Ruby is an object-oriented scripting language that is growing in popularity. It's simple and clean and useful in many domains. One domain that demonstrates the simplicity of the language is in network application development.

The Daytime protocol server is shown in Listing 12.3. Ruby provides numerous classes for networking development; the one illustrated here supports TCP server sockets (`TCPserver`). At line 4, we create our server socket and bind it to the Daytime protocol server (identified by the string "`daytime`"). At line 12, we await an incoming connection using the **`accept`** method. When one arrives, we emit the current time to the client at line 19 using the **`write`** method. Finally, the socket is closed at line 23 using the **`close`** method.

Listing 12.3 Daytime Protocol Server in the Ruby Language (on the CD-ROM at `./source/ch12/dayserv.rb`)

```
 1:    require 'Socket'
 2:
 3:    # Create a new TCP Server using port 13
 4:    servsock = TCPserver::new("daytime")
 5:
 6:    # Debug data - emit the server socket info
 7:    print("server address : ", servsock.addr::join(":"),"\n")
 8:
 9:    while true
10:
11:     # Await a connection from a client socket
12:     clisock = servsock::accept
13:
14:     # Emit some debugging data on the peer
15:     print("accepted ", clisock.peeraddr::join(":"),"\n")
16:     print(clisock, " is accepted\n")
17:
18:     # Emit the time through the socket to the client
19:     clisock.write( Time::new )
20:     clisock.write("\n" )
```

```

    # Close the client connection
    clisock.close

  end
```

The Sockets API is also useful in other types of languages, such as functional ones. The scheme language is LISP-like in syntax, but it easily integrates the functionality of the Sockets API.

In Listing 12.4 we illustrate the Daytime protocol client in the Scheme language. Lines 2 and 3 define two global constants using in the client. At line 5, the stream-client procedure is created. We create our Socket at lines 6 and 7 of the stream type using the **`socket-connect`** procedure. We provide our previously defined host and port values to identify to whom we should connect. This is bound to the `sock` variable using the `let` expression. Having a connected socket, we read from the socket at line 8 using another `let` expression. The return value of **`read-string`** is bound to result, which is then printed at line 9 using **`write-string`**. We emit a newline at line 10 and then close the socket using the **`close-socket`** procedure at line 11. The client is started at line 16 by calling the defined procedure **`stream-client`**.

ON THE CD

Listing 12.4 Daytime Protocol Client in the Scheme Language (on the CD-ROM at `./source/ch12/daycli.scm`)

```
 1:   ; Define a couple of server constants
 2:   (define host "localhost")
 3:   (define port 13)
 4:
 5:   (define (stream-client)
 6:    (let ((sock (socket-connect protocol-family/internet
 7:           socket-type/stream host port)))
 8:     (let ((result (read-string 100 (socket:inport sock))))
 9:      (write-string result)
10:      (newline)
11:      (close-socket sock) ) ) )
12:
13:   ;
14:   ; Invoke the stream client
15:   ;
16:   (stream-client)
```

Space permitting, we could explore Sockets applications in a multitude of other applications (such as Python, Perl, C++, Java, and Tcl) [Jones03]. The key is that Sockets aren't just a C language construct, but are useful in many languages.

SUMMARY

In this chapter, we've provided a quick tour of Sockets programming in C. We investigated the Sockets programming paradigm, covering the basic elements of networking such as hosts, interfaces, protocols, and ports. The Sockets API was explored in a sample server and client in C and then in detail looking at the functions of the API. Finally, use of the Sockets API was discussed from a multilanguage perspective, illustrating its applicability to non-C language scenarios.

SOCKETS PROGRAMMING APIS

```
#include <sys/types.h>
#include <sys/socket.h>
#include <unistd.h>
int socket( int domain, int type, int protocol );
int bind( int sock, struct sockaddr *addr, int addrLen );
int listen( int sock, int backlog );
int accept( int sock, struct sockaddr *addr, int *addrLen );
int connect( int sock, (struct sockaddr *)servaddr, int addrLen );
int send( int sock, const void *msg, int len, unsigned int flags );
int recv( int sock, void *buf, int len, unsigned int flags );
int sendto( int sock, const void *msg, int len,
unsigned int flags,
const struct sockaddr *to, int tolen );
int recvfrom( int sock, void *buf, int len,
 unsigned int flags,
 struct sockaddr *from, int *fromlen );
int getsockopt( int sock, int level, int optname,
  void *optval, socklen_t *optlen );
int setsockopt( int sock, int level, int optname,
  const void *optval, socklen_t optlen );
int close( int sock );
struct sockaddr_in {
    int16_t sin_family;
        uint16_t sin_port;
        struct in_addr sin_addr;
        char sin_zero[8];
};
struct in_addr {
        uint32_t s_addr;
};
```

```
#include <netdb.h>
struct hostent *gethostbyname( const char *name );
int getsockname( int sock, struct sockaddr *name,
                 socklen_t *namelen );
int getpeername( int sock, struct sockaddr *name,
                 socklen_t *namelen );
    struct hostent {
        char *h_name;
        char **h_aliases;
        int h_addrtype;
        int h_length;
        char **h_addr_list;
    }
    #define h_addr h_addr_list[0]
```

REFERENCES

[Jones03] Jones, M. Tim, *BSD Sockets Programming from a Multilanguage Perspective*, M. Tim Jones, Charles River Media, 2003.

RESOURCES

scsh—The Scheme Shell at *http://www.scsh.net.*

The Ruby Language at *http://www.ruby-lang.org/en/*

Stevens, W. Richard, *Unix Network Programming—Networking APIs: Sockets and XTI Volume 1*, W. Richard Stevens, Prentice Hall PTR, 1998.

13 GNU/Linux Process Model

In This Chapter

- Creating Processes with `fork()`
- Review of Process-related API Functions
- Raising and Catching Signals
- Available Signals and Their Uses
- GNU/Linux Process-related Commands

INTRODUCTION

In this chapter, we'll introduce the GNU/Linux process model. We'll define elements of a process, how processes communicate with each other, and how to control and monitor them. First, we'll do a quick review of fundamental APIs and then follow up with a more detailed review, complete with sample applications that illustrate each technique.

GNU/LINUX PROCESSES

GNU/Linux presents two fundamental types of processes. These are *kernel threads* and *user processes*. We'll focus our attention here to user processes (those created by `fork` and `clone`). Kernel threads are created within the kernel context via the `kernel_thread()` function.

When a subprocess is created (via **`fork`**), a new child task is created with a copy of the memory used by the original parent task. This memory is separate between the two processes. Any variables present when the `fork` takes place are available to the child. But after the **`fork`** completes, any changes that the parent makes to a variable are not seen by the child. This is important to consider when using the `fork` API function.

When a new task is created, the memory space used by the parent isn't actually copied to the child. Instead, both the parent and child reference the same memory space, with the memory pages marked as copy-on-write. When any of the processes attempt to write to the memory, a new set of memory pages is created for the process that is private to it alone. In this way, creating a new process is an efficient mechanism, with copying of the memory space deferred until writes take place. In the default case, the child process inherits open file descriptors, the memory image, and CPU state (such as the PC and assorted registers).

Certain elements are not copied from the parent and instead are created specifically for the child. We'll look at examples of these in the following sections. What's relevant to understand at this stage is that a process can create subprocesses (known as *children*) and generally control them.

WHIRLWIND TOUR OF PROCESS APIS

As we defined previously, we can create a new process with the **`fork`** or **`clone`** API function. But in fact, we create a new process every time we execute a command or start a program. Consider the simple program shown in Listing 13.1.

Listing 13.1 First Process Example (on the CD-ROM at `./source/ch13/process.c`)

```
#include <stdio.h>
#include <unistd.h>
#include <sys/types.h>

int main()
{
  pid_t myPid;
  pid_t myParentPid;
  gid_t myGid;
  uid_t myUid;

```

```
    myPid = getpid();
    myParentPid = getppid();
    myGid = getgid();
    myUid = getuid();

    printf( "my process id is %d\n", myPid );

    printf( "my parent's process id is %d\n", myParentPid );

    printf( "my group id is %d\n", myGid );

    printf( "my user id is %d\n", myUid );

    return 0;
}
```

Every process in GNU/Linux has a unique identifier called a process ID (or pid). Every process also has a parent (except for the `init` process). In Listing 13.1, we use the **`getpid()`** function to get the current process ID and the **`getppid()`** function to retrieve the process's parent ID. Then we grab the group ID and the user ID using **`getuid()`** and **`getgid()`**.

If we were to compile and then execute this application, we'd see the following:

```
$ ./process
my process id is 10932
my parent's process id is 10795
my group id is 500
my user id is 500
$
```

We see our process ID is 10932, and our parent is 10795 (our bash shell). If we execute the application again, we see:

```
$ ./process
my process id is 10933
my parent's process id is 10795
my group id is 500
my user id is 500
$
```

Note that our process ID has changed, but all other values have remained the same. This is expected, as the only thing we've done is created a new process that performs its I/O and then exits. Each time a new process is created, a new pid is allocated to it.

Creating a Subprocess with fork

Let's now move on to the real topic of this chapter, creating new processes within a given process. The **fork** API function is the most common method to achieve this.

The **fork** call is an oddity when you consider what is actually occurring. When the **fork** API function returns, the split occurs, and the return value from fork identifies in which context the process is running. Consider the following code snippet:

```
pid_t pid;
...
pid = fork();
if (pid > 0) {
  /* Parent context, child is pid */
} else if (pid == 0) {
  /* Child context */
} else {
  /* Parent context, error occurred, no child created */
}
```

We see here three possibilities from the return of the **fork** call. When the return value of **fork** is greater than zero, then we're in the parent context and the value represents the pid of the child. When the return value is zero, then we're in the child process's context. Finally, any other value (less than zero) represents an error and is performed within the context of the parent.

Let's now look at a sample application of fork (shown in Listing 13.2). This working example illustrates the **fork** call, identifying the contexts. At line 11, we call fork to split our process into parent and child. Both the parent and child emit some text to standard-out in order to see each execution. Note that a shared variable (role) is updated by both parent and child and emitted at line 45.

ON THE CD

Listing 13.2 Working Example of the **fork** Call (on the CD-ROM at ./source/ch13/smplfork.c)

```
1:     #include <sys/types.h>
2:     #include <unistd.h>
3:     #include <errno.h>
4:
5:     int main()
6:     {
7:       pid_t ret;
8:       int   status, i;
9:       int   role = -1;
10:
```

```
  ret = fork();

  if (ret > 0) {

    printf("Parent: This is the parent process (pid %d)\n",
             getpid());

    for (i = 0 ; i < 10 ; i++) {
      printf("Parent: At count %d\n", i);
      sleep(1);
    }

    ret = wait( &status );

    role = 0;

  } else if (ret == 0) {

    printf("Child: This is the child process (pid %d)\n",
             getpid());

    for (i = 0 ; i < 10 ; i++) {
      printf("Child: At count %d\n", i);
      sleep(1);
    }

    role = 1;

  } else {

    printf("Parent: Error trying to fork() (%d)\n", errno);

  }

  printf("%s: Exiting...\n",
          ((role == 0) ? "Parent" : "Child"));

  return 0;
}
```

The output of the application shown in Listing 13.2 is shown in the following. We see that the child is started and in this case immediately emits some output (its pid and the first `count` line). The parent and the child then switch off from the GNU/Linux scheduler, each sleeping for one second and emitting a new `count`.

```
# ./smplfork
Child: This is the child process (pid 11024)
Child: At count 0
Parent: This is the parent process (pid 11023)
Parent: At count 0
Parent: At count 1
Child: At count 1
Parent: At count 2
Child: At count 2
Parent: At count 3
Child: At count 3
Parent: At count 4
Child: At count 4
Parent: At count 5
Child: At count 5
Child: Exiting...
Parent: Exiting...
#
```

At the end, we see the `role` variable used to emit the role of the process (parent or child). In this case, while the `role` variable was shared between the two processes, once the write occurs, the memory is split, and each process has its own variable, independent of the other. How this occurs is really unimportant. What's important to note is that each process has a copy of its own set of variables.

Synchronizing with the Creator Process

One element of Listing 13.2 was ignored, but we'll dig into it now. At line 23, the **`wait`** function was called within the context of the parent. The **`wait`** function suspends the parent until the child exits. If the **`wait`** function is not called by the parent and the child exits, the child becomes what is known as a "zombie" process (neither alive nor dead). It can be problematic to have these processes lying around (due to the resources that they waste), so handling child `exit` is necessary. Note that if the parent exits first, the children that have been spawned are inherited by the `init` process.

Another way to avoid zombie processes is to tell the parent to ignore child exit signals when they occur. This can be achieved using the **`signal`** *API function, which we explore in the next section, "Catching a Signal." In any case, once the child has stopped, any system resources that were used by the process are immediately released.*

The first two methods that we'll discuss for synchronizing the exit of a child process are the **wait** and **waitpid** API functions. The **waitpid** API function provides greater control over the wait process; here we'll look exclusively at the **wait** API function.

The **wait** function suspends the caller (in this case, the parent) awaiting the **exit** of the child. Once the child exits, the integer value reference (passed to wait) is filled in with the particular exit status. Sample use of the **wait** function, including parsing of the successful status code, is shown in the following code snippet:

```
int status;
pid_t pid;
...
pid = wait( &status );
if ( WIFEXITED(status) ) {
  printf( "Process %d exited normally\n", pid );
}
```

The wait function can set other potential status values, which we'll investigate in the "wait" section, later in this chapter.

Catching a Signal

A signal is fundamentally an asynchronous callback for processes in GNU/Linux. We can register to receive a signal when an event occurs for a process or register to ignore signals when a default action exists. GNU/Linux supports a variety of signals, which we'll cover later. Signals are an important topic here in process management because they allow processes to communicate with one another.

To catch a signal, we provide a signal handler for the process (a kind of callback function) and the signal that we're interested in for this particular callback. Let's now look at an example of registering for a signal. In this example, we'll register for the SIGINT signal. This particular signal is used to identify that a Ctrl+C was received.

Our main program in Listing 13.3 (lines 14–24) begins with registering our callback function (also known as the "signal handler"). We use the **signal** API function to register our handler (at line 17). We specify first the signal of interest and then the handler function that will react to the signal. At line 21, we pause, which suspends the process until a signal is received.

Our signal handler is shown at Listing 13.3 at lines 6–12. We simply emit a message to stdout and then flush it to ensure that it's been emitted. We return from our signal handler, which allows our main function to continue from the **pause** call and **exit**.

Listing 13.3 Registering for Catching a Signal (on the CD-ROM at `./source/ch13/sigcatch.c`)

```
#include <stdio.h>
#include <sys/types.h>
#include <signal.h>
#include <unistd.h>

void catch_ctlc( int sig_num )
{
  printf( "Caught Control-C\n" );
  fflush( stdout );

  return;
}

int main()
{

  signal( SIGINT, catch_ctlc );

  printf("Go ahead, make my day.\n");

  pause();

  return 0;
}
```

Raising a Signal

In the previous example, we illustrated a process receiving a signal. We can also have a process send a signal to another process using the **`kill`** API function. The **`kill`** API function takes a process ID (to whom the signal is to be sent) and the signal to send.

Let's look at a simple example of two processes communicating via a signal. This will use the classic parent/child process creation via **`fork`** (see Listing 13.4).

At lines 8–13, we declare our signal handler. This handler is very simple, as shown, and simply emits some text to `stdout` indicating that the signal was received, in addition to the process context (identified by the pid).

Our `main` (lines 15–61) is a simple parent/child **`fork`** example. Our parent context (starting at line 25) installs our signal handler and then pauses (awaiting the receipt of a signal). It then continues by awaiting the `exit` of the child process.

The child context (starting at line 39) sleeps for one second (allowing the parent context to execute and install its signal handler) and then raises a signal. Note that we use the **kill** API function (**line 47**) to direct the signal to the parent process ID (via **getppid**). The signal we use is SIGUSR1, which is a user-definable signal. Once the signal has been raised, the child sleeps another two seconds and then exits.

Listing 13.4 Raising a Signal from a Child to a Parent Process (on the CD-ROM at ./source/ch13/raise.c)

```
1:   #include <stdio.h>
2:   #include <sys/types.h>
3:   #include <sys/wait.h>
4:   #include <unistd.h>
5:   #include <signal.h>
6:   #include <errno.h>
7:
8:   void usr1_handler( int sig_num )
9:   {
10:
11:    printf( "Parent (%d) got the SIGUSR1\n", getpid() );
12:
13:  }
14:
15:  int main()
16:  {
17:    pid_t ret;
18:    int   status;
19:    int   role = -1;
20:
21:    ret = fork();
22:
23:    if (ret > 0) {                /* Parent Context */
24:
25:      printf( "Parent: This is the parent process (pid %d)\n",
26:               getpid() );
27:
28:      signal( SIGUSR1, usr1_handler );
29:
30:      role = 0;
31:
32:      pause();
33:
34:      printf( "Parent: Awaiting child exit\n" );
35:      ret = wait( &status );
36:
37:    } else if (ret == 0) {        /* Child Context */
38:
```

```
        printf( "Child: This is the child process (pid %d)\n",
                getpid() );

        role = 1;

        sleep( 1 );

        printf( "Child: Sending SIGUSR1 to pid %d\n", getppid() );
        kill( getppid(), SIGUSR1 );

        sleep( 2 );

      } else {                          /* Parent Context - Error */

        printf( "Parent: Error trying to fork() (%d)\n", errno );

      }

      printf( "%s: Exiting...\n",
               ((role == 0) ? "Parent" : "Child") );

      return 0;
    }
```

While this was probably self-explanatory, looking at its output can be beneficial to understanding exactly what's going on. The output for the application shown in Listing 13.4 is as follows:

```
$ ./raise
Child: This is the child process (pid 14960)
Parent: This is the parent process (pid 14959)
Child: Sending SIGUSR1 to pid 14959
Parent (14959) got the SIGUSR1
Parent: Awaiting child exit
Child: Exiting...
Parent: Exiting...
$
```

We see that the child performs its first `printf` first (the `fork` gave control of the CPU to the child first). The child then sleeps, allowing the parent to perform its first `printf`, install the signal handler, and then pause awaiting a signal. Now that the parent has suspended, the child can then execute again (once the 1-second sleep has finished). It emits its message, indicating that the signal is being raised, and then raises the signal using the **`kill`** API function. The parent then performs the `printf` within the signal handler (in the context of the parent process as shown by the process ID) and then suspends again awaiting child `exit` via the **`wait`** API function. The child process can then execute again, and once the 2-second sleep has finished, it exits, releasing the parent from the `wait` call so that it too can exit.

It's fairly simple to understand, but it's a powerful mechanism for coordination and synchronization between processes. The entire thread is shown graphically in Figure 13.1. This illustrates the coordination points that exist within our application (shown as dashed horizontal lines from the child to the parent).

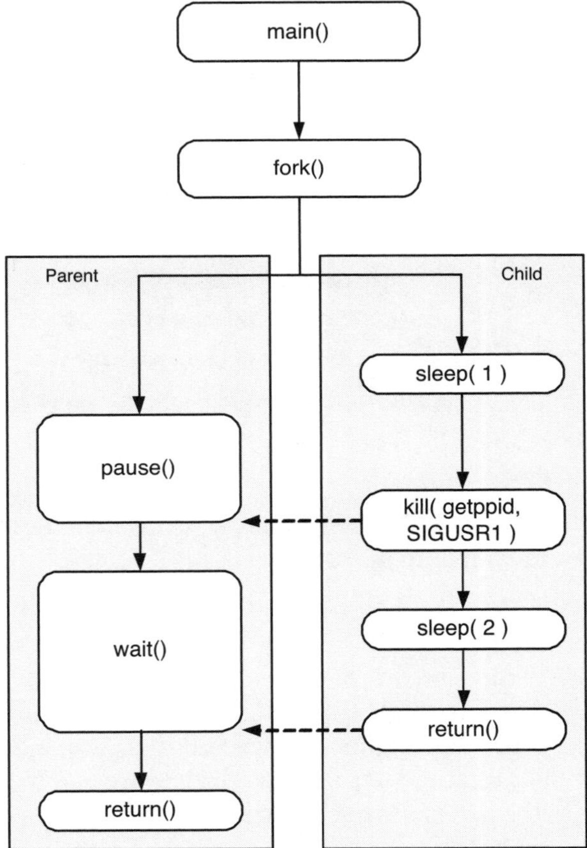

FIGURE 13.1 Graphical illustration of Listing 13.4.

If we're raising a signal to ourselves (same process), we could also use the **`raise`** *API function. This takes the signal to be raised but no process ID argument (because it's automatically* **`getpid`***).*

TRADITIONAL PROCESS API

We've looked at a number of different API functions that relate to the GNU/Linux process model. Let's now dig further into these functions (and others) and explore them in greater detail. Table 13.1 provides a list of the functions that we'll explore in the remainder of this section, including their uses.

TABLE 13.1 Traditional Process and Related APIs

API Function	Use
`fork`	Create a new child process
`wait`	Suspend execution until a child processes exits
`waitpid`	Suspend execution until a specific child process exits
`signal`	Install a new signal handler
`pause`	Suspend execution until a signal is caught
`kill`	Raise a signal to a specified process
`raise`	Raise a signal to the current process
`exec`	Replace the current process image with a new process image
`exit`	Cause normal program termination of the current process

We'll address each of these functions in detail in the remainder of this chapter, illustrated in sample applications.

fork

The **fork** API function provides the means to create a new child subprocess from an existing parent process. The new child process is identical to the parent process in almost every way. Some differences include the process ID (a new ID for the child) and that the parent process ID is set to the parent. File locks and signals that are pending to the parent are not inherited by the child process. The prototype for the **fork** function is defined as follows:

```
pid_t fork( void );
```

The **fork** API function takes no arguments and returns a pid (process identifier). The **fork** call has a very unique structure in that the `return` value identifies the context in which the process is running. If the `return` value is zero, then the current process is the newly created child process. If the `return` value is greater than zero, then the current process is the parent, and the `return` value represents the process ID of the child. This is illustrated in the following snippet:

```
#include <sys/types.h>
#include <unistd.h>
#include <errno.h>
...
pid_t ret;
ret = fork();
if        ( ret > 0 ) {
  /* Parent Process */
  printf( "My pid is %d and my child's is %d\n",
               getpid(), ret );
} else if ( ret == 0 ) {
  /* Child Process */
  printf( "My pid is %d and my parent's is %d\n",
               getpid(), getppid() );
} else {
  /* Parent Process — error */
  printf( "An error occurred in the fork (%d)\n", errno );
}
```

Within the **`fork()`** call, the process is duplicated, and then control is returned to the unique process (parent and child). If the `return` value of **`fork`** is less than zero, then an error has occurred. The `errno` value will represent either `EAGAIN` or `ENOMEM`. Both errors arise from a lack of available memory.

The **`fork`** API function is very efficient in GNU/Linux because of its unique implementation. Rather than copy the page tables for the memory when the `fork` takes place, the parent and child share the same page tables but are not permitted to write to them. When a `write` takes place to one of the shared page tables, the page table is copied for the writing process so that it has its own copy. This is called "copy-on-write" in GNU/Linux and permits the `fork` to take place very quickly. Only as writes occur to the share data memory does the segregation of the page tables take place.

wait

The purpose of the **`wait`** API function is to suspend the calling process until a child process (created by this process) exits or until a signal is delivered. If the parent isn't currently waiting on the child to exit, the child exits, and the child process becomes a zombie process.

The **`wait`** function provides an asynchronous mechanism as well. If the child process exits before the parent has had a chance to call **`wait`**, then the child becomes a zombie but then is freed once `wait` is called. The **`wait`** function, in this case, returns immediately.

The prototype for the **wait** function is defined as:

```
pid_t wait( int *status );
```

The **wait** function returns the pid of the child that exited, or –1 if an error occurred. The `status` variable (whose reference is passed into **wait** as its only argument) returns status information about the child `exit`. This variable can be evaluated using a number of macros. These macros are listed in Table 13.2.

TABLE 13.2 Macro Functions to Evaluate **wait** Status

Macro	Description
WIFEXITED	Nonzero if the child exited normally
WEXITSTATUS	Returns the `exit` status of the child
WIFSIGNALED	Returns `true` if child exited due to a signal that wasn't caught by the child
WTERMSIG	Returns the signal number that caused the child to exit (relevant only if `WIFSIGNALED` was `true`)

The general form of the status evaluation macro is demonstrated in the following code snippet:

```
pid = wait( &status );
if        ( WIFEXITED(status) ) {
  printf( "Child exited normally with status %d\n",
              WEXITSTATUS(status) );
} else if ( WIFSIGNALED(status) ) {
  printf( "Child exited by signal with status %d\n",
              WTERMSIG(status) );
}
```

In some cases, we're not interested in the **exit** status of our child processes. In the signal API function discussion, we'll look at a way to ignore this status so that `wait` is not necessary to be called by the parent to avoid child zombie processes.

waitpid

While the **wait** API function suspends the parent until a child exits (any child), the **waitpid** API function suspends until a specific child exits. The **waitpid** function provides some other capabilities, which we'll explore here. The **waitpid** function prototype is defined as:

```
pid_t waitpid( pid_t pid, int *status, int options );
```

The return value for **waitpid** is the process identifier for the child that exited. The return value can also be zero if the options argument was set to WNOHANG and no child process has exited (returns immediately).

The arguments to **waitpid** are a pid value, a reference to a return status, and a set of options. The pid value can be a child process ID or other values that provide different behaviors. Table 13.3 lists the possible pid values for **waitpid**.

TABLE 13.3 Pid Arguments for waitpid

Value	Description
>0	Suspend until the child identified by the pid value has exited
0	Suspend until any child exits whose group ID matches that of the calling process
−1	Suspend until any child exits (identical to the wait function)
<−1	Suspend until any child exits whose group ID is equal to the absolute value of the pid argument

The status argument for **waitpid** is identical to the wait function, except that two new status macros are possible (see Table 13.4). These macros are seen only if the WUNTRACED option is specified.

TABLE 13.4 Extended Macro Functions for waitpid

Macro	Description
WIFSTOPPED	Returns true if the child process is currently stopped.
WSTOPSIG	Returns the signal that caused the child to stop (relevant only if WIFSTOPPED was nonzero).

The final argument to **waitpid** is the options argument. Two options are available: WNOHANG and WUNTRACED. WNOHANG, as we discussed, avoids suspension of the parent process and returns only if a child has exited. The WUNTRACED option returns for children that have been stopped and not yet reported.

Let's now look at some examples of the **waitpid** function. In the first code snippet, we'll **fork** off a new child process and then await it explicitly (rather than the **wait** method that waits for any child).

```
pid_t child_pid, ret;
int status;
...
child_pid = fork();
if (child_pid == 0) {
  // Child process...
} else if (child_pid > 0) {
  ret = waitpid( child_pid, &status, 0 );
  /* Note ret should equal child_pid on success */
  if ( WIFEXITED(status) ) {
    printf( "Child exited normally with status %d\n",
                  WEXITSTATUS(status) );
  }
}
```

In this example, we **fork** off our child and then use **waitpid** with the child's process ID. Note here that we can use the status macro functions that were defined with `wait` (as demonstrated with `WIFEXITED`). If we didn't want to wait for the child, we could specify `WNOHANG` as an option. This requires us to call **waitpid** periodically to handle the child `exit`:

```
ret = waitpid( child_pid, &status, WNOHANG );
```

The following line awaits a child process exiting the defined group. Note that we negate the group ID in the call to **waitpid**. Also notable is passing `NULL` as the status reference. In this case, we're not interested in getting the child's **exit** status. In any case, the `return` value is the process ID for the child process that exited.

```
pid_t group_id;
...
ret = waitpid( -group_id, NULL, 0 );
```

signal

The **signal** API function allows us to install a signal handler for a process. The signal handler passed to the **signal** API function has the form:

```
void signal_handler( int signal_number );
```

Once installed, the function is called for the process when the particular signal is raised to the process. The prototype for the **signal** API function is defined as:

```
sighandler_t signal( int signum, sighandler_t handler );
```

where the `sighandler_t typedef` is:

```
typedef void (*sighandler_t)(int);
```

The **signal** function returns the previous signal handler that was installed, which allows the new handler to chain the older handlers together (if necessary).

A process can install handlers to catch signals, and it can also define that signals should be ignored (SIG_IGN). To ignore a signal for a process, the following code snippet can be used:

```
signal( SIGCHLD, SIG_IGN );
```

Once this particular code is executed, it is not necessary for a parent process to wait for the child to **exit** using **wait** or **waitpid**.

Signal handlers for a process can be of three different types. They can be ignored (via SIG_IGN), the default handler for the particular signal type (SIG_DFL), or a user-defined handler (installed via **signal**).

A large number of signals exist for GNU/Linux. They are provided in Table 13.4–13.18 with their meanings. The signals are split into four groups, based upon default action for the signal.

The first group (terminate) lists the signals whose default action is to terminate the process. The second group (ignore) lists the signals for which the default action is to ignore the signal. The third group (core) lists those signals whose action is to both terminate the process and perform a core dump (generate a core dump file). And finally, the fourth group (stop) stops the process (suspend, rather than terminate).

TABLE 13.5 GNU/Linux Signals That Default to Terminate

Signal	Description
SIGHUP	Hang up—commonly used to restart a task
SIGINT	Interrupt from the keyboard
SIGKILL	Kill signal
SIGUSR1	User-defined signal
SIGUSR2	User-defined signal
SIGPIPE	Broken pipe (no reader for write)
SIGALRM	Timer signal (from API function alarm)
SIGTERM	Termination signal
SIGPROF	Profiling timer expired

TABLE 13.6 GNU/Linux Signals That Default to Ignore

Signal	Description
SIGCHLD	Child stopped or terminated
SIGCLD	Same as SIGCHLD
SIGURG	Urgent data on a socket

TABLE 13.7 GNU/Linux Signals That Default to Stop

Signal	Description
SIGSTOP	Stop process
SIGTSTP	Stop initiated from TTY
SIGTTIN	Background process has TTY input
SIGTTOU	Background process has TTY output

TABLE 13.8 GNU/Linux Signals That Default to Core Dump

Signal	Description
SIGQUIT	quit signal from keyboard
SIGILL	Illegal instruction encountered
SIGTRAP	Trace or breakpoint trap
SIGABRT	Abort signal (from API function abort)
SIGIOT	IOT trap, same as SIGABRT
SIGBUS	Bus error (invalid memory access)
SIGFPE	Floating-point exception
SIGSEGV	Segment violation (invalid memory access)

It's important to note that the SIGSTOP and SIGKILL signals cannot be ignored or caught by the application. One other signal not categorized above is the SIGCONT signal, which is used to continue a process if it was previously stopped.

GNU/Linux also supports 32 real-time signals (of POSIX 1003.1-2001). The signals are numbered from 32 (SIGRTMIN) up to 63 (SIGRTMAX) and can be sent using the `sigqueue` API function. The receiving process must use `sigaction` to install the signal handler (discussed later in this chapter) in order to collect other data provided in this signaling mechanism.

Let's now look at a simple application that installs a signal handler at the parent, which is inherited by the child (see Listing 13.4). In this listing, we first declare a signal handler (lines 8–13) that will be installed by the parent prior to the `fork` (at line 21). Installing the handler prior to the `fork` means that child will inherit this signal handler as well.

After the **`fork`** (at line 23), the parent and child context emit an identification string to `stdout` and then call the **`pause`** API function (which suspends each process until a signal is received). When a signal is received, the signal handler will print out the context in which it caught the signal (via **`getpid`**) and then either exit (child process) or await the **`exit`** of the child (parent process).

ON THE CD

Listing 13.5 Signal Demonstration with a Parent and Child Process (on the CD-ROM at `./source/ch13/sigtest.c`)

```
1:      #include <stdio.h>
2:      #include <sys/types.h>
3:      #include <sys/wait.h>
4:      #include <unistd.h>
5:      #include <signal.h>
6:      #include <errno.h>
7:
8:      void usr1_handler( int sig_num )
9:      {
10:
11:       printf( "Process (%d) got the SIGUSR1\n", getpid() );
12:
13:     }
14:
15:     int main()
16:     {
17:       pid_t ret;
18:       int   status;
19:       int   role = -1;
20:
21:       signal( SIGUSR1, usr1_handler );
22:
23:       ret = fork();
24:
25:       if (ret > 0) {                  /* Parent Context */
26:
27:         printf( "Parent: This is the parent process (pid %d)\n",
```

```
                    getpid() );

        role = 0;

        pause();

        printf( "Parent: Awaiting child exit\n" );
        ret = wait( &status );

    } else if (ret == 0) {          /* Child Context */

        printf( "Child: This is the child process (pid %d)\n",
                getpid() );

        role = 1;

        pause();

    } else {                        /* Parent Context — Error */

        printf( "Parent: Error trying to fork() (%d)\n", errno );

    }

    printf( "%s: Exiting...\n",
            ((role == 0) ? "Parent" : "Child") );

    return 0;
}
```

Let's now look at the sample output for this application to better understand what happens. Note that neither the parent nor the child raises any signals to each other. We'll take care of sending the signal at the command line, using the **kill** command.

```
# ./sigtest &
[1] 20152
# Child: This is the child process (pid 20153)
Parent: This is the parent process (pid 20152)

# kill -10 20152
Process (20152) got the SIGUSR1
Parent: Awaiting child exit
```

```
# kill -10 20153
Process (20153) got the SIGUSR1
Child: Exiting...
Parent: Exiting...
#
```

We begin by running the application (called `sigtest`) and placing it in the background (via the `&` symbol). We see the expected outputs from the child and parent processes identifying that the `fork` has occurred and that both processes are now active and awaiting signals at the respective **pause** calls. We use the **kill** command with the signal of interest (–10, or `SIGUSR1`) and the process identifier to which to send the signal. In this case, we send the first `SIGUSR1` to the parent process (20152). The parent immediately identifies receipt of the signal via the signal handler, but note that it executes within the context of the parent process (as identified by the process ID of 20152). The parent then returns from the **pause** function and awaits the exit of the child via the wait function. We then send another `SIGUSR1` signal to the child using the **kill** command. In this case, we direct the **kill** command to the child by its process ID (20153). The child also indicates receipt of the signal by the signal handler and in its own context. The child then exits and permits the parent to return from the **wait** function and exit also.

Despite the simplicity of the signals mechanism, it can be quite a powerful method to communicate with processes in an asynchronous fashion.

pause

The **pause** function is used to suspend the calling process until a signal is received. Once the signal is received, the calling process returns from the `pause` function, permitting it to continue. The prototype for the **pause** API function is:

```
int pause( void );
```

If the process has installed a signal handler for the signal that was caught, then the **pause** function returns after the signal handler has been called and returns.

kill

The **kill** API function is used to raise a signal to a process or set of processes. A return of zero indicates that the signal was successfully sent, otherwise –1 is returned. The **kill** function prototype is:

```
int kill( pid_t pid, int sig_num );
```

The `sig_num` argument represents the signal to send. The `pid` argument can be a variety of different values (as shown in Table 13.9).

TABLE 13.9 Values of `pid` Argument for **`kill`** Function

`pid`	**Description**
0	Signal sent to the process defined by `pid`
0	Signal sent to all processes within the process group
1	Signal sent to all processes (except for the `init` process)
–1	Signal sent to all processes within the process group defined by the absolute value of `pid`

Some simple examples of the **`kill`** function are now explored. We can send a signal to ourselves using the following code snippet:

```
kill( getpid(), SIGHUP );
```

The process group allows us to collect a set of processes together that can be signaled together as a group. API functions such as **`getpgrp`** (get process group) and **`setpgrp`** (set process group) can be used to read and set the process group identifier. We can send a signal to all processes within a defined process group as

```
kill( 0, SIGUSR1 );
```

or to another process group as

```
pid_t group;
...
kill( -group, SIGUSR1 );
```

We can also mimic the behavior of sending to the current process group by identifying the group and then passing the negative of this value to signal:

```
pid_t group = getpgrp();
...
kill( -group, SIGUSR1 );
```

Finally, we can send a signal to all processes (except for `init`) using the –1 `pid` identifier. This of course requires that we have permission to do this.

```
kill( -1, SIGUSR1 );
```

raise

The **raise** API function can be used to send a specific signal to the current process (the process context in which the **raise** function is called). The prototype for the **raise** function is:

```
int raise( int sig_num );
```

The **raise** function is a constrained version of the **kill** API function that targets only the current process (**getpid()**).

exec Variants

The **fork** API function provided a mechanism to split an application into separate parent and child processes, sharing the same code but potentially serving different roles. The **exec** family of functions replaces the current process image altogether.

> *Once the* **exec** *function replaces the current process, its* `pid` *is the same as the creating process.*

The prototypes for the variants of **exec** are provided here:

```
int execl( const char *path, const char *arg, ... );
int execlp( const char *path, const char *arg, ... );
int execle( const char *path, const char *arg, ...,
              char * const envp[] );
int execv( const char *path, char *const argv[] );
int execvp( const char *file, char *const argv[] );
int execve( const char *filename, char *const argv[],
          char *const envp[] );
```

One of the notable differences between these functions is that one set takes a list of parameters (`arg0`, `arg1`, and so on) and the other takes an `argv` array. The `path` argument specifies the program to run, and the remaining parameters specify the arguments to pass to the program.

The **exec** commands permit the current process context to be replaced with the program (or command) specified as the first argument. Let's look at a quick example of **execcl** to achieve this:

```
execl( "/bin/ls", "ls", "-la", NULL );
```

This command replaces the current process with the `ls` image (list directory). We specify the command to execute as the first argument (including its path). The second argument is the command again (recall that `arg0` of the `main` program call is the name of the program). The third argument is an option that we pass to `ls`, and finally, we identify the end of our list with a `NULL`. Invoking an application that performs this command results in an `ls -la`.

The important item to note here is that the current process context is replaced by the command requested via **`execl`**. Therefore, when the preceding command is successfully executed, it will never return.

One additional item to note is that **`execl`** includes the absolute path to the command. If we had executed **`execlp`** instead, the full path would not have been required because the parent's `PATH` definition is used to find the command.

One interesting example of **`execlp`** is its use in creating a simple shell (on top of an existing shell). We'll support only simple commands within this shell (those that take no arguments). See Listing 13.6 for an example.

ON THE CD

Listing 13.6 Simple Shell Interpreter Using **`execlp`** (on the CD-ROM at `./source/ch13/simpshell.c`)

```
#include <sys/types.h>
#include <sys/wait.h>
#include <unistd.h>
#include <stdio.h>
#include <stdlib.h>
#include <string.h>

#define MAX_LINE        80

int main()
{
  int status;
  pid_t childpid;
  char cmd[MAX_LINE+1];
  char *sret;

  while (1) {

    printf("mysh>");

    sret = fgets( cmd, sizeof(cmd), stdin );

    if (sret == NULL) exit(-1);

    cmd[ strlen(cmd)-1] = 0;

    if (!strncmp(cmd, "bye", 3)) exit(0);
```

```
28:
29:         childpid = fork();
30:
31:         if (childpid == 0) {
32:
33:           execlp( cmd, cmd, NULL );
34:
35:         } else if (childpid > 0) {
36:
37:           waitpid( childpid, &status, 0 );
38:
39:         }
40:
41:         printf("\n");
42:
43:       }
44:
45:       return 0;
46:     }
```

Our simple shell interpreter is built around the simple parent/child `fork` application. The parent forks off the child (at line 29) and then awaits completion. The child takes the command read from the user (at line 21) and executes this using **`execlp`** (line 33). We simply specify the command as the command to execute and also include it for `arg0` (second argument). The `NULL` terminates the argument list, in this case no arguments are passed for the command. The child process never returns, but its **`exit`** status is recognized by the parent at the **`waitpid`** function (line 37).

As the user types in commands, they are executed via **`execlp`**. Typing in the command `bye` causes the application to exit.

Since no arguments are passed to the command (via **`execlp`**), the user may type in only commands and no arguments. Any arguments that are provided are simply ignored by the interpreter.

A sample execution of this application is shown here:

```
$ ./simpshell
mysh>date
Sat Apr 24 13:47:48 MDT 2004
mysh>ls
simpshell    simpshell.c
mysh>bye
$
```

We see that after executing our shell, the prompt is displayed, indicating that commands can be entered. The `date` command is entered first, which provides the current date and time. Next, we do an `ls`, which gives us the contents of the current directory. Finally, we exit the shell using the `bye` internal command.

Let's look at one final `exec` variant as a way to explore the argument and environment aspects of a process. The **`execve`** variant allows an application to provide a command with a list of command-line arguments (as a vector) as well as an environment for the new process (as a vector of environment variables). Let's look back at the **`execve`** prototype:

```
int execve( const char *filename, char *const argv[],
            char *const envp[] );
```

The `filename` argument is the program to execute (which must be a binary executable or a script (that includes the `#!` interpreter spec at the top of the file). The `argv` argument is an array of arguments for the command (with the first argument being the command itself (same as the `filename` argument). Finally, the `envp` argument is an array of key/value strings containing environment variables. Consider the following simple example that retrieves the environment variables through the `main` function (on the CD-ROM at `./source/ch13/sigenv.c`).

```
#include <stdio.h>
#include <unistd.h>
int main( int argc, char *argv[], char *envp[] )
{
  int ret;
  char *args[]={ "ls", "-la", NULL };
  ret = execve( "/bin/ls", args, envp );
  fprintf( stderr, "execve failed\n" );
  return 0;
}
```

The first item to note in this example is the `main` function definition. We use a variant that passes in a third parameter that lists the environment for the process. This can also be gathered by the program using the special `environ` variable, which has the definition:

```
extern char *environ[];
```

POSIX systems do not support the `envp` *argument to* `main`*, so it's best to use the* `environ` *variable.*

We specify our argument vector (`args`), which contains our command name and arguments, terminated by a `NULL`. This is provided as the argument vector to **`execve`**, along with the environment (passed in through the `main` function). This particular example simply performs an `ls` operation (by replacing the process with the `ls` command). Note also that we provide the `-la` option.

We could also specify our own environment similar to the `args` vector. For example, the following specifies a new environment for the process:

```
char *envp[] = { "PATH=/bin", "FOO=99", NULL };
...
ret = execve( command, args, envp );
```

The `envp` variable provides the set of variables that define the environment for the newly created process.

alarm

The **`alarm`** API function can be very useful to time out other functions. The **`alarm`** function works by raising a `SIGALRM` signal once the number of seconds passed to `alarm` has expired. The function prototype for **`alarm`** is:

```
unsigned int alarm( unsigned int secs );
```

The user passes in the number of seconds to wait before sending the `SIGALRM` signal. The **`alarm`** function returns zero if no alarm was previously scheduled; otherwise, it returns the number of seconds pending on the previous alarm.

Here's an example of **`alarm`** to kill the current process if the user isn't able to enter a password in a reasonable amount of time (see Listing 13.7). At line 18, we install our signal handler for the `SIGALRM` signal. The signal handler is for the `wakeup` function (lines 6–9), which simply raises the `SIGKILL` signal. This will terminate the application. We then emit the message to enter the password within three seconds and try to read the password from the keyboard (`stdin`). If the `read` call succeeds, we disable the alarm (by calling **`alarm`** with an argument of zero). The else portion of the test (line 30) would check the user password and continue. If the alarm timed out, a `SIGALRM` would be generated, resulting in a `SIGKILL` signal, which would terminate the program.

Listing 13.7 Example Use of `alarm` and Signal Capture (on the CD-ROM at `./source/ch13/alarm.c`)

```
1:     #include <stdio.h>
2:     #include <unistd.h>
3:     #include <signal.h>
4:     #include <string.h>
5:
6:     void wakeup( int sig_num )
```

```
{
  raise(SIGKILL);
}

#define MAX_BUFFER      80

int main()
{
  char buffer[MAX_BUFFER+1];
  int ret;

  signal( SIGALRM, wakeup );

  printf("You have 3 seconds to enter the password\n");

  alarm(3);

  ret = read( 0, buffer, MAX_BUFFER );

  alarm(0);

  if (ret == -1) {

  } else {

    buffer[strlen(buffer)-1] = 0;
    printf("User entered %s\n", buffer);

  }

}
```

exit

The **exit** API function terminates the calling process. The argument passed to **exit** is returned to the parent process as the status of the parent's **wait** or **waitpid** call. The function prototype for **exit** is:

```
void exit( int status );
```

The process calling exit also raises a SIGCHLD to the parent process and frees the resources allocated by the process (such as open file descriptors). If the process had registered a function with atexit or on_exit, these would be called (in the reverse order to their registration).

This call is very important because it indicates success or failure to the shell environment. Scripts that rely on a program's `exit` status can behave improperly if the application does not provide an adequate status. This call provides that linkage to the scripting environment. Returning 0 to the script indicates a `TRUE` or `SUCCESS` indication.

POSIX Signals

Before we end our discussion of process-related functions, let's take a quick look at the POSIX signal APIs. The POSIX-compliant signals were introduced first in BSD and provide a portable API over the use of the signal API function. Let's now have a look at a multiprocess application that uses the **`sigaction`** function to install a signal handler. The **`sigaction`** API function has the following prototype:

```
#include <signal.h>
int sigaction( int signum,
               const struct sigaction *act,
               struct sigaction *oldact );
```

`signum` is the signal for which we're installing the handler, `act` specifies the action to take for `signum`, and `oldact` is used to store the previous action. The **`sigaction`** structure contains a number of elements that can be configured:

```
struct sigaction {
    void (*sa_handler)( int );
    void (*sa_sigaction)( int, siginfo_t *, void * );
    sigset_t sa_mask;
    int sa_flags;
};
```

The `sa_handler` is a traditional signal handler that accepts a single argument (and `int` representing the signal). The `sa_sigaction` is a more refined version of a signal handler. The first `int` argument is the signal, and the third `void*` argument is a context variable (provided by the user). The second argument (`siginfo_t`) is a special structure that provides more detailed information about the signal that was generated:

```
siginfo_t {
  int        si_signo;    /* Signal number */
  int        si_errno;    /* Errno value */
  int        si_code;     /* Signal code */
  pid_t      si_pid;      /* Pid of signal sending process */
  uid_t      si_uid;      /* User id of signal sending process */
```

```
    int         si_status;    /* Exit value or signal */
    clock_t     si_utime;     /* User time consumed */
    clock_t     si_stime;     /* System time consumed */
    sigval_t    si_value      /* Signal value */
    int         si_int;       /* POSIX.1b signal */
    void *      si_ptr        /* POSIX.1b signal */
    void *      si_addr       /* Memory location which caused fault */
    int         si_band;      /* Band Event */
    int         si_fd;        /* File Descriptor */
}
```

One of the interesting items to note from `siginfo_t` is that with this API, we can identify the source of the signal (`si_pid`). The `si_code` field can be used to identify how the signal was raised. For example, if its value was `SI_USER`, then it was raised by a `kill`, `raise`, or `sigsend` API function. If `SI_KERNEL`, then it was raised by the kernel. `SI_TIMER` indicates that a timer expired and resulted in the signal generation.

The `si_signo`, `si_errno`, and `si_code` are set for all signals. The `si_addr` field (indicating the memory location where the fault occurred) is set for `SIGILL`, `SIGFPE`, `SIGSEGV`, and `SIGBUS`. The `sigaction` main page identifies which fields are relevant for which signals.

The `sa_flags` argument of **`sigaction`** allows a modification of the behavior of **`sigaction`** function. For example, if we provide `SA_SIGINFO`, then the **`sigaction`** will use the `sa_sigaction` field to identify the signal handler instead of `sa_handler`. Flag `SA_ONESHOT` can be used to restore the signal handler to the prior state after the signal handler has been called once. The `SA_NOMASK` (or `SA_NODEFER`) flag can be used to not inhibit the reception of the signal while in the signal handler (use with care).

Our example function is provided in Listing 13.8. The only real difference we see here from other examples is that **`sigaction`** is used at line 49 to install our signal handler. We create a **`sigaction`** structure at line 42 and then initialize it with our function at line 48 and also identify that we're using the new **`sigaction`** handler via the `SA_SIGINFO` flag at line 47. When our signal finally fires (at line 34 in the parent process), our signal handler emits the originating pid at line 12 (using the `si_pid` field of the `siginfo` reference).

ON THE CD

Listing 13.8 Simple Application Illustrating `sigaction` for Signal Installation (on the CD-ROM at `./source/ch13/posixsig.c`)

```
1:      #include <sys/types.h>
2:      #include <sys/wait.h>
3:      #include <signal.h>
4:      #include <stdio.h>
5:      #include <unistd.h>
6:      #include <errno.h>
```

```

static int stopChild = 0;

void sigHandler( int sig, siginfo_t *siginfo, void *ignore )
{
  printf("Got SIGUSR1 from %d\n", siginfo->si_pid);
  stopChild=1;

  return;
}

int main()
{
  pid_t ret;
  int   status;
  int   role = -1;

  ret = fork();

  if (ret > 0) {

    printf("Parent: This is the parent process (pid %d)\n",
            getpid());

    /* Let the child init */
    sleep(1);

    kill( ret, SIGUSR1 );

    ret = wait( &status );

    role = 0;

  } else if (ret == 0) {

    struct sigaction act;

    printf("Child: This is the child process (pid %d)\n",
            getpid());

    act.sa_flags = SA_SIGINFO;
    act.sa_sigaction = sigHandler;
    sigaction( SIGUSR1, &act, 0 );

```

```
        printf("Child Waiting...\n");
        while (!stopChild);

        role = 1;

      } else {

        printf("Parent: Error trying to fork() (%d)\n", errno);

      }

      printf("%s: Exiting...\n",
              ((role == 0) ? "Parent" : "Child"));

      return 0;
    }
```

The `sigaction` function provides a more advanced mechanism for signal handling, in addition to greater portability. For this reason, `sigaction` should be used over `signal`.

SYSTEM COMMANDS

In this section, we'll look at a few of the GNU/Linux commands that work with the previously mentioned API functions. We'll look below at commands that permit us to inspect the process list and send a signal to a process or to an entire process group.

ps

The `ps` command provides a snapshot in time of the current set of processes active on a given system. The `ps` command takes a large variety of options; we'll explore a few here.

In the simplest form, we can simply type `ps` at the keyboard to see a subset of the processes that are active:

```
$ ps
  PID TTY          TIME CMD
22001 pts/0    00:00:00 bash
22186 pts/0    00:00:00 ps
$
```

First, we see our bash session (our own process) and our `ps` command process (every command in GNU/Linux is executed within its own subprocess). We could see all of the processes running using the `-a` option (list shortened for brevity):

```
$ ps -a
  PID TTY          TIME CMD
    1 ?        00:00:05 init
    2 ?        00:00:00 keventd
    3 ?        00:00:00 kapmd
    4 ?        00:00:00 ksoftirqd_CPU0
...
22001 pts/0    00:00:00 bash
22074 ?        00:00:00 sendmail
22189 pts/0    00:00:00 ps
$
```

In this example, we see a number of other processes including the mother-of-all-processes (`init`, process ID 1) and assorted kernel threads. If we wanted to see only those processes that were associated with our user, we could accomplish this with the `–User` option:

```
$ ps –User mtj
  PID TTY          TIME CMD
22000 ?        00:00:00 sshd
22001 pts/0    00:00:00 bash
22190 pts/0    00:00:00 ps
$
```

Another very useful option is `-H`, which tells us the process hierarchy. In the next example, we'll request all processes for user `mtj` but then also request their hierarchy (parent/child relationships):

```
$ ps –User mtj -H
  PID TTY          TIME CMD
22000 ?        00:00:00 sshd
22001 pts/0    00:00:00   bash
22206 pts/0    00:00:00     ps
#
```

Here we see that our base process is an `sshd` session (since we're connected to this server via the secure shell). This is the parent of my bash session, which in turn is the parent of the `ps` command that we just executed.

The ps command can be very useful, especially when we're interested in finding our process identifiers to kill a process or send it a signal.

top

The top command is related to ps, but top runs in real time and lists the activity of the processes for the given CPU. In addition to the process list, we can also see statistics about the CPU (number of processes, number of zombies, memory used, and so on). We're obviously in need of a memory upgrade here (only 4MB free). This sample list has again been shortened for brevity.

```
 19:27:49  up 79 days, 10:04,  2 users,  load average: 0.00, 0.00, 0.00
47 processes: 44 sleeping, 3 running, 0 zombie, 0 stopped
CPU states:   0.0% user   0.1% system   0.0% nice   0.0% iowait  99.8%
idle
Mem:   124984k av,  120892k used,    4092k free,       0k shrd,
52572k buff
                     79408k actv,       4k in_d,     860k in_c
Swap:  257032k av,    5208k used,  251824k free
37452k cached
  PID USER      PRI  NI  SIZE  RSS SHARE STAT %CPU %MEM   TIME CPU
COMMAND
22226 mtj        15   0  1132 1132   868 R     0.1  0.9   0:00   0 top
    1 root       15   0   100   76    52 S     0.0  0.0   0:05   0 init
    2 root       15   0     0    0     0 SW    0.0  0.0   0:00   0
keventd
    3 root       15   0     0    0     0 RW    0.0  0.0   0:00   0 kapmd
    4 root       34  19     0    0     0 SWN   0.0  0.0   0:00   0
ksoftirqd_CPU0
...
 1708 root       15   0   196    4     0 S     0.0  0.0   0:00   0 login
 1709 root       15   0   284    4     0 S     0.0  0.0   0:00   0 bash
22001 mtj        15   0  1512 1512  1148 S     0.0  1.2   0:00   0 bash
```

The rate of sampling can also be adjusted for top, in addition to a number of other options (see the top man page for more details).

kill

The kill command, like the kill API function, allows us to send a signal to a process. We can also use it to list the signals that are relevant for the given processor architecture. For example, if we'd like to see the signals that are available for the given processor, we'd use the -l option:

```
# kill -l
 1) SIGHUP       2) SIGINT       3) SIGQUIT      4) SIGILL
 5) SIGTRAP      6) SIGABRT      7) SIGBUS       8) SIGFPE
 9) SIGKILL     10) SIGUSR1     11) SIGSEGV     12) SIGUSR2
13) SIGPIPE     14) SIGALRM     15) SIGTERM     17) SIGCHLD
18) SIGCONT     19) SIGSTOP     20) SIGTSTP     21) SIGTTIN
22) SIGTTOU     23) SIGURG      24) SIGXCPU     25) SIGXFSZ
26) SIGVTALRM   27) SIGPROF     28) SIGWINCH    29) SIGIO
30) SIGPWR      31) SIGSYS      33) SIGRTMIN    34) SIGRTMIN+1
35) SIGRTMIN+2  36) SIGRTMIN+3  37) SIGRTMIN+4  38) SIGRTMIN+5
39) SIGRTMIN+6  40) SIGRTMIN+7  41) SIGRTMIN+8  42) SIGRTMIN+9
43) SIGRTMIN+10 44) SIGRTMIN+11 45) SIGRTMIN+12 46) SIGRTMIN+13
47) SIGRTMIN+14 48) SIGRTMIN+15 49) SIGRTMAX-14 50) SIGRTMAX-13
51) SIGRTMAX-12 52) SIGRTMAX-11 53) SIGRTMAX-10 54) SIGRTMAX-9
55) SIGRTMAX-8  56) SIGRTMAX-7  57) SIGRTMAX-6  58) SIGRTMAX-5
59) SIGRTMAX-4  60) SIGRTMAX-3  61) SIGRTMAX-2  62) SIGRTMAX-1
63) SIGRTMAX
#
```

For a running process, we could send a signal as follows. In this example, we'll send the `SIGSTOP` signal to the process identified by the pid `23000`.

```
# kill -s SIGSTOP 23000
```

This places the process in the `STOPPED` state (not running). We could start the process up again by giving it the `SIGCONT` signal, as:

```
# kill -s SIGCONT 23000
```

Like the **`kill`** API function, we can signal an entire process group by providing a pid of 0. Similarly, all processes within the process group can be sent a signal by sending the negative of the process group.

SUMMARY

This chapter explored the traditional process API provided in GNU/Linux. We investigated process creation with **`fork`**, validating the status return of **`fork`**, and various process-related API functions such as **`getpid`** (get process ID) and **`getppid`** (get parent process ID). We then looked at process support functions such as `wait` and **`waitpid`** and the signal mechanism that permits processes to communicate with one another. Finally, we looked at a number of GNU/Linux commands that allow us to review active processes and also the commands to signal them.

REFERENCES

GNU/Linux `signal` and `sigaction` main pages.

API SUMMARY

```
#include <sys/types.h>
#include <unistd.h>
#include <sys/wait.h>
#include <signal.h>
pid_t fork( void );
pid_t wait( int *status );
pid_t waitpid( pid_t pid, int *status, int options );
sighandler_t signal( int signum, sighandler_t handler );
int pause( void );
int kill( pid_t pid, int sig_num );
int raise( int sig_num );
int execl( const char *path, const char *arg, ... );
int execlp( const char *path, const char *arg, ... );
int execle( const char *path, const char *arg, ...,
              char * const envp[] );
int execv( const char *path, char *const argv[] );
int execvp( const char *file, char *const argv[] );
int execve( const char *filename, char *const argv[],
          char *const envp[] );
unsigned int alarm( unsigned int secs );
void exit( int status );
int sigaction( int signum,
           const struct sigaction *act,
           struct sigaction *oldact );
```

14 POSIX Threads (Pthreads) Programming

In This Chapter

- Threads and Processes
- Creating Threads
- Synchronizing Threads
- Communicating Between Threads
- POSIX Signals API
- Threaded Application Development Topics

INTRODUCTION

Multithreaded applications are a useful paradigm for system development because they offer many facilities not available to traditional GNU/Linux processes. In this chapter, we'll explore pthreads programming and the functionality provided by the pthreads API.

The 2.4 GNU/Linux kernel POSIX thread library was based upon the Linux-Threads implementation (introduced in 1996), which was built on the existing GNU/Linux process model. The 2.6 kernel utilizes the new Native POSIX Thread Library, or NPTL (introduced in 2002), which is a higher performance implementation with numerous advantages over the older component. For example, NPTL provides real thread groups (within a process), compared to one thread per process in the prior model. We'll outline those differences when it's useful to know.

To know which pthreads library is being used, issue the following command:

```
$ getconf GNU_LIBPTHREAD_VERSION
```

This will provide either LinuxThreads or NPTL, each with a version number.

What's a Thread?

To define a thread, let's look back at Linux processes to understand their makeup. Both processes and threads have control flows and can run concurrently, but they differ in some very distinct ways. Threads, for example, share data, where processes explicitly don't. When a process is forked (recall from Chapter 12, "Introduction to Sockets Programming"), a new process is created with its own globals and stack (see Figure 14.1). When a thread is created, the only new element created is a stack that is unique for the thread (see Figure 14.2). The code and global data are common between the threads. This is advantageous, but the shared nature of threads can also be problematic. We'll investigate this later in the chapter.

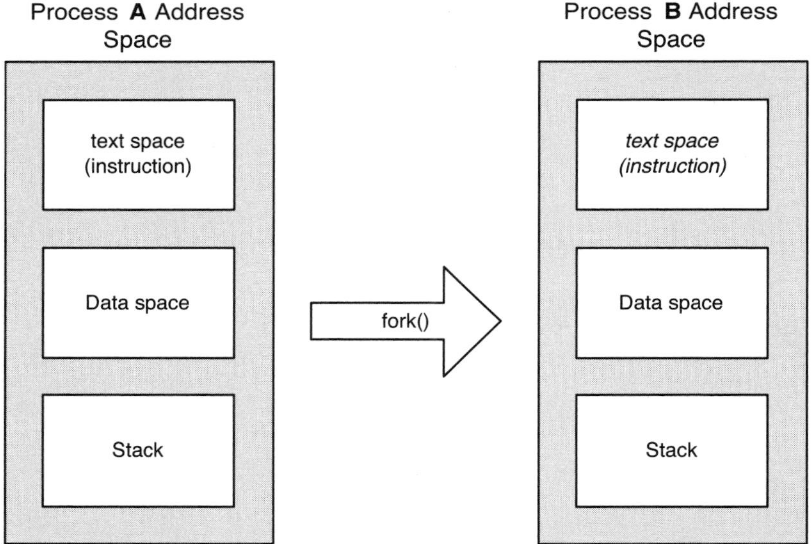

FIGURE 14.1 Forking a new process.

A GNU/Linux process can create and manage numerous threads. Each thread is identified by a thread identifier that is unique for every thread in a system. Each thread also has its own stack (as shown in Figure 14.2) and also a unique context (program counter, save registers, and so forth). But since the data space is shared by

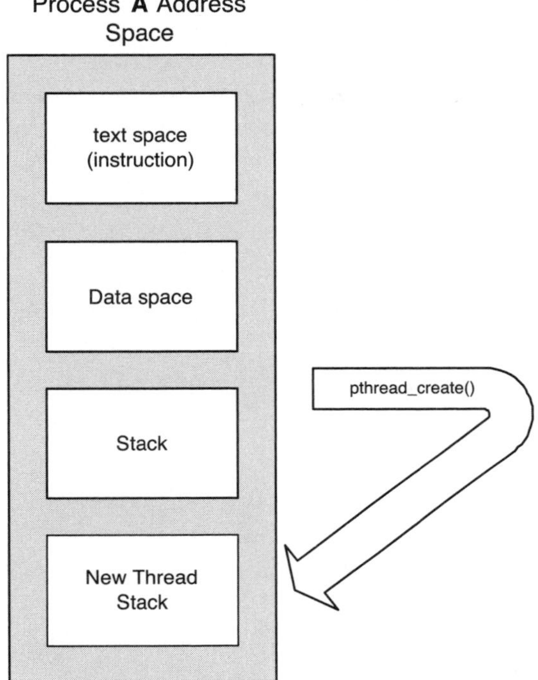

FIGURE 14.2 Creating a new thread.

threads, they share more than just user data. For example, file descriptors for open files or sockets are shared also. Therefore, when a multithreaded application uses a socket or file, the access to the resource must be protected against multiple accesses. We'll look at methods for achieving that in this chapter.

While writing multithreaded applications can be easier in some ways than traditional process-based applications, there are problems to understand. The shared data aspect of threads is probably the most difficult to design around, but it is also powerful and can lead to simpler applications with higher performance. The key is to strongly consider shared data while developing threaded applications. Another important consideration is that serious multithreaded application development should utilize the 2.6 kernel rather than the 2.4 kernel (given the new NPTL threads implementation).

Thread Function Basics

The APIs that we've discussed thus far follow a fairly uniform model of returning –1 when an error occurs, with the actual error value in the `errno` process variable. The threads API returns 0 on success but a positive value to indicate an error.

The Pthreads API

While the pthreads API is comprehensive, it's quite easy to understand and use. We'll now explore the pthreads API, looking at the basics of thread creation through the specialized communication and synchronization methods that are available.

All multithreaded programs must make the `pthread` function prototypes and symbols available for use. This is accomplished by including the `pthread` standard header, as:

```
#include <pthread.h>
```

NOTE

The examples that follow are written for brevity, and in some cases, return values are not checked. To avoid debugging surprises, you are strongly encouraged to check all system call return values and never assume that a function is successful.

Thread Basics

All multithreaded applications must create threads and ultimately destroy them. This is provided in two functions by the pthreads API:

```
int pthread_create( pthread_t *thread,
                        pthread_attr_t *attr,
                void *(*start_routine)(void *), void *arg );
int pthread_exit( void *retval );
```

The **`pthread_create`** function permits the creation of a new thread, while **`pthread_exit`** allows a thread to terminate itself. There also is a function to permit one thread to terminate another, but we'll investigate that later.

To create a new thread, we call **`pthread_create`** and associate our **`pthread_t`** object with a function (`start_routine`). This function represents the top level code that will be executed within the thread. We can optionally provide a set of attributes via `pthread_attr_t` (via **`pthread_attr_init`**). Finally, the fourth argument (`arg`) is an optional argument that is passed to the thread upon creation.

Let's now look at a short example of thread creation (see Listing 14.1). In our `main` function, we first create a `pthread_t` object at line 10. This object represents our new thread. We call **`pthread_create`** at line 12 and provide the `pthread_t` object (which will be filled in by the **`pthread_create`** function) in addition to our function that contains the code for the thread (argument 3, `myThread`). A zero return indicates successful creation of the thread.

ON THE CD

Listing 14.1 Creating a Thread with `pthread_create` (on the CD-ROM at `./source/ch14/ptcreate.c`)

```
#include <pthread.h>
#include <stdlib.h>
#include <stdio.h>
#include <string.h>
#include <errno.h>

int main()
{
  int ret;
  pthread_t mythread;

  ret = pthread_create( &mythread, NULL, myThread, NULL );

  if (ret != 0) {
    printf( “Can’t create pthread (%s)\n”, strerror( errno ) );
    exit(-1);
  }

  return 0;
}
```

The **`pthread_create`** function returns zero if successful, otherwise a nonzero value is returned. Now let's look at the thread function itself, which will also demonstrate our **`pthread_exit`** function (see Listing 14.2). Our thread simply emits a message to `stdout` that it ran and then terminated at line 6 with **`pthread_exit`**.

ON THE CD

Listing 14.2 Terminating a Thread with `pthread_exit` (on the CD-ROM at `./source/ch14/ptcreate.c`)

```
1:   void *myThread( void *arg )
2:   {
3:     printf(“Thread ran!\n”);
4:
5:     /* Terminate the thread */
6:     pthread_exit( NULL );
7:   }
```

Our thread didn't use the `void` pointer argument, but this could be used to provide the thread with a specific personality, passed in at creation (see argument four of line 12 in Listing 14.1). The argument could represent a scalar value or a structure containing a variety of elements. The `exit` value presented to **`pthread_exit`** must not be of local scope, otherwise it won't exist once the thread is destroyed. The **`pthread_exit`** function does not return.

The startup cost for new threads is minimal in the new NPTL implementation, compared to the older LinuxThreads. In addition to significant improvements and optimizations in the NPTL, the allocation of thread memory structures is improved (thread data structures and thread local storage are now provided on the local thread stack).

Thread Management

Before we dig in to thread synchronization and coordination, let's look at a couple of miscellaneous thread functions that can be of use. The first is the **`pthread_self`** function, which can be used by a thread to retrieve its unique identifier. Recall in **`pthread_create`** that a `pthread_t` object reference was passed in as the first argument. This permits the thread creator to know the identifier for the thread just created. The thread itself can also retrieve this identifier by calling **`pthread_self`**.

```
pthread_t pthread_self( void );
```

Consider the updated thread function in Listing 14.3, which illustrates retrieving the `pthread_t` handle. At line 5, we call **`pthread_self`** to grab the handle and then emit it to `stdout` at line 7 (converting it to an `int`).

Listing 14.3 Retrieving the `pthread_t` Handle with **`pthread_self`** (on the CD-ROM at `./source/ch14/ptcreate.c`)

```
 1:     void *myThread( void *arg )
 2:     {
 3:       pthread_t pt;
 4:
 5:       pt = pthread_self();
 6:
 7:       printf("Thread %x ran!\n", (int)pt );
 8:
 9:       pthread_exit( NULL );
10:     }
```

Most applications require some type of initialization, but with threaded applications, the job can be difficult. The **`pthread_once`** function allows a developer to create an initialization routine that is invoked for a multithreaded application only once (even though multiple threads may attempt to invoke it).

The **pthread_once** function requires two objects: a pthread_once_t object (that has been preinitialized with **pthread_once_init**) and an initialization function. Consider the partial example in Listing 14.4. The first thread to call **pthread_once** will invoke the initialization function (initialize_app), but subsequent calls to **pthread_once** will result in no calls to initialize_app.

Listing 14.4 Providing a Single-use Initialization Function with **pthread_once**

```
#include <pthread.h>

pthread_once_t my_init_mutex = pthread_once_init;

void initialize_app( void )
{
  /* Single-time init here */
}

void *myThread( void *arg )
{
  ...

  pthread_once( &my_init_mutex, initialize_app );

  ...
}
```

The number of threads in LinuxThreads was a compile-time option (1000), whereas NPTL supports a dynamic number of threads. NPTL can support up to 2 billion threads on an IA-32 system [Drepper and Molnar03].

Thread Synchronization

The ability to synchronize threads is an important aspect of multithreaded application development. We'll look at a number of methods, but first we'll look at the most basic method, the ability for the creator thread to wait for the created thread to finish (otherwise known as a join). This activity is provided by the **pthread_join** API function. When called, the pthread_join call suspends the calling thread until a join is complete. When the join is done, the caller receives the joined thread's termination status as the return from **pthread_join**. The **pthread_join** function (somewhat equivalent to the wait function for processes) has the following prototype:

```
int pthread_join( pthread_t th, void **thread_return );
```

The `th` argument is the thread to which we wish to join. This argument is returned from `pthread_create` or passed via the thread itself via `pthread_self`. The `thread_return` can be `NULL`, which means we'll not capture the return status of the thread. Otherwise, the return value from the thread is stored in `thread_return`.

A thread is automatically joinable when using the default attributes of **`pthread_create`**. *If the attribute for the thread is defined as detached, then the thread can't be joined (because it's detached from the creating thread).*

To join with a thread, we must have the thread's identifier, which is retrieved from the **`pthread_create`** function. Let's look at a complete example (see Listing 14.5).

In this example, permit the creation of five distinct threads by calling **`pthread_create`** within a loop (lines 18–23) and storing the resulting thread identifiers in a `pthread_t` array (line 16). Once the threads are created, we begin the `join` process, again in a loop (lines 25–32). The **`pthread_join`** returns zero on success, and upon success, the `status` variable is emitted (note that this value is returned at line 8 within the thread itself).

Listing 14.5 Joining Threads with **`pthread_join`** (on the CD-ROM at `./source/ch14/ptjoin.c`)

```
 1:     #include <pthread.h>
 2:     #include <stdio.h>
 3:
 4:     void *myThread( void *arg )
 5:     {
 6:       printf( "Thread %d started\n", (int)arg );
 7:
 8:       pthread_exit( arg );
 9:     }
10:
11:     #define MAX_THREADS     5
12:
13:     int main()
14:     {
15:       int ret, i, status;
16:       pthread_t threadIds[MAX_THREADS];
17:
18:       for (i = 0 ; i < MAX_THREADS ; i++) {
19:         ret = pthread_create( &threadIds[i], NULL, myThread,
                                  (void *)i );
20:         if (ret != 0) {
```

```
      printf( “Error creating thread %d\n”, (int)threadIds[i] );
    }
  }

  for (i = 0 ; i < MAX_THREADS ; i++) {
    ret = pthread_join( threadIds[i], (void **)&status );
    if (ret != 0) {
      printf( “Error joining thread %d\n”, (int)threadIds[i] );
    } else {
      printf( “Status = %d\n”, status );
    }
  }

  return 0;
}
```

The **pthread_join** function suspends the caller until the requested thread has been joined. In many cases, we simply don't care about the thread once it's created. In these cases, we can identify this by detaching the thread. The creator or the thread itself can detach itself. We can also specify that the thread is detached when we create the thread (as part of the attributes). Once a thread is detached, it can never be joined. The **pthread_detach** function has the following prototype:

```
int pthread_detach( pthread_t th );
```

Let's now look at the process of detaching the thread within the thread itself (see Listing 14.6). Recall that a thread can identify its own identifier by calling **thread_self**.

Listing 14.6 Detaching a Thread from Within with **pthread_detach**

```
1:  void *myThread( void *arg )
2:  {
3:    printf( “Thread %d started\n”, (int)arg );
4:
5:    pthread_detach( pthread_self() );
6:
7:    pthread_exit( arg );
8:  }
```

At line 5, we simply call **pthread_detach**, specifying the thread identifier by calling **pthread_self**. When this thread exits, all resources are immediately freed (as it's detached and will never be joined by another thread). The **pthread_detach** function returns zero on success, nonzero if an error occurs.

GNU/Linux automatically places a newly created thread into the joinable state. This is not the case in other implementations, which can default to detached.

Thread Mutexes

A mutex is a variable that permits threads to implement critical sections. These sections enforce exclusive access to variables by threads, which if left unprotected would result in data corruption. This topic is discussed in detail in Chapter 16, "Synchronization with Semaphores."

Let's start by reviewing the mutex API, and then we'll illustrate the problem being solved. To create a mutex, we simply declare a variable that represents our mutex and initialize it with a special symbolic constant. The mutex is of type `pthread_mutex_t` and demonstrated as:

```
pthread_mutex_t myMutex = PTHREAD_MUTEX_INITIALIZER
```

As shown here, the initialization makes this mutex a fast mutex. The mutex initializer can actually be of one of three types, as shown in Table 14.1.

TABLE 14.1 Mutex Initializers

Type	Description
`PTHREAD_MUTEX_INITIALIZER`	Fast Mutex
`PTHREAD_RECURSIVE_MUTEX_INITIALIZER_NP`	Recursive Mutex
`PTHREAD_ERRORCHECK_MUTEX_INITIALIZER_NP`	Error-checking Mutex

The recursive mutex is a special mutex that allows the mutex to be locked several times (without blocking), as long as it's locked by the same thread. Even though the mutex can be locked multiple times without blocking, the thread must unlock the mutex the same number of times that it was locked. The error-checking mutex can be used to help find errors when debugging. Note that the `_NP` suffix for recursive and error-checking mutexes indicates that it's not portable.

Now that we have a mutex, we can lock and unlock it to create our critical section. This is done with the **`pthread_mutex_lock`** and **`pthread_mutex_unlock`** API functions. Another function called **`pthread_mutex_trylock`** can be used to try to lock a mutex, but it won't block if the mutex is already locked. Finally, we can destroy an existing mutex using **`pthread_mutex_destroy`**. These have the prototype:

```
int pthread_mutex_lock( pthread_mutex_t *mutex );
int pthread_mutex_trylock( pthread_mutex_t *mutex );
int pthread_mutex_unlock( pthread_mutex_t *mutex );
int pthread_mutex_destroy( pthread_mutex_t *mutex );
```

All functions return zero on success or a nonzero error code. All errors returned from **pthread_mutex_lock** and **pthread_mutex_unlock** are assertable (not recoverable). Therefore, we'll use the return of these functions to abort our program.

Locking a thread is the means by which we enter a critical section. Once our mutex is locked, we can safely enter the section without having to worry about data corruption or multiple access. To exit our critical section, we unlock the semaphore and we're done. The following code snippet illustrates a simple critical section:

```
pthread_mutex_t cntr_mutex = PTHREAD_MUTEX_INITIALIZER;
...
assert( pthread_mutex_lock( &cntr_mutex ) == 0 );
/* Critical Section */
/* Increment protected counter */
counter++;
/* Critical Section */
  assert( pthread_mutex_unlock( &cntr_mutex ) == 0 );
```

A critical section is a section of code that can be executed by at most one process at a time. The critical section exists to protect shared resources from multiple access.

The **pthread_mutex_trylock** operates under the assumption that if we can't lock our mutex, there's something else that we should do instead of blocking on the **pthread_mutex_lock** call. This call is demonstrated as:

```
ret = pthread_mutex_trylock( &cntr_mutex );
if (ret == EBUSY) {
  /* Couldn't lock, do something else */
} else if (ret == EINVAL) {
  /* Critical error */
  assert(0);
} else {
  /* Critical Section */
  ret = thread_mutex_unlock( &cntr_mutex );
}
```

Finally, to destroy our mutex, we simply provide it to the **pthread_mutex_destroy** function. The **pthread_mutex_destroy** function will succeed only if no thread currently has the mutex locked. If the mutex is locked, the function will fail and return the EBUSY error code. The **pthread_mutex_destroy** call is demonstrated with the following snippet:

```
ret = pthread_mutex_destroy( &cntr_mutex );
if (ret == EBUSY) {
  /* Mutex is locked, can't destroy */
} else {
  /* Mutex was destroyed */
}
```

Let's now look at an example that ties these functions together to illustrate why mutexes are important in multithreaded applications. We'll build on our previous applications that provide a basic infrastructure for task creation and joining. Consider the example in Listing 14.7. At line 4, we create our mutex and initialize it as a fast mutex. In our thread, our job is to increment the protVariable counter some number of times. This occurs for each thread (here we create 10), so we'll need to protect the variable from multiple access. We place our variable increment within a critical section by first locking the mutex and then, after incrementing the protected variable, unlocking it. This ensures that each task has sole access to the resource when the increment is performed and protects it from corruption. Finally, at line 52, we destroy our mutex using the **pthread_mutex_destroy** API function.

ON THE CD

Listing 14.7 Protecting a Variable in a Critical Section with Mutexes (on the CD-ROM at ./source/ch14/ptmutex.c)

```
1:     #include <pthread.h>
2:     #include <stdio.h>
3:
4:     pthread_mutex_t cntr_mutex = PTHREAD_MUTEX_INITIALIZER;
5:
6:     long protVariable = 0L;
7:
8:     void *myThread( void *arg )
9:     {
10:      int i, ret;
11:
12:      for (i = 0 ; i < 10000 ; i++) {
13:
14:        ret = pthread_mutex_lock( &cntr_mutex );
15:
16:        assert( ret == 0 );
17:
18:        protVariable++;
19:
```

```
    ret = pthread_mutex_unlock( &cntr_mutex );

    assert( ret == 0 );

  }

  pthread_exit( NULL );
}

#define MAX_THREADS     10

int main()
{
  int ret, i;
  pthread_t threadIds[MAX_THREADS];

  for (i = 0 ; i < MAX_THREADS ; i++) {
    ret = pthread_create( &threadIds[i], NULL, myThread, NULL );
    if (ret != 0) {
      printf( "Error creating thread %d\n", (int)threadIds[i] );
    }
  }

  for (i = 0 ; i < MAX_THREADS ; i++) {
    ret = pthread_join( threadIds[i], NULL );
    if (ret != 0) {
      printf( "Error joining thread %d\n", (int)threadIds[i] );
    }
  }

  printf( "The protected variable value is %ld\n", protVariable );

  ret = pthread_mutex_destroy( &cntr_mutex );

  if (ret != 0) {
    printf( "Couldn't destroy the mutex\n");
  }

  return 0;
}
```

When using mutexes, it's important to minimize the amount of work done in the critical section to what really needs to be done. Since other threads will block until a mutex is unlocked, minimizing the critical section time can lead to better performance.

Thread Condition Variables

Now that we have mutexes out of the way, let's explore condition variables. A condition variable is a special thread construct that allows a thread to wake up another thread based upon a condition. While mutexes provide a simple form of synchronization (based upon the lock status of the mutex), condition variables are a means for one thread to wait for an event and another to signal it that the event has occurred. An event can mean anything here. A thread blocks on a mutex but can wait on any condition variable. Think of them as wait queues, which is exactly what the implementation does in GNU/Linux.

Consider this problem of a thread awaiting a particular condition being met. With only mutexes, the thread would have to poll to acquire the mutex, check the condition, and then release the mutex if there was no work to do (the condition wasn't met). That kind of busy looping can lead to poor performing applications and should therefore be avoided.

The pthreads API provides a number of functions supporting condition variables. These functions provide condition variable creation, waiting, signaling, and destruction. The condition variable API functions are presented below:

```
int pthread_cond_wait( pthread_cond_t *cond,
                       pthread_mutex_t *mutex );
int pthread_cond_timedwait( pthread_cond_t *cond,
                            pthread_mutex_t *mutex,
                            const struct timespec *abstime );
int pthread_cond_signal( pthread_cond_t *cond );
int pthread_cond_broadcast( pthread_cond_t *cond );
int pthread_cond_destroy( pthread_cond_t *cond );
```

To create a condition variable, we simply create a variable of type `pthread_cond_t`. We initialize this by setting it to `PTHREAD_COND_INITIALIZER` (similar to mutex creation and initialization). This is demonstrated as:

```
pthread_cond_t recoveryCond = PTHREAD_COND_INITIALIZER;
```

Condition variables require the existence of a mutex that is associated with them, which we create as before:

```
pthread_mutex_t recoveryMutex = PTHREAD_MUTEX_INITIALIZER;
```

Now let's look at a thread awaiting a condition. In this example, let's say we have a thread whose job is to warn of overload conditions. Work comes in on a queue, with an accompanying counter identifying the amount of work to do. When the amount of work exceeds a certain value (MAX_NORMAL_WORKLOAD), then our thread should wake up and perform a recovery. Our fault thread for synchronizing with the alert thread is illustrated as:

```
/* Fault Recovery Thread Loop */
while ( 1 ) {
  assert( pthread_mutex_lock( &recoveryMutex ) == 0);
  while (workload < MAX_NORMAL_WORKLOAD) {
    pthread_cond_wait( &recoveryCond, &recoveryMutex );
  }
  /*————————*/
  /* Recovery Code. */
  /*————————*/
  assert( pthread_mutex_unlock( &recoveryMutex ) == 0);
}
```

This is the standard pattern when dealing with condition variables. We start by locking the mutex, entering **pthread_cond_wait**, and upon waking up from our condition, unlocking the mutex. The mutex must be locked first because upon entry to **pthread_cond_wait**, the mutex is automatically unlocked. When we return from **pthread_cond_wait**, the mutex has been reacquired, meaning that we'll need to unlock it afterward. The mutex is necessary here to handle race conditions that exist in this call sequence. To ensure that our condition is met, we loop around the **pthread_cond_wait**, and if the condition is not satisfied (in this case, our workload is normal), then we reenter the **pthread_cond_wait** call. Note that since the mutex is locked upon return from **pthread_cond_wait**, we don't need to call **pthread_mutex_lock** here.

Now let's look at the signal code. This is considerably simpler than that code necessary to wait for the condition. Two possibilities exist for signaling: sending a single signal, or broadcasting to all waiting threads.

The first case is signaling one thread. In either case, we first lock the mutex before calling the signal function and then unlock when we're done. To signal one thread, we call the **pthread_cond_signal** function, as:

```
pthread_mutex_lock( &recoveryMutex );
pthread_cond_signal( &recoveryCond );
pthread_mutex_unlock( &recovery_Mutex );
```

Once the mutex is unlocked, exactly one thread is signaled and allowed to execute. Each function returns zero on success or an error code. If our architecture supports multiple threads for recovery, we could instead use the **pthread_cond_broadcast**. This function awakes all threads currently awaiting the condition. This is demonstrated as:

```
pthread_mutex_lock( &recoveryMutex );
pthread_cond_broadcast( &recoveryCond );
pthread_mutex_unlock( &recovery_Mutex );
```

Once the mutex is unlocked, the series of threads is then permitted to perform recovery (though one by one since they're dependent upon the mutex).

The pthreads API also supports a version of timed-wait for a condition variable. This function, **pthread_cond_timedwait**, allows the caller to specify an absolute time representing when to give up and return to the caller. The return value will be ETIMEDOUT, to indicate that the function returned because of a time-out rather than a successful return. The following code snippet illustrates its use:

```
struct timeval currentTime;
struct timespec  expireTime;
int ret;
...
assert( pthread_mutex_lock( &recoveryMutex ) == 0);
gettimeofday( &currentTime );
expireTime.tv_sec = currentTime.tv_sec + 1;
expireTime.tv_nsec = currentTime.tv_usec * 1000;
ret = 0;
while ((workload < MAX_NORMAL_WORKLOAD) && (ret != ETIMEDOUT) {
  ret = pthread_cond_timedwait( &recoveryCond, &recoveryMutex,
                &expireTime );
}
if (ret == ETIMEDOUT) {
  /* Timeout – perform timeout processing */
} else {
  /* Condition met – perform condition recovery processing */
}
assert( pthread_mutex_unlock( &recoveryMutex ) == 0);
```

The first item to note is the generation of a timeout. We use the gettimeofday function to get the current time and then add one second to it in the timespec structure. This will be passed to **pthread_cond_timedwait** to identify the time at which we desire a timeout if the condition has not been met. In this case, which is very similar to the standard **pthread_cond_wait** example, we check in our loop that the **pthread_cond_timedwait** function has not returned ETIMEDOUT. If it has, we exit our loop and then check again to perform timeout processing. Otherwise, we perform our standard condition processing (recovery for this example) and then reacquire the mutex.

The final function to note here is **pthread_cond_destroy**. We simply pass the condition variable to the function, as:

```
pthread_mutex_destroy( &recoveryCond );
```

It's important to note that in the GNU/Linux implementation, no resources are actually attached to the condition variable, so this function simply checks to see if any threads are currently pending on the condition variable.

Let's now look at a complete example that brings together all of the elements discussed above for condition variables. In this example, we'll illustrate condition variables in the context of producers and consumers. We'll create a producer thread that creates work and then *N* consumer threads that operate on the (simulated) work.

Our first listing (Listing 14.8) shows the main program. This listing is similar to our previous examples of creating and then joining threads, with a few changes. We create two types of threads in this listing. At lines 18–21, we create a number of consumer threads, and at line 24, we create a single producer thread. We'll look at these shortly. After creation of the last thread, we join the producer thread (resulting in a suspend of the main application until it has completed). We then wait for the work to complete (as identified by a simple counter, `workCount`). We want to allow the consumer threads to complete their work, so we wait until this variable is zero, indicating that all work is consumed.

The block of code at lines 33–36 shows joins for the consumer threads, with one interesting change. In this example, the consumer threads never quit, so we cancel them here using the **pthread_cancel** function. This function has the prototype:

```
int pthread_cancel( pthread_t thread );
```

This permits us to terminate another thread when we're done with it. In this example, we've produced the work that we need the consumers to work on, so we cancel each thread in turn (line 34). Finally, we destroy our `condition` variable and mutex at lines 37 and 38, respectively.

ON THE CD

Listing 14.8 Producer/Consumer Example Initialization and `main` (on the CD-ROM at `./source/ch14/ptcond.c`)

```
1:     #include <pthread.h>
2:     #include <stdio.h>
3:
4:     pthread_mutex_t cond_mutex = PTHREAD_MUTEX_INITIALIZER;
```

```
pthread_cond_t condition = PTHREAD_COND_INITIALIZER;

int workCount = 0;

#define MAX_CONSUMERS   10

int main()
{
  int i;
  pthread_t consumers[MAX_CONSUMERS];
  pthread_t producer;

  /* Spawn the consumer thread */
  for ( i = 0 ; i < MAX_CONSUMERS ; i++ ) {
    pthread_create( &consumers[i], NULL,
                      consumerThread, NULL );
  }

  /* Spawn the single producer thread */
  pthread_create( &producer, NULL,
                      producerThread, NULL );

  /* Wait for the producer thread */
  pthread_join( producer, NULL );

  while ((workCount > 0));

  /* Cancel and join the consumer threads */
  for ( i = 0 ; i < MAX_CONSUMERS ; i++ ) {
    pthread_cancel( consumers[i] );
  }

  pthread_mutex_destroy( &cond_mutex );
  pthread_cond_destroy( &condition );

  return 0;
}
```

Next, let's look at the producer thread function (Listing 14.9). The purpose of the producer thread is to produce work, simulated by incrementing the `workCount` variable. A nonzero `workCount` indicates that work is available to do. We loop for a number of times to create work, as is shown at lines 8–22. As shown in the condition variable sample, we first lock our mutex at line 10 and then create work to do (increment `workCount`). We then notify the awaiting consumer (worker) threads at line 14 using the **pthread_cond_broadcast** function. This will notify any awaiting consumer threads that work is now available to do. Next, at line 15, we unlock the mutex, allowing the consumer threads to lock the mutex and perform their work.

At lines 20–22, we simply do some busy work to allow the kernel to schedule another task (thereby avoiding synchronous behavior, for illustration purposes).

When all of the work has been produced, we permit the producer thread to exit (which will be joined in our `main` function at line 28 of Listing 14.8).

ON THE CD

Listing 14.9 Producer Thread Example for Condition Variables (on the CD-ROM at `./source/ch14/ptcond.c`)

```
 1:   void *producerThread( void *arg )
 2:   {
 3:     int i, j, ret;
 4:     double result=0.0;
 5:
 6:     printf(“Producer started\n”);
 7:
 8:     for ( i = 0 ; i < 30 ; i++ ) {
 9:
10:       ret = pthread_mutex_lock( &cond_mutex );
11:       if (ret == 0) {
12:         printf( “Producer: Creating work (%d)\n”, workCount );
13:         workCount++;
14:         pthread_cond_broadcast( &condition );
15:         pthread_mutex_unlock( &cond_mutex );
16:       } else {
17:         assert(0);
18:       }
19:
20:       for ( j = 0 ; j < 60000 ; j++ ) {
21:         result = result + (double)random();
22:       }
23:
24:     }
25:
26:     printf(“Producer finished\n”);
27:
28:     pthread_exit( NULL );
29:   }
```

Now let's look at the consumer thread (see Listing 14.10). Our first task is to detach ourselves (line 5), since we won't ever join with the creating thread. Then we go into our work loop (lines 9–22) to process the workload. We first lock the condition mutex at line 11 and then wait for the condition to occur at line 12. We then check to make sure that the condition is true (there's work to do) at line 14. Note that since we're broadcasting to threads, we may not have work to do for every thread, so we test before we assume that work is available.

Once we've completed our work (in this case, simply decrementing the work count at line 15), we release the mutex at line 19 and wait again for work at line 11. Note that since we cancel our thread, we'll never see the `printf` at line 23, nor will we exit the thread at line 25. The **`pthread_cancel`** function terminates the thread so that the thread does not terminate normally.

Listing 14.10 Consumer Thread Example for Condition Variables (on the CD-ROM at `./source/ch14/ptcond.c`)

```
1:   void *consumerThread( void *arg )
2:   {
3:     int ret;
4:
5:     pthread_detach( pthread_self() );
6:
7:     printf( “Consumer %x: Started\n”, pthread_self() );
8:
9:     while( 1 ) {
10:
11:      assert( pthread_mutex_lock( &cond_mutex ) == 0);
12:      assert( pthread_cond_wait( &condition, &cond_mutex ) == 0 );
13:
14:      if (workCount) {
15:        workCount–;
16:        printf( “Consumer %x: Performed work (%d)\n”,
17:                  pthread_self(), workCount );
18:      }
19:      assert( pthread_mutex_unlock( &cond_mutex ) == 0);
20:
21:    }
22:
23:    printf( “Consumer %x: Finished\n”, pthread_self() );
24:
25:    pthread_exit( NULL );
26:  }
```

Let's look at this application in action. For brevity, we'll show only the first 30 lines emitted, but this will give you a good indication of how the application behaves (see Listing 14.11). We can see the consumer threads starting up, the producer starting, and then work being created and consumed in turn.

Listing 14.11 Application Output for Condition Variable Application

```
# ./ptcond
Consumer 4082cd40: Started
Consumer 4102ccc0: Started
Consumer 4182cc40: Started
Consumer 42932bc0: Started
Consumer 43132b40: Started
Consumer 43932ac0: Started
Consumer 44132a40: Started
Consumer 449329c0: Started
Consumer 45132940: Started
Consumer 459328c0: Started
Producer started
Producer: Creating work (0)
Producer: Creating work (1)
Consumer 4082cd40: Performed work (1)
Consumer 4102ccc0: Performed work (0)
Producer: Creating work (0)
Consumer 4082cd40: Performed work (0)
Producer: Creating work (0)
Producer: Creating work (1)
Producer: Creating work (2)
Producer: Creating work (3)
Producer: Creating work (4)
Producer: Creating work (5)
Consumer 4082cd40: Performed work (5)
Consumer 4102ccc0: Performed work (4)
Consumer 4182cc40: Performed work (3)
Consumer 42932bc0: Performed work (2)
Consumer 43132b40: Performed work (1)
Consumer 43932ac0: Performed work (0)
Producer: Creating work (0)
```

The design of multithreaded applications follows a small number of patterns (or models). The master/servant model is common where a single master doles out work to a collection of servants. The pipeline model splits work up into stages where one or more threads make up each of the work phases.

BUILDING THREADED APPLICATIONS

Building pthread-based applications is very simple. All that's necessary is to specify the pthreads library during compilation as:

```
gcc -pthread threadapp.c -o threadapp -lpthread
```

This will link our application with the `pthread` library, making the `pthread` functions available for use. Note also that we specify the `-pthread` option, which adds support for multithreading to the application (such as reentrancy). The option also ensures that certain global system variables (such as `errno`) are provided on a per-thread basis.

One topic that's important to discuss in multithreaded applications is that of reentrancy. Consider two threads, each of which uses the `strtok` function. The `strtok` function uses an internal buffer for token processing of a string. This internal buffer can be used by only one user at a time, which is fine in the process world (forked processes), but in the thread world runs into problems. If each thread attempts to call `strtok`, then the internal buffer is corrupted, leading to undesirable (and unpredictable) behavior. To fix this, rather than use an internal buffer, a thread-supplied buffer could be used instead. This is exactly what happens with the thread-safe version of `strtok`, called `strtok_r`. The suffix `_r` indicates that the function is thread safe.

SUMMARY

Multithreaded application development is a powerful model for the development of high-performance software systems. GNU/Linux provides the POSIX pthreads API for a standard and portable programming model. In this chapter, we explored the standard thread creation, termination, and synchronization functions. This includes the basic synchronization using a join, but also more advanced coordination using mutexes and condition variables. Finally, building pthread applications was investigated, along with some of the pitfalls that can be encountered (such as reentrancy) and how to deal with them. The GNU/Linux 2.6 kernel (using NPTL) provides a closer POSIX implementation and more efficient IPC and kernel support, than the prior LinuxThreads version.

REFERENCES

[Drepper and Molnar 2003] Drepper, Ulrich and Molnar, Ingo. (2003) *The Native POSIX Thread Library for Linux*. Red Hat, Inc.

API SUMMARY

```
#include <pthread.h>
int pthread_create( pthread_t *thread,
                    pthread_attr_t *attr,
                    void *(*start_routine)(void *), void *arg );
int pthread_exit( void *retval );
pthread_t pthread_self( void );
int pthread_join( pthread_t th, void **thread_return );
int pthread_detach( pthread_t th );
int pthread_mutex_lock( pthread_mutex_t *mutex );
int pthread_mutex_trylock( pthread_mutex_t *mutex );
int pthread_mutex_unlock( pthread_mutex_t *mutex );
int pthread_mutex_destroy( pthread_mutex_t *mutex );
int pthread_cond_wait( pthread_cond_t *cond,
                       pthread_mutex_t *mutex );
int pthread_cond_timedwait( pthread_cond_t *cond,
                            pthread_mutex_t *mutex,
                            const struct timespec *abstime );
int pthread_cond_signal( pthread_cond_t *cond );
int pthread_cond_broadcast( pthread_cond_t *cond );
int pthread_cancel( pthread_t thread );
```

15 IPC with Message Queues

In This Chapter

INTRODUCTION

The topic of inter-process communication is an important one because it allows us the ability to build systems out of numerous communicating asynchronous processes. This is beneficial because we can naturally segment the functionality of a large system into a number of distinct elements. Because GNU/Linux processes utilize independent memory spaces, a function in one process cannot call another in a different process. Message queues provide one means to permit communication and coordination between processes. In this chapter, we'll review the message queue model (which conforms to the SystemV UNIX model), as well as explore some sample code that utilizes the message queue API.

QUICK OVERVIEW OF MESSAGE QUEUES

Let's begin by taking a whirlwind tour of the POSIX-compliant message queue API. We'll look at code examples that illustrate creating a message queue, configuring its size, sending and receiving a message, and then removing the message queue. Once we've had a taste of the message queue API, we'll dive in deeper in the following sections.

Using the message queue API requires that the function prototypes and symbols be available to the application. This is done by including the `msg.h` header file as:

```
#include <sys/msg.h>
```

We'll first introduce a common header file that defines some common information needed for the writer and reader of the message. We define our system-wide queue ID (111) at line 3. This isn't the best way to define the queue, but later on we'll look at a way to define a unique system ID. Lines 5–10 define our message type, with the required long type at the head of the structure (line 6).

Listing 15.1 Common Header File Used by the Sample Applications (on the CD-ROM at `./source/ch15/common.h`)

```
1:     #define MAX_LINE        80
2:
3:     #define MY_MQ_ID        111
4:
5:     typedef struct {
6:             long  type;               // Msg Type (> 0)
7:             float fval;               // User Message
8:             unsigned int uival;       // User Message
9:             char strval[MAX_LINE+1];  // User Message
10:    } MY_TYPE_T;
```

Creating a Message Queue

To create a message queue, we use the **msgget** API function. This function takes a message queue ID (a unique identifier, or key, within a given host) and another argument identifying the message flags. The flags, in the queue create example (see Listing 15.2) specify that a queue is to be created (`IPC_CREAT`) as well as the access permissions of the message queue (read/write permission for system, user, and group).

*The result of the **msgget** function is a handle, which is similar to a file descriptor, pointing to the message queue with the particular ID.*

ON THE CD

Listing 15.2 Creating a Message Queue with **msgget** (on the CD-ROM at `./source/ch15/mqcreate.c`)

```
1:  #include <stdio.h>
2:  #include <sys/msg.h>
3:  #include "common.h"
4:
5:  int main()
6:  {
7:    int msgid;
8:
9:    /* Create the message queue with the id MY_MQ_ID */
10:   msgid = msgget( MY_MQ_ID, 0666 | IPC_CREAT );
11:
12:   if (msgid >= 0) {
13:
14:     printf( "Created a Message Queue %d\n", msgid );
15:
16:   }
17:
18:   return 0;
19: }
```

Upon creating the message queue at line 10 (in Listing 15.2), we get a return integer that represents a handle for the message queue. This message queue ID can be used in subsequent message queue calls to send or receive messages.

Configuring a Message Queue

When we create a message queue, some of the details of the process that created the queue are automatically stored with it (for permissions) as well as a default queue size in bytes (16KB). We can adjust this size using the **`msgctl`** API function. Listing 15.3 illustrates reading the defaults for the message queue, adjusting the queue size, and then configuring the queue with the new set.

ON THE CD

Listing 15.3 Configuring a Message Queue with **msgctl** (on the CD-ROM at `./source/ch15/mqconf.c`)

```
#include <stdio.h>
#include <sys/msg.h>
#include "common.h"

int main()
{
```

```
 7:     int msgid, ret;
 8:     struct msqid_ds buf;
 9:
10:     /* Get the message queue for the id MY_MQ_ID */
11:     msgid = msgget( MY_MQ_ID, 0 );
12:
13:     /* Check successful completion of msgget */
14:     if (msgid >= 0) {
15:
16:       ret = msgctl( msgid, IPC_STAT, &buf );
17:
18:       buf.msg_qbytes = 4096;
19:
20:       ret = msgctl( msgid, IPC_SET, &buf );
21:
22:       if (ret == 0) {
23:
24:         printf( "Size successfully changed for queue %d.\n", msgid );
25:
26:       }
27:
28:     }
29:
30:     return 0;
31:   }
```

First, at line 11, we get the message queue ID using **msgget**. Note that the second argument here is 0 because we're not creating the message queue, just retrieving its ID. We use this at line 16 to get the current queue data structure using the `IPC_STAT` command and our local buffer (for which the function will fill in the defaults). We adjust the queue size at line 18 (by modifying the `msg_qbytes` field of the structure) and then write it back at line 20 using the **msgctl** API function with the `IPC_SET` command. We could also modify the user or group ID of the message queue or its mode. We'll discuss these in more detail later.

Writing a Message to a Message Queue

Now let's look at actually sending a message through a message queue. A message within the context of a message queue has only one constraint. The object that's being sent must include a long variable at its head that defines the message type. We'll discuss this more later, but it's simply a way to differentiate messages that have been loaded onto a queue (and also how those messages can be read from the queue). The general structure for a message is:

```
typedef struct {
  long type;
  char message[80];
} MSG_TYPE_T;
```

In this example (`MSG_TYPE_T`), we have our required `long` at the head of the message, followed by the user-defined message (in this case, a string of 80 characters).

To send a message to a message queue (see Listing 15.4), we use the **`msgsnd`** API function. Following a similar pattern to our previous examples, we first identify the message queue ID using the **`msgget`** API function (line 11). Once this is known, we can send a message to it. Next, we initialize our message at lines 16–19. This includes specifying the mandatory type (must be greater than 0), a floating-point value (`fval`) and unsigned int value (`uival`), and a character string (`strval`). To send this message, we call the **`msgsnd`** API function. The arguments for this function are the message queue ID (`qid`), our message (a reference to `myObject`), the size of the message we're sending (the size of `MY_TYPE_T`), and finally a set of message flags (for now, `0`, but we'll investigate more later).

Listing 15.4 Sending a Message with **`msgsnd`**. (on the CD-ROM at `./source/ch15/mqsend.c`)

```
 1:    #include <sys/msg.h>
 2:    #include <stdio.h>
 3:    #include “common.h”
 4:
 5:    int main()
 6:    {
 7:      MY_TYPE_T myObject;
 8:      int qid, ret;
 9:
10:      /* Get the queue ID for the existing queue */
11:      qid = msgget( MY_MQ_ID, 0 );
12:
13:      if (qid >= 0) {
14:
15:        /* Create our message with a message queue type of 1 */
16:        myObject.type = 1L;
17:        myObject.fval = 128.256;
18:        myObject.uival = 512;
19:        strncpy( myObject.strval, “This is a test.\n”, MAX_LINE );
20:
21:        /* Send the message to the queue defined by the queue ID */
22:        ret = msgsnd( qid, (struct msgbuf *)&myObject,
23:                       sizeof(MY_TYPE_T), 0 );
```

```
        if (ret != -1) {

          printf( "Message successfully sent to queue %d\n", qid );

        }

      }

      return 0;
    }
```

That's it! This message is now held in the message queue, and at any point in the future, it can be read (and consumed) by the same or a different process.

Reading a Message from a Message Queue

Now that we have a message in our message queue, let's look at reading that message and displaying its contents. We retrieve the ID of the message queue using **msgget** at line 12 and then use this as the target queue from which to read using the **msgrcv** API function at lines 16–17. The arguments to **msgrcv** are first the message queue ID (`qid`), the message buffer into which our message will be copied (`myObject`), the size of the object (`sizeof(MY_TYPE_T)`), the message type that we want to read (`1`), and the message flags (`0`). Note that when we sent our message (in Listing 15.4), we specified our message type as `1`. We use this same value here to read the message from the queue. Had we used another value, the message would not have been read. More on this subject in the "`msgrcv`" section later in this chapter.

ON THE CD

Listing 15.5 Reading a Message with `msgrcv` (on the CD-ROM at `./source/ch15/mqrecv.c`)

```
 1:    #include <sys/msg.h>
 2:    #include <stdio.h>
 3:    #include "common.h"
 4:
 5:    int main()
 6:    {
 7:      MY_TYPE_T myObject;
 8:      int qid, ret;
 9:
10:      qid = msgget( MY_MQ_ID, 0 );
11:
12:      if (qid >= 0) {
13:
```

```
    ret = msgrcv( qid, (struct msgbuf *)&myObject,
                  sizeof(MY_TYPE_T), 1, 0 );

    if (ret != -1) {

      printf( "Message Type: %ld\n", myObject.type );
      printf( "Float Value:  %f\n",  myObject.fval );
      printf( "Uint Value:   %d\n",  myObject.uival );
      printf( "String Value: %s\n",  myObject.strval );

    }

  }

  return 0;
}
```

The final step in our application in Listing 15.5 is to emit the message read from the message queue. We use our object type to access the fields in the structure and simply emit them with `printf`.

Removing a Message Queue

As a final step, let's look at how we can remove a message queue (and any messages that may be held on it). We use the **`msgctl`** API function for this purpose with the command of `IPC_RMID`. This is illustrated in Listing 15.6.

ON THE CD

Listing 15.6 Removing a Message Queue with `msgctl` (on the CD-ROM at `./source/ch15/mqdel.c`)

```
#include <stdio.h>
#include <sys/msg.h>
#include "common.h"

int main()
{
  int   msgid, ret;

  msgid = msgget( MY_MQ_ID, 0 );

  if (msgid >= 0) {

    /* Remove the message queue */
    ret = msgctl( msgid, IPC_RMID, NULL );

    if (ret != -1) {
```

```

            printf( "Queue %d successfully removed.\n", msgid );

        }

    }

    return 0;
}
```

In Listing 15.6, we first identify the message queue ID using **msgget** and then use this with **msgctl** to remove the message queue. Any messages that happened to be on the message queue when **msgctl** was called would be immediately removed.

That does it for our whirlwind tour. In the next section, we'll dig deeper into message queue API and look at some of the behaviors of the commands that weren't covered already.

The Message Queue API

Let's now dig into the message queue API and investigate each of the functions in more detail. For a quick review, Table 15.1 provides the API functions and their purposes.

TABLE 15.1 Message Queue API Functions and Uses

API Function	Uses
msgget	Create a new message queue
	Get a message queue ID
msgsnd	Send a message to a message queue
msgrcv	Receive a message from a message queue
msgctl	Get the info about a message queue
	Set the info for a message queue
	Remove a message queue

Figure 15.1 graphically illustrates the message queue API functions and their relationship in the process.

We'll address these functions now in detail, identifying each of the uses with descriptive examples.

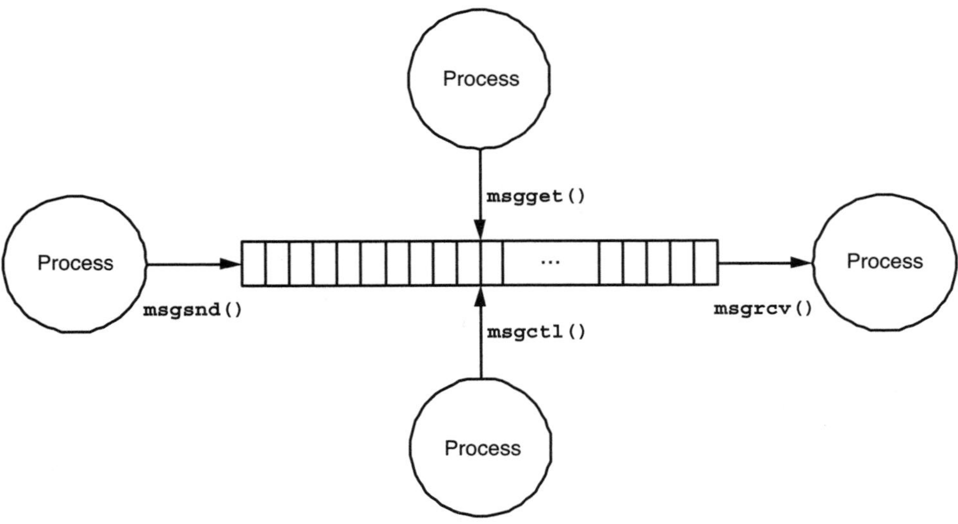

FIGURE 15.1 Message queue API functions.

msgget

The **msgget** API function serves two basic roles: to create a message queue or to get the identifier of a message queue that already exists. The result of the **msgget** function (unless an error occurs) is the message queue identifier (used by all other message queue API functions). The prototype for the **msgget** function is defined as follows:

```
int msgget( key_t key, int msgflag );
```

The `key` argument defines a system-wide identifier that uniquely identifies a message queue. `key` must be a nonzero value or the special symbol `IPC_PRIVATE`. The `IPC_PRIVATE` variable simply tells the **msgget** function that no key is provided and to simply make one up. The problem with this is that no other process can then find the message queue, but for local message queues (private queues), this method works fine.

The `msgflag` argument allows the user to specify two distinct parameters: a command and an optional set of access permissions. Permissions replicate those found as modes for the file creation functions (see Table 15.2). The command can take three forms. The first is simply `IPC_CREAT`, which instructs **msgget** to create a new message queue (or return the ID for the queue if it already exists). The second includes two commands (IPC_CREAT | IPC_EXCL), which request that the message queue be created, but if it already exists, the API function should fail and return an error response (EEXIST). The third possible command argument is simply 0. This form tells **msgget** that the message queue identifier for an existing queue is being requested.

TABLE 15.2 Message Queue Permissions for the **msgget** msgflag Argument

Symbol	Value	Meaning
S_IRUSR	0400	User has read permission
S_IWUSR	0200	User has write permission
S_IRGRP	0040	Group has read permission
S_IWGRP	0020	Group has write permission
S_IROTH	0004	Other has read permission
S_IWOTH	0002	Other has write permission

Let's look at a few examples of the **msgget** function to create message queues or access existing ones. Assume in the following code snippets that msgid is an int value (int msgid). Let's start by creating a private queue (no key is provided).

```
msgid = msgget( IPC_PRIVATE, IPC_CREAT | 0666 );
```

If the **msgget** API function fails, -1 is returned with the actual error value provided within the process's errno variable.

Let's now say that we want to create a message queue with a key value of 0x111. We also want to know if the queue already exists, so we'll use the IPC_EXCL in this example:

```
// Create a new message queue
msgid = msgget( 0x111, IPC_CREAT | IPC_EXCL | 0666 );
 if (msgid == -1) {
   printf("Queue already exists...\n");
 } else {
   printf("Queue created...\n");
 }
```

An interesting question you've probably asked yourself now is how can you coordinate the creation of queues using IDs that may not be unique? What happens if someone already used the 0x111 key? Luckily, there's a way to create keys in a system-wide fashion that ensures uniqueness. The `ftok` system function provides the means to create system-wide unique keys using a file in the filesystem and a number. As the file (and its path) will by default be unique in the filesystem, a unique key can be created easily. Let's look at an example of using `ftok` to create a unique key. Assume that the file with path `/home/mtj/queues/myqueue` exists.

```
key_t myKey;
int   msgid;
// Create a key based upon the defined path and number
myKey = ftok( "/home/mtj/queues/myqueue", 0 );
msgid = msgget( myKey, IPC_CREAT | 0666 );
```

This will create a key for this path and number. Each time `ftok` is called with this path and number, the same key will be generated. Therefore, it provides a useful way to generate a key based upon a file in the filesystem.

One last example is getting the message queue ID of an existing message queue. The only difference in this example is that we provide no command, only the key:

```
msgid = msgget( 0x111, 0 );
if (msgid == -1) {
  printf("Queue doesn't exist...\n");
}
```

The `msgflags` (second argument to **`msgget`**) is zero in this case, which indicates to this API function that an existing message queue is being sought.

One final note on message queues is the default settings that are given to a message queue when it is created. The configuration of the message queue is noted in the parameters shown in Table 15.3. Note that there's no way to change these defaults within **`msgget`**. In the next section, we'll look at some of the parameters that can be changed and their effects.

The user can override the `msg_perm.uid`, `msg_perm.gid`, `msg_perm.mode`, and `msg_qbytes` directly. More on this topic in the next section.

msgctl

The **`msgctl`** API function provides three distinct features for message queues. The first is the ability to read the current set of message queue defaults (via the `IPC_STAT` command). The second is the ability to modify a subset of the defaults (via `IPC_SET`). Finally, the ability to remove a message queue is provided (via `IPC_RMID`). The `msgctl` prototype function is defined as:

```
#include <sys/msg.h>
int msgctl( int msgid, int cmd, struct msqid_ds *buf );
```

TABLE 15.3 Message Queue Configuration and Defaults in `msgget`

Parameter	**Default Value**
`msg_perm.cuid`	Effective user ID of the calling process (creator)
`msg_perm.uid`	Effective user ID of the calling process (owner)
`msg_perm.cgid`	Effective group ID of the calling process (creator)
`msg_perm.gid`	Effective group ID of the calling process (owner)
`msg_perm.mode`	Permissions (lower 9 bits of `msgflag`)
`msg_qnum`	0 (Number of messages in the queue)
`msg_lspid`	0 (Process ID of last `msgsnd`)
`msg_lrpid`	0 (Process ID of last `msgrcv`)
`msg_stime`	0 (last `msgsnd` time)
`msg_rtime`	0 (Last `msgrcv` time)
`msg_ctime`	Current time (last change time)
`msg_qbytes`	Queue size in bytes (system limit)—(16KB)

Let's start by looking at **`msgctl`** as a means to remove a message queue from the system. This is the simplest use of **`msgctl`** and can be demonstrated very easily. In order to remove a message queue, all that's needed is the message queue identifier that is returned by **`msgctl`**.

While a system-wide unique key is required to create a message queue, only the message queue ID (returned from **`msgget`***) is required to configure a queue, send a message from a queue, receive a message from a queue, or remove a queue.*

Let's look at an example of message queue removal using **`msgctl`**. We first get the message queue identifier using **`msgget`** and then use this ID in our call to **`msgctl`**.

```
int msgid, ret;
...
msgid = msgget( QUEUE_KEY, 0 );
if (msgid != -1) {
  ret = msgctl( msgid, IPC_RMID, NULL );
      if (ret == 0) {
    // queue was successfully removed.
  }
}
```

If any processes are currently blocked on a **msgsnd** or **msgrcv** API function, those functions will return with an error (-1) with the errno process variable set to EIDRM. The process performing the IPC_RMID must have adequate permissions to remove the message queue. If permissions do not allow the removal, an error return is generated with an errno variable set to EPERM.

Now let's look at IPC_STAT (read configuration) and IPC_SET (write configuration) commands together for **msgctl**. In the previous section, we identified the range of parameters that make up the configuration and status parameters. Now let's look at which of the parameters can be directly manipulated or used by the application developer. Table 15.4 lists the parameters that can be updated once a message queue has been created.

TABLE 15.4 Message Queue Parameters That May Be Updated

Parameter	Description
msg_perm.uid	Message queue user owner
msg_perm.gid	Message queue group owner
msg_perm.mode	Permissions (see Table 15.2)
msg_qbytes	Size of message queue in bytes

Changing these parameters is a very simple process. The process should be that the application first reads the current set of parameters (via IPC_STAT) and then modifies the parameters of interest before writing them back out (via IPC_SET). See Listing 15.7 for an illustration of this process.

Listing 15.7 Setting All Possible Options in **msgctl** (on the CD-ROM at ./source/ch15/mqrdset.c)

```
#include <stdio.h>
#include <sys/msg.h>
#include <unistd.h>
#include <sys/types.h>
#include <errno.h>
#include "common.h"

int main()
{
  int msgid, ret;
  struct msqid_ds buf;

```

```
13:     /* Get the message queue for the id MY_MQ_ID */
14:     msgid = msgget( MY_MQ_ID, 0 );
15:
16:     /* Check successful completion of msgget */
17:     if (msgid >= 0) {
18:
19:       ret = msgctl( msgid, IPC_STAT, &buf );
20:
21:       buf.msg_perm.uid = geteuid();
22:       buf.msg_perm.gid = getegid();
23:       buf.msg_perm.mode = 0644;
24:       buf.msg_qbytes = 4096;
25:
26:       ret = msgctl( msgid, IPC_SET, &buf );
27:
28:       if (ret == 0) {
29:
30:         printf( "Parameters successfully changed.\n");
31:
32:       } else {
33:
34:         printf( "Error %d\n", errno );
35:
36:       }
37:
38:     }
39:
40:     return 0;
41:   }
```

At line 14, we get our message queue identifier, and then we use this at line 19 to retrieve the current set of parameters. At line 21, we set the `msg_perm.uid` (effective user ID) with the current effective user ID using the `geteuid()` function. Similarly, we set the `msg_perm.gid` (effective group ID) at line 22 using the `getegid()` function. At line 23 we set the mode, and at line 24 we set the maximum queue size (in bytes). In this case we set it to 4KB. We now take this structure and set the parameters for the current message queue using the **`msgctl`** API function. This is done with the `IPC_SET` command in **`msgctl`**.

When setting the `msg_perm.mode` *(permissions), it's important to note that this is traditionally defined as an octal value. Note at line 23 of Listing 15.7 that a leading zero is shown, indicating that the value is octal. If, for example, a decimal value of* `666` *were provided instead of octal* `0666`*, permissions would be invalid, and therefore undesirable behavior would result. For this reason, it can be beneficial to use the symbols as shown in Table 15.2.*

We can also use the **msgctl** API function to identify certain message queue-specific parameters, such as the number of messages currently on the message queue. Listing 15.8 illustrates the collection and printing of the accessible parameters.

Listing 15.8 Reading Current Message Queue Settings (on the CD-ROM at `./source/ch15/mqstats.c`)

```
#include <stdio.h>
#include <sys/msg.h>
#include <unistd.h>
#include <sys/types.h>
#include <time.h>
#include “common.h”

int main()
{
  int msgid, ret;
  struct msqid_ds buf;

  /* Get the message queue for the id MY_MQ_ID */
  msgid = msgget( MY_MQ_ID, 0 );

  /* Check successful completion of msgget */
  if (msgid >= 0) {

    ret = msgctl( msgid, IPC_STAT, &buf );

    if (ret == 0) {

      printf( “Number of messages queued: %ld\n”,
               buf.msg_qnum );
      printf( “Number of bytes on queue : %ld\n”,
               buf.msg_cbytes );
      printf( “Limit of bytes on queue  : %ld\n”,
               buf.msg_qbytes );

      printf( “Last message writer (pid): %d\n”,
               buf.msg_lspid );
      printf( “Last message reader (pid): %d\n”,
               buf.msg_lrpid );

      printf( “Last change time         : %s”,
               ctime(&buf.msg_ctime) );

      if (buf.msg_stime) {
```

```
39:              printf( "Last msgsnd time         : %s",
40:                        ctime(&buf.msg_stime) );
41:            }
42:            if (buf.msg_rtime) {
43:              printf( "Last msgrcv time         : %s",
44:                        ctime(&buf.msg_rtime) );
45:            }
46:
47:          }
48:
49:        }
50:
51:        return 0;
52:      }
```

Listing 15.8 begins as most other message queue examples, with the collection of the message queue ID from **msgget**. Once we have our ID, we use this to collect the message queue structure using **msgctl** and the command IPC_STAT. We pass in a reference to the msqid_ds structure, which is filled in by the **msgctl** API function. We then emit the information collected in lines 23–45.

At lines 23–24, we emit the number of messages that are currently enqueued on the message queue (msg_qnum). The current total number of bytes that are enqueued is identified by msg_cbytes (lines 25–26), and the maximum number of bytes that may be enqueued is defined by msg_qbytes (lines 27–28).

We can also identify the last reader and writer process pids (lines 30–33). These refer to the effective process ID of the calling process that called **msgrcv** or **msgsnd.**

The msg_ctime element refers to the last time the message queue was changed (or when it was created). It's in standard time_t format, so we pass msg_ctime to ctime to grab the ASCII text version of the calendar date and time. We do the same for msg_stime (last **msgsnd** time) and msg_rtime (last **msgrcv** time). Note that in the case of msg_stime and msg_rtime, we emit the sting dates only if their values are nonzero. If the values are zero, there have been no **msgrcv** or **msgsnd** API functions called.

msgsnd

The **msgsnd** API function allows a process to send a message to a queue. As we saw in the introduction, the message is purely user defined except that the first element in the message must be a long word for the type field. The function prototype for the **msgsnd** function is defined as:

```
int msgsnd( int msgid, struct msgbuf *msgp, size_t msgsz,
                        int msgflg );
```

The msgid argument is the message queue ID (returned from the **msgget** function). The msgbuf represents the message to be sent; at a minimum it is a long value representing the message type. The msgsz argument identifies the size of the msgbuf passed in to **msgsend**, in bytes. Finally, the msgflag argument allows the developer to alter the behavior of the **msgsnd** API function.

The **msgsnd** function has some default behavior that the developer should consider. If insufficient room exists on the message queue to write the message, the process will be blocked until sufficient room exists. Otherwise, if room exists, the call succeeds immediately with a zero return to the caller.

Since we've already looked at some of the standard uses of **msgsnd**, let's look at some of the more specialized cases. The blocking behavior is desirable in most cases as it can be the most efficient. In some cases, we may want to try to send a message, and if we're unable (due to the insufficient space on the message queue), do something else. Let's look at this example in the following code snippet:

```
ret = msgsnd( msgid, (struct msgbuf *)&myMessage,
                      sizeof(myMessage), IPC_NOWAIT );
if (ret == 0) {
  // Message was successfully enqueued
} else {
  if (errno == EAGAIN) {
    // Insufficient space, do something else...
  }
}
```

The IPC_NOWAIT symbol (passed in as the msgflags) tells the **msgsnd** API function that if insufficient space exists, don't block but instead return immediately. We know this because an error was returned (indicated by the -1 return value), and the errno variable was set to EAGAIN. Otherwise, with a zero return, the message was successfully enqueued on the message queue for the receiver.

While a message queue should not be deleted while processes pend on **msgsnd**, a special error return is surfaced when this occurs. If a process is currently blocked on a msgsnd and the message queue is deleted, then a -1 value is returned with an errno value set to EIDRM.

One final item to note on **msgsnd** is the parameters that are modified when the **msgsnd** API call finishes. Table 15.3 lists the entire structure, but the items modified after successful completion of the **msgsnd** API function are listed in Table 15.5.

Note that the msg_stime *is the time that the message was enqueued and not the time that the* msgsnd *API function was called. This can be important if the* **msgsnd** *function blocks (due to a full message queue).*

TABLE 15.5 Structure Updates after Successful **msgsnd** Completion

Parameter	Update
msg_lspid	Set to the process ID of the process that called msgsnd
msg_qnum	Incremented by one
msg_stime	Set to the current time

msgrcv

Let's now focus on the last function in the message queue API. The **msgrcv** API function provides the means to read a message from the queue. The user provides a message buffer (filled in within **msgrcv**) and the message type of interest. The function prototype for **msgrcv** is defined as:

```
ssize_t msgrcv( int msgid, struct msgbuf *msgp, size_t msgsz,
                long msgtyp, int msgflg );
```

The arguments passed to **msgrcv** include the msgid (message queue identifier received from **msgget**), a reference to a message buffer (msgp), the size of the buffer (msgsz), the message type of interest (msgtyp), and finally a set of flags (msgflag). The first three arguments are self-explanatory, so let's concentrate on the latter two: msgtyp and msgflag.

The msgtyp argument (message type) specifies to **msgrcv** the messages to be received. Each message within the queue contains a message type. The msgtyp argument to **msgrcv** defines that only those types of messages are sought. If no messages of that type are found, the calling process blocks until a message of the desired type is enqueued. Otherwise, the first message of the given type is returned to the caller. The caller could provide a 0 as the msgtyp, which tells **msgrcv** to ignore the message type and return the first message on the queue. One exception to the message type request is discussed with msgflg.

The msgflg argument allows the caller to alter the default behavior of the **msgrcv** API function. As with **msgsnd**, we can instruct **msgrcv** not to block if no messages are waiting on the queue. This is done also with the IPC_NOWAIT flag. We discussed the use of msgtyp with a 0 and nonzero value, but what if we were interested in any flag except a certain one? This can be accomplished by setting msgtyp with the undesired message type and setting the flag MSG_EXCEPT within msgflg. Finally, the use of flag MSG_NOERROR instructs **msgrcv** to ignore the size check of the incoming message and the available buffer passed from the user and simply truncate the message if the user buffer isn't large enough. All of the options for msgtyp are described in Table 15.6, and options for msgflg are shown in Table 15.7.

TABLE 15.6 `msgtyp` Arguments for **`msgrcv`**

`msgtyp`	**Description**
0	Read the first message available on the queue.
>0	If the `msgflg MSG_EXCEPT` is set, read the first message on the queue not equal to the `msgtyp`. Otherwise, if `MSG_EXCEPT` is not set, read the first message on the queue with the defined `msgtyp`.
<0	The first message on the queue that is less than or equal to the absolute value of `msgtyp` is returned.

TABLE 15.7 `msgflg` Arguments for `msgrcv`

Flag	**Description**
`IPC_NOWAIT`	Return immediately if no messages awaiting of the given msgtyp (no blocking).
`MSG_EXCEPT`	Return first message available other than `msgtyp`.
`MSG_NOERROR`	Truncate the message if user buffer isn't of sufficient size.

When a message is read from the queue, the internal structure representing the queue is automatically updated as shown in Table 15.8.

*Note that `msg_rtime` is the time that the message was dequeued and not the time that the `msgrcv` API function was called. This can be important if the **`msgrcv`** function blocks (due to an empty message queue).*

TABLE 15.8 Structure Updates after Successful `msgsnd` Completion

Parameter	**Update**
`msg_lrpid`	Set to the process ID of process calling **`msgrcv`**.
`msg_qnum`	Decremented by one.
`msg_rtime`	Set to the current time.

Let's now look at some examples to illustrate **msgrcv** and the use of msgtyp and msgflg options. The most common use of **msgrcv** is to read the next available message from the queue:

```
ret = msgrcv( msgid, (struct msgbuf *)&buf, sizeof(buf), 0, 0 );
if (ret != -1) {
  printf("Message of type %ld received\n", *(long *)&buf );
}
```

Note that we check specifically for a return value that's not -1. We do this because **msgrcv** actually returns the number of bytes read. If the return is -1, errno contains the error that occurred.

If we desired not to block on the call, we could do this very simply as:

```
ret = msgrcv( msgid, (struct msgbuf *)&buf, sizeof(buf),
                    0, IPC_NOWAIT );
if (ret != -1) {
  printf("Message of type %ld received\n", *(long *)&buf );
} else if (errno == EAGAIN) {
  printf("Message unavailable\n");
}
```

With the presence of an error return from **msgrcv** and errno set to EAGAIN, it's understood that no messages were available for read. This isn't actually an error, just an indication that no messages were available in the nonblocking scenario.

Message queues permit multiple writers and readers to the same queue. These could be the same process, but very likely each is a different process. Let's say that we have a process that manages only a certain type of message. We identify this particular message by its message type. In the next example, we provide a snippet from a process whose job it is to manage only messages of type 5.

```
ret = msgrcv( msgid, (struct msgbuf *)&buf, sizeof(buf), 5, 0 );
```

Any message sent of type 5 would be received by the process executing this code snippet. To manage all other message types (other than 5), we could use the MSG_EXCEPT flag to receive these. Take for example:

```
ret = msgrcv( msgid, (struct msgbuf *)&buf, sizeof(buf),
                  5, MSG_EXCEPT );
```

Any message received on the queue other than type 5 will be read using this line. If only messages of type 5 are available, this function blocks until a message not of type 5 is enqueued.

One final note on **msgrcv** is what happens if a process is blocked on a queue that is removed. The removal is permitted to occur, and the process blocked on the queue receives an error return with the errno set to EIDRM (as with blocked **msgsnd** calls). It's therefore important to fully recognize the error returns that are possible.

USER UTILITIES

GNU/Linux provides the ipcs command to explore IPC assets from the command line. The ipcs utility provides information on message queues as well as semaphores and shared memory segments. We'll look at its use for message queues here.

The general form of the ipcs utility for message queues is:

```
# ipcs -q
```

This presents all of the message queues that are visible to the process. Let's start by creating a message queue (as was done in Listing 15.1):

```
# ./mqcreate
Created a Message Queue 819200
# ipcs -q

—— Message Queues ——
key        msqid      owner      perms      used-bytes   messages
0x0000006f 819200     mtj        666        0            0
```

We see our newly created queue (key 0x6f, or decimal 111). If we send a message to the message queue (such as was illustrated with Listing 15.4), we see the following:

```
# ./mqsend
Message successfully sent to queue 819200
# ipcs -q

—— Message Queues ——
key        msqid      owner      perms      used-bytes   messages
0x0000006f 819200     mtj        666        96           1
```

We see now that a message is contained on the queue that occupies 96 bytes. We could also take a deeper look at the queue by specifying the message queue ID. This is done with ipcs using the -i option:

```
# ipcs -q -i 819200
Message Queue msqid=819200
```

```
uid=500 gid=500 cuid=500       cgid=500       mode=0666
cbytes=96       qbytes=16384    qnum=1  lspid=22309     lrpid=0
send_time=Sat Mar 27 18:59:34 2004
rcv_time=Not set
change_time=Sat Mar 27 18:58:43 2004
```

We're now able to review the structure representing the message queue (as defined in Table 15.3). The ipcs utility can be very useful to view Snapshots of message queues for application debugging.

We can also delete queues from the command line using the ipcrm command. To delete our previously created message queue, we'd simply use the ipcrm command as follows:

```
$ ipcrm -q 819200
$
```

As with the message queue API functions, we pass the message queue ID as the indicator of the message queue to remove.

SUMMARY

In this chapter, we introduced the message queue API and its application of interprocess communication. We began with a whirlwind tour of the API and then detailed each of the functions, including the behavioral modifiers (msgflg arguments). Finally, we reviewed the ipcs utility and demonstrated its use as a debugging tool as well as the ipcrm command for removing message queues from the command line.

MESSAGE QUEUE APIS

```
    #include <sys/types.h>
    #include <sys/ipc.h>
    #include <sys/msg.h>
    int msgget( key_t key, int msgflg );
    int msgctl( int msgid, int cmd, struct msqid_ds *buf );
    int msgsnd( int msgid, structu msgbuf *msgp, size_t msgsz,
                int msgflg );
size_t msgrcv( int msgid, struct msgbuf *msgp, size_t msgsz,
               long msgtyp, int msgflg );
```

16 Synchronization with Semaphores

In This Chapter

- Introduction to GNU/Linux Semaphores
- Discussion of Binary and Counting Semaphores
- Creating and Configuring Semaphores
- Acquiring and Releasing Semaphores
- Single Semaphores or Semaphore Arrays
- The `ipcs` and `ipcrm` Utilities for Semaphores

INTRODUCTION

In this chapter, we'll explore the topic of semaphores. GNU/Linux provides both binary and counting semaphores using the same POSIX compliant API function set. We'll also investigate semaphores in GNU/Linux and their similarities with some of the other IPC (inter-process communication) mechanisms.

SEMAPHORE THEORY

Let's first go through a quick review of semaphore theory. A semaphore is nothing more than a variable that is protected. It provides a means to restrict access to a resource that is shared amongst two or more processes. Two operations are permitted, commonly called acquire and release. The acquire operation allows a process to take the semaphore, and if it has already been acquired, then the process blocks until it's available. If a process has the semaphore, it can release it, which allows other processes to acquire it. The process of releasing a semaphore automatically wakes up the next process awaiting it on the acquire operation. Consider the simple example in Figure 16.1.

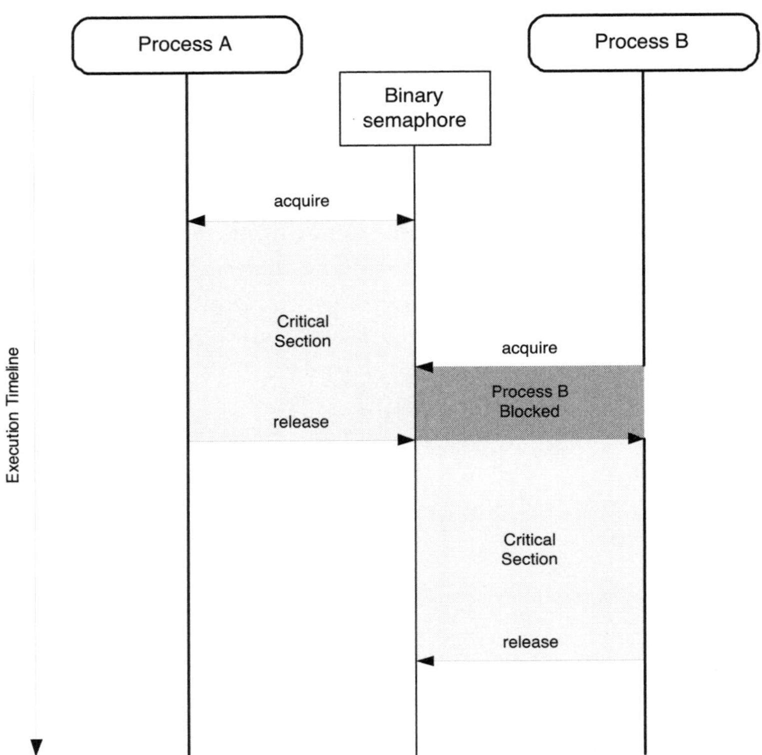

FIGURE 16.1 Simple binary semaphore example with two processes.

As shown in Figure 16.1, two processes are both vying for the single semaphore. Process A performs the acquire first and therefore is provided with the semaphore. The period in which the semaphore is owned by the process is commonly called a *critical section*. The critical section can be performed by only one process, therefore the need for the coordination provided by the semaphore. While Process A has the semaphore, Process B is not permitted to perform its critical section.

Note that while Process A is in its critical section, Process B attempts to acquire the semaphore. As the semaphore has already been acquired by Process A, Process B is placed into a blocked state. When Process A finally releases the semaphore, it is then granted to Process B, which is allowed to enter its critical section. Process B at a later time releases the semaphore, making it available for acquisition.

Semaphores commonly represent a point of synchronization in a system. For example, a semaphore could represent access to a shared resource. Only when the process has access to the semaphore can it access the shared resources. This ensures that only one process will have access to the shared resource at a time, thus providing coordination between two or more users of the resource.

Semaphores were invented by Edsger Dijkstra for the T.H.E. operating system. Originally, the semaphore operations were defined as P and V. The P stands for the Dutch "Proberen," or to test, and the V for "Verhogen," or to increment.

Edsger Dijkstra used the train analogy to illustrate the critical section. Imagine two parallel train tracks that for a short duration merge into a single track. The single track is the shared resource and is also the critical section. The semaphore ensures that only one train is permitted on the shared track at a time. Not having the semaphore could have disastrous results on two trains trying to use the shared track at the same time. The effect on software is just as treacherous.

Types of Semaphores

Semaphores come in two basic varieties. The first are *binary semaphores*, as illustrated in Figure 16.1. The binary semaphore represents a single resource, and therefore when one process has acquired it, others are blocked until it is released.

The other style is the *counting semaphore*, which is used to represent shared resources in quantities greater than one. Consider a pool of buffers. A counting semaphore could represent the entire set of buffers by setting its value to the number of buffers available. Each time a process requires a buffer, it acquires the semaphore, which decrements its value. When the semaphore value reaches zero, processes are blocked until it becomes nonzero. When a semaphore is released, the semaphore value is increased, thus permitting other processes to acquire a semaphore (and associated buffer). This is the one use for a counting semaphore (see Figure 16.2).

In the counting semaphore example, each process requires two resources before being able to perform its desired activities. In this example, the value of the counting semaphore is 3, which means that only one process will be permitted to fully operate at a time. Process A acquires its two resources first, which means that Process B blocks until Process A releases at least one of its resources.

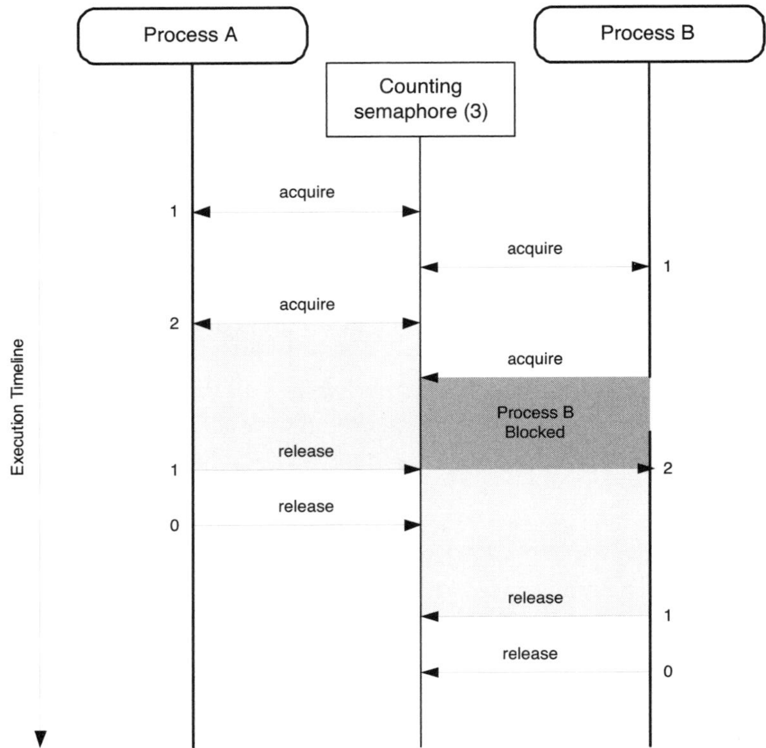

FIGURE 16.2 Counting semaphore example with two processes.

QUICK OVERVIEW OF GNU/LINUX SEMAPHORES

Let's start our discussion with a whirlwind tour of the GNU/Linux semaphore API. We'll look at code examples illustrating each of the API capabilities such as creating a new semaphore, finding a semaphore, acquiring a semaphore, releasing a semaphore, configuring a semaphore, and finally, removing a semaphore. Once we've finished the quick overview, we'll dig deeper into the semaphore API.

Semaphores in GNU/Linux are actually semaphore arrays. A single semaphore can represent an array of 64 semaphores. This unique feature of GNU/Linux permits atomic operations over numerous semaphores at the same time. In the early discussions of GNU/Linux semaphores, we'll explore single semaphore uses. In the detailed discussions that follow, we'll look at the more complex semaphore array examples.

Using the semaphore API requires that the function prototypes and symbols be available to the application. This is done by including the following three header files:

```
#include <sys/types.h>
#include <sys/ipc.h>
#include <sys/sem.h>
```

Creating a Semaphore

To create a semaphore (or get an existing semaphore), we use the **semget** API function. This function takes a semaphore key, a semaphore count, and a set of flags. The count represents the number of semaphores in the set. In this case, we'll specify the need for one semaphore. The semaphore flags, argument 3 as shown in Listing 16.1, specify that the semaphore is to be created (`IPC_CREAT`). We also specify the read/write permissions to use (in this case `0666` for read/write for the user, group, and system in octal). An important item to consider is that when a semaphore is created, its value is zero. This suits us for this example, but we'll investigate later how to initialize the semaphore's value.

Listing 16.1 demonstrates creating a semaphore. In the following examples, we'll use the key `MY_SEM_ID` to represent our globally unique semaphore. At line 10, we use the `semget` with our semaphore key, semaphore set count, and command (with read/write permissions).

ON THE CD

Listing 16.1 Creating a Semaphore with **semget** (on the CD-ROM at `./source/ch16/semcreate.c`)

```
 1:     #include <stdio.h>
 2:     #include <sys/sem.h>
 3:     #include "common.h"
 4:
 5:     int main()
 6:     {
 7:       int semid;
 8:
 9:       /* Create the semaphore with the id MY_SEM_ID */
10:       semid = semget( MY_SEM_ID, 1, 0666 | IPC_CREAT );
11:
12:       if (semid >= 0) {
13:
14:         printf( "semcreate: Created a semaphore %d\n", semid );
15:
16:       }
```

```

    return 0;
  }
```

Upon completion of this simple application, a new globally available semaphore would be available with a key identified by `MY_SEM_ID`. Any process in the system could use this semaphore.

Getting and Releasing a Semaphore

Now let's look at an application that attempts to acquire an existing semaphore and also another application that releases it. Recall that our previously created semaphore (in Listing 16.1) was initialized with a value of zero. This is identical to a binary semaphore already having been acquired.

Listing 16.2 illustrates an application acquiring our semaphore. The GNU/Linux semaphore API is a little more complicated than many semaphore APIs, but it is POSIX compliant and therefore important for porting to other UNIX-like operating systems.

Listing 16.2 Getting a Semaphore with **semop**

```
#include <stdio.h>
#include <sys/sem.h>
#include <stdlib.h>
#include "common.h"

int main()
{
  int semid;
  struct sembuf sb;

  /* Get the semaphore with the id MY_SEM_ID */
  semid = semget( MY_SEM_ID, 1, 0 );

  if (semid >= 0) {

    sb.sem_num = 0;
    sb.sem_op = -1;
    sb.sem_flg = 0;

    printf( "semacq: Attempting to acquire semaphore %d\n", semid );

```

```
22:        /* Acquire the semaphore */
23:        if ( semop( semid, &sb, 1 ) == -1 ) {
24:
25:          printf( "semacq: semop failed.\n" );
26:          exit(-1);
27:
28:        }
29:
30:        printf( "semacq: Semaphore acquired %d\n", semid );
31:
32:      }
33:
34:      return 0;
35:    }
```

We begin by identifying the semaphore identifier with **semget** at line 12. If this is successful, we build our semaphore operations structure (identified by the `sembuf` structure). This structure contains the semaphore number, the operation to be applied to the semaphore, and a set of operation flags. Since we have only one semaphore, we use the semaphore number zero to identify it. To acquire the semaphore, we specify an operation of -1. This subtracts one from the semaphore, but only if it's greater than zero to begin with. If the semaphore is already zero, the operation (and the process) will block until the semaphore value is incremented.

With our `sembuf` created (variable `sb`), we use this with the API function **semop** to acquire the semaphore. We specify our semaphore identifier, our `sembuf` structure, and then the number of `sembufs` that were passed in (in this case, one). This implies that we can provide an array of `sembufs`, which we'll investigate later. As long as the semaphore operation can finish (semaphore value is nonzero), then it returns with success (a non –1 value). This means that the process performing the **semop** has acquired the semaphore.

Let's now look at a release example. In this example, we'll demonstrate the **semop** API function from the perspective of releasing the semaphore.

In many cases, the release would follow the acquire in the same process. This usage allows synchronization between two processes. The first process attempts to acquire the semaphore and then blocks when it's not available. The second process, knowing that another process is sitting blocked on the semaphore, releases it, allowing the process to continue. This provides a lock-step operation between the processes and is practical and useful.

Listing 16.3 Releasing a Semaphore with **semop** (on the CD-ROM at ./source/ch16/semrel.c)

```
1:  #include <stdio.h>
2:  #include <sys/sem.h>
3:  #include <stdlib.h>
4:  #include "common.h"
5:
6:  int main()
7:  {
8:    int semid;
9:    struct sembuf sb;
10:
11:   /* Get the semaphore with the id MY_SEM_ID */
12:   semid = semget( MY_SEM_ID, 1, 0 );
13:
14:   if (semid >= 0) {
15:
16:     printf( "semrel: Releasing semaphore %d\n", semid );
17:
18:     sb.sem_num = 0;
19:     sb.sem_op  = 1;
20:     sb.sem_flg = 0;
21:
22:     /* Release the semaphore */
23:     if (semop( semid, &sb, 1 ) == -1) {
24:
25:       printf("semrel: semop failed.\n");
26:       exit(-1);
27:
28:     }
29:
30:     printf( "semrel: Semaphore released %d\n", semid );
31:
32:   }
33:
34:   return 0;
35: }
```

At line 12 of Listing 16.3, we first identify our semaphore of interest using the **semget** API function. Having our semaphore identifier, we build our `sembuf` structure to release the semaphore at line 23 using the **semop** API function. In this example, our `sem_op` element is 1 (compared to the –1 in Listing 16.2). In this example, we're releasing the semaphore, which means that we're making it nonzero (and thus available).

It's important to note the symmetry the `sembuf` *uses in Listings 16.2 and 16.3. To acquire the semaphore, we subtract 1 from its value. To release the semaphore, we add 1 to its value. When the semaphore's value is 0, it's unavailable, forcing any processing attempting to acquire it to block. An initial value of 1 for the semaphore defines it as a binary semaphore. If the semaphore value is greater than 0, it can be considered a counting semaphore.*

Let's now look at a sample application of each of the functions discussed thus far. Listing 16.4 illustrates execution of Listing 16.1 `semcreate`, Listing 16.2 `semacq`, and Listing 16.3 `semrel`.

Listing 16.4 Execution of the Sample Semaphore Applications

```
1:     # ./semcreate
2:     semcreate: Created a semaphore 1376259
3:     # ./semacq &
4:     [1] 12189
5:     semacq: Attempting to acquire semaphore 1376259
6:     # ./semrel
7:     semrel: Releasing semaphore 1376259
8:     semrel: Semaphore released 1376259
9:     # semacq: Semaphore acquired 1376259
10:
11:    [1]+  Done                    ./semacq
12:    #
```

At line 1, we create the semaphore. We emit the identifier associated with this semaphore, `1376259` (which is shown at line 2). Next, at line 3, we perform the `semacq` application, which acquires the semaphore. We run this in the background (identified by the trailing `&` symbol) because this application will immediately block as the semaphore is unavailable. At line 4, we see the creation of the new subprocess (where `[1]` represents the number of subprocesses and `12189` is its process ID, or pid). The `semacq` application prints out its message, indicating that it's attempting to acquire the semaphore, but then it blocks. We then execute the `semrel` application to release the semaphore (line 6). We see two messages from this application; the first at line 7 indicates that it is about to release the semaphore, and then at line 8, we see that it was successful. Immediately thereafter, we see the `semacq` application was able to acquire the newly released semaphore, given its output at line 9. Finally, at line 11, we see the `semacq` application subprocess finish. Since it unblocked (based upon the presence of its desired semaphore), the `semacq`'s main function reached its return, and thus the process finished.

Configuring a Semaphore

While there are a number of elements that can be configured for a semaphore, let's look here specifically at reading and writing the value of the semaphore (the current count).

In the first example, Listing 16.5, we'll demonstrate reading the current value of the semaphore. We achieve this using the **semctl** API function.

ON THE CD

Listing 16.5 Retrieving the Current Semaphore Count (on the CD-ROM at ./source/ch16/semcrd.c)

```
1:     #include <stdio.h>
2:     #include <sys/sem.h>
3:     #include <stdlib.h>
4:     #include "common.h"
5:
6:     int main()
7:     {
8:       int semid, cnt;
9:
10:      /* Get the semaphore with the id MY_SEM_ID */
11:      semid = semget( MY_SEM_ID, 1, 0 );
12:
13:      if (semid >= 0) {
14:
15:        /* Read the current semaphore count */
16:        cnt = semctl( semid, 0, GETVAL );
17:
18:        if (cnt != -1) {
19:
20:          printf("semcrd: current semaphore count %d.\n", cnt);
21:
22:        }
23:
24:      }
25:
26:      return 0;
27:    }
```

Reading the semaphore count is performed at line 16. We specify the semaphore identifier, the index of the semaphore (0), and the command (GETVAL). Note that the semaphore is identified by an index because it could represent an array of semaphores (rather than one). The return value from this command is either –1 for error or the count of the semaphore.

We can configure a semaphore with a count using a similar mechanism (as shown in Listing 16.6).

Listing 16.6 Setting the Current Semaphore Count

```
1:   #include <stdio.h>
2:   #include <sys/sem.h>
3:   #include <stdlib.h>
4:   #include "common.h"
5:
6:   int main()
7:   {
8:     int semid, ret;
9:
10:    /* Get the semaphore with the id MY_SEM_ID */
11:    semid = semget( MY_SEM_ID, 1, 0 );
12:
13:    if (semid >= 0) {
14:
15:      /* Read the current semaphore count */
16:      ret = semctl( semid, 0, SETVAL, 6 );
17:
18:      if (ret != -1) {
19:
20:        printf( "semcrd: semaphore count updated.\n" );
21:
22:      }
23:
24:    }
25:
26:    return 0;
27:  }
```

As with retrieving the current semaphore value, we can set this value using the **semctl** API function. The difference here is that along with the semaphore identifier (semid) and semaphore index (0), we specify the set command (SETVAL) and a value. In this example (line 16 of Listing 16.6), we're setting the semaphore value to 6. Setting the value to 6, as shown here, changes the binary semaphore to a counting semaphore. Six semaphore acquires would be permitted before an acquiring process would block.

Removing a Semaphore

Removing a semaphore is also performed through the **semctl** API function. After retrieving the semaphore identifier (line 10 in Listing 16.7), we use this to remove the semaphore using the **semctl** API function and the IPC_RMID command (at line 14).

Listing 16.7 Removing a Semaphore

```
 1:    #include <stdio.h>
 2:    #include <sys/sem.h>
 3:    #include "common.h"
 4:
 5:    int main()
 6:    {
 7:      int semid, ret;
 8:
 9:      /* Get the semaphore with the id MY_SEM_ID */
10:      semid = semget( MY_SEM_ID, 1, 0 );
11:
12:      if (semid >= 0) {
13:
14:        ret = semctl( semid, 0, IPC_RMID);
15:
16:        if (ret != -1) {
17:
18:          printf( "Semaphore %d removed.\n", semid );
19:
20:        }
21:
22:      }
23:
24:      return 0;
25:    }
```

As you can probably see, the semaphore API probably is not the simplest that you've used before.

That's it for our whirlwind tour; next we'll explore the semaphore API in greater detail and look at some of its other capabilities.

THE SEMAPHORE API

As we noted before, the semaphore API handles not only the case of managing a single semaphore, but also groups (or arrays) of semaphores. We'll investigate their use in this section. For a quick review, Table 16.1 shows the API functions and describes their uses. In the following discussion, we'll continue to use the term *semaphore*, but this could refer instead to a semaphore array.

TABLE 16.1 Semaphore API Functions and Their Uses

API Function	Uses
semget	Create a new semaphore
	Get an existing semaphore
semop	Acquire or release a semaphore
semctl	Get info about a semaphore
	Set info about a semaphore
	Remove a semaphore

We'll address each of these functions in the following sections using both simple examples (single semaphore) and the more complex uses (semaphore arrays).

semget

The **semget** API function serves two fundamental roles. Its first use is in the creation of new semaphores. The second use is identifying an existing semaphore. In both cases, the response from **semget** is a semaphore identifier (a simple integer value representing the semaphore). The prototype for the **semget** API function is defined as:

```
int semget( key_t key, int nsems, int semflg );
```

The key argument specifies a system-wide identifier that uniquely identifies this semaphore. The key must be nonzero or the special symbol IPC_PRIVATE. The IPC_PRIVATE variable tells **semget** that no key is provided and to simply make one up. Since no key exists, there's no way for other processes to know about this semaphore. Therefore, it's a private semaphore for this particular process.

The developer can create a single semaphore (with an nsems value of 1) or multiple semaphores. If we're using **semget** to get an existing semaphore, this value can simply be zero.

Finally, the semflg argument allows the developer to alter the behavior of the **semget** API function. The semflg argument can take on three basic forms, depending upon what is desired by the caller. In the first form, the developer desires to create a new semaphore. In this case, the semflg argument must be the IPC_CREAT value OR'd with the permissions (see Table 16.2). The second form also provides for semaphore creation, but with the constraint that if the semaphore already exists, an error is generated. This second form requires the semflg argument to be set to (IPC_CREAT | IPC_EXCL) along with the permissions. If the second form is used and the semaphore already exists, the call will fail –1 return) with `errno` set to `EEXIST`. The third form takes a zero for `semflg` and identifies that an existing semaphore is being requested.

TABLE 16.2 Semaphore Permissions for the `semget semflg` Argument

Symbol	Value	Meaning
S_IRUSR	0400	User has read permission
S_IWUSR	0200	User has write permission
S_IRGRP	0040	Group has read permission
S_IWGRP	0020	Group has write permission
S_IROTH	0004	Other has read permission
S_IWOTH	0002	Other has write permission

Let's now look at a few examples of **`semget`**, used in each of the three scenarios defined above. In the examples shown below, assume `semid` is an `int` value, and `mySem` is of type `key_t`. In the first example, we'll create a new semaphore (or access an existing one) of the private type.

```
semid = semget( IPC_PRIVATE, 1, IPC_CREAT | 0666 );
```

Once the **`semget`** call completes, our semaphore identifier is stored in `semid`. Otherwise, if an error occurred, a –1 would be returned. Note that in this example (using `IPC_PRIVATE`), `semid` is all we have to identify this semaphore. If `semid` were somehow lost, there would be no way to find this semaphore again.

In the next example, we'll create a semaphore using a system-wide unique key value (`0x222`). We also desire that if the semaphore already exists, we don't simply get its value, but instead fail the call. Recall that this is provided by the `IPC_EXCL` command, as:

```
// Create a new semaphore
semid = semget( 0x222, 1, IPC_CREAT | IPC_EXCL | 0666 );
if ( semid == -1 ) {
  printf( "Semaphore already exists, or error\n" );
} else {
  printf( "Semaphore created (id %d)\n", semid );
}
```

If we didn't want to rely on the fact that 0x222 may not be unique in our system, we could use the **ftok** system function. This function provides the means to create a new unique key in the system. It does this by using a known file in the filesystem and an integer number. The file in the filesystem will be unique by default (considering its path). Therefore, by using the unique file (and integer), it's an easy task to then create a unique system-wide value. Let's look at an example of the use of **ftok** to create a unique key value. We'll assume for this example that our file and path are defined as /home/mtj/semaphores/mysem.

```
key_t mySem;
int   semid;
// Create a key based upon the defined path and number
myKey = ftok( "/home/mtj/semaphores/mysem", 0 );
semid = semget( myKey, 1, IPC_CREAT | IPC_EXCL | 0666 );
```

Note that each time **ftok** is called with those parameters, the same key is generated (which is why this method works at all!). As long as each process that needs access to the semaphore knows about the file and number, the key can be recalculated and then used to identify the semaphore.

In the examples discussed thus far, we've created a single semaphore. We can create an array of semaphores by simply specifying an nsems value greater than one, such as:

```
semarrayid = semget( myKey, 10, IPC_CREAT | 0666 );
```

The result is a semaphore array being created that consists of 10 semaphores. The return value (semarrayid) represents the entire set of semaphores. We'll look at how individual semaphores can be addressed in the **semctl** and **semop** discussions.

In our last example of **semget**, we'll simply use it to get the semaphore identifier of an existing semaphore. In this example, we specify our key value and no command:

```
semid = semget( 0x222, 0, 0 );
if ( semid == -1 ) {
  printf( "Semaphore does not exist...\n" );
}
```

One final note on semaphores is that, just like message queues, a set of defaults is provided to the semaphore as it's created. The parameters that are defined are shown in Table 16.3. Later on in the discussion of **semctl**, we'll demonstrate how some of the parameters can be changed.

TABLE 16.3 Semaphore Internal Values

Parameter	Default Value
sem_perm.cuid	Effective User ID of the calling process (creator)
sem_perm.uid	Effective User ID of the calling process (owner)
sem_perm.cgid	Effective Group ID of the calling process (creator)
sem_perm.gid	Effective Group ID of the calling process (owner)
sem_perm.mode	Permissions (lower 9 bits of semflg)
sem_nsems	Set to the value of nsems
sem_otime	Set to zero (last semop time)
sem_ctime	Set to the current time (create time)

The process can override some of these parameters. We'll explore this later in our discussion of semctl.

semctl

The **semctl** API function provides a number of control operations on semaphores or semaphore arrays. Example functionality ranges from setting the value of the semaphore (as shown in Listing 16.6) to removing a semaphore or semaphore array (see Listing 16.7). We'll look at these and other examples in this section.

The function prototype for the **semctl** call is shown below:

```
int semctl( int semid, int semnum, int cmd, ... );
```

The first argument defines the semaphore identifier, the second, the semaphore number of interest, the third the command to be applied, and then potentially another argument (usually defined as a union). The operations that can be performed are shown in Table 16.4.

We'll now look at some examples of each of these operations in **semctl**, focusing on semaphore array examples where applicable. In our first example, we'll illustrate the setting of a semaphore value and then returning its value. In this example, we first set the value of the semaphore to 10 (using the command SETVAL) and then read it back out using GETVAL. Note that the semnum argument (argument 2) defines an individual semaphore. Later on, we'll look at the semaphore array case with GETALL and SETALL.

TABLE 16.4 Operations That Can Be Performed Using `semctl`

Command	Description	Fourth Argument
GETVAL	Return the semaphore value.	
SETVAL	Set the semaphore value.	int
GETPID	Return the pid that last operated on the semaphore (semop).	
GETNCNT	Return the number of processes awaiting the defined semaphore to increase in value.	int
GETZCNT	Return the number of processes awaiting the defined semaphore to become zero.	int
GETALL	Return the value of each semaphore in a semaphore array.	u_short*
SETALL	Set the value of each semaphore in a semaphore array.	u_short*
IPC_STAT	Return the effective user, group, and permission for a semaphore.	struct semid_ds*
IPC_SET	Set the effective user, group, and permissions for a semaphore.	struct semid_ds*
IPC_RMID	Remove the semaphore or semaphore-array.	

```
int semid, ret, value;
...
/* Set the semaphore to 10 */
ret = semctl( semid, 0, SETVAL, 10 );
...
/* Read the semaphore value (return value) */
value = semctl( semid, 0, GETVAL );
```

The `GETPID` command allows us to identify the last process that performed a **`semop`** on the semaphore. The process identifier is the return value, and argument 4 is not used in this case.

```
int semid, pid;
...
pid = semctl( semid, 0, GETPID );
```

If no **semop** has been performed on the semaphore, the return value will be zero.

To identify the number of semaphores that are currently awaiting a semaphore to increase in value, we can use the GETNCNT command. We can also identify the number of processes that are awaiting the semaphore value to become zero using GETZCNT. Both of these commands are illustrated below for the semaphore numbered zero:

```
int semid, count;
/* How many processes are awaiting this semaphore to increase */
count = semctl( semid, 0, GETNCNT );
/* How many processes are awaiting this semaphore to become zero */
count = semctl( semid, 0, GETZCNT );
```

Now let's look at an example of some semaphore array operations. Listing 16.8 illustrates both the SETVAL and GETVAL commands with **semctl**.

In this example, we'll create a semaphore array of 10 semaphores. The creation of the semaphore array is performed at lines 20–21 using the **semget** API function. Note that since we're going to create and remove the semaphore array within this same function, we use no key and instead use the IPC_PRIVATE key. Our MAX_SEMAPHORES symbolic defines the number of semaphores that we'll create, and we finally specify that we're creating the semaphore array (IPC_CREAT) with the standard permissions.

Next, we initialize our semaphore value array (lines 26–30). While this is not a traditional example, we initialize each semaphore to one plus its semnum (so semaphore zero has a value of one, semaphore one has a value of two, and so on). We do this so that we can inspect the value array later and know what we're looking at. At line 33, we set our arg.array parameter to the address of the array (sem_array). Note that we're using the semun union, which defines some commonly used types for semaphores. In this case, we use the unsigned short field to represent an array of semaphore values.

At line 36, we use the **semctl** API function and the SETALL command to set the semaphore values. We provide our semaphore identifier, semnum as zero (unused in this case), the SETALL command, and finally our semun union. Upon return of this API function, the semaphore array identified by semid has the values as defined in sem_array.

Next, we explore the GETALL command, which is used to retrieve the array of values for the semaphore array. We first set our arg.array to a new array (just to avoid reusing our existing array that has the contents that we're looking for), at line 41. At line 44, we call **semctl** again with our semid, zero for semnum (unused here, again), the GETALL command, and our semun union.

To illustrate what we've read, we loop through the `sem_read_array` and emit each value for each semaphore index within the semaphore array (lines 49–53).

While `GETALL` allows us to retrieve the entire semaphore array in one call, we could have performed the same action using the `GETVAL` command, calling **`semctl`** for each semaphore of the array. This is illustrated at lines 56–62. This also applies to the `SETVAL` command to mimic the `SETALL` behavior.

Finally, at line 65, we use the **`semctl`** API function with the `IPC_RMID` command to remove the semaphore array.

Listing 16.8 Creating and Manipulating Semaphore Arrays (on the CD-ROM at `./source/ch16/semall.c`)

```
 1:     #include <stdio.h>
 2:     #include <sys/types.h>
 3:     #include <sys/sem.h>
 4:     #include <errno.h>
 5:
 6:     #define MAX_SEMAPHORES  10
 7:
 8:     int main()
 9:     {
10:       int i, ret, semid;
11:       unsigned short sem_array[MAX_SEMAPHORES];
12:       unsigned short sem_read_array[MAX_SEMAPHORES];
13:
14:       union semun {
15:         int val;
16:         struct semid_ds *buf;
17:         unsigned short *array;
18:       } arg;
19:
20:       semid = semget( IPC_PRIVATE, MAX_SEMAPHORES,
21:                         IPC_CREAT | 0666 );
22:
23:       if (semid != -1) {
24:
25:         /* Initialize the sem_array */
26:         for ( i = 0 ; i < MAX_SEMAPHORES ; i++ ) {
27:
28:           sem_array[i] = (unsigned short)(i+1);
29:
30:         }
31:
32:         /* Update the arg union with the sem_array address */
33:         arg.array = sem_array;
```

```
    /* Set the values of the semaphore-array */
    ret = semctl( semid, 0, SETALL, arg );
    if (ret == -1) printf("SETALL failed (%d)\n", errno);
    /* Update the arg union with another array for read */
    arg.array = sem_read_array;
    /* Read the values of the semaphore array */
    ret = semctl( semid, 0, GETALL, arg );
    if (ret == -1) printf("GETALL failed (%d)\n", errno);
    /* print the sem_read_array */
    for ( i = 0 ; i < MAX_SEMAPHORES ; i++ ) {
      printf("Semaphore %d, value %d\n", i, sem_read_array[i] );
    }
    /* Use GETVAL in a similar manner */
    for ( i = 0 ; i < MAX_SEMAPHORES ; i++ ) {
      ret = semctl( semid, i, GETVAL );
      printf("Semaphore %d, value %d\n", i, ret );
    }
    /* Delete the semaphore */
    ret = semctl( semid, 0, IPC_RMID );
  } else {
    printf("Could not allocate semaphore (%d)\n", errno);
  }
  return 0;
}
```

Executing this application (called `semall`) produces the output shown in Listing 16.9. Not surprisingly, the `GETVAL` emits identical output as that shown for the `GETALL`.

Listing 16.9 Output from the `semall` Application Shown in Listing 16.8

```
# ./semall
Semaphore 0, value 1
Semaphore 1, value 2
Semaphore 2, value 3
Semaphore 3, value 4
Semaphore 4, value 5
Semaphore 5, value 6
Semaphore 6, value 7
Semaphore 7, value 8
Semaphore 8, value 9
Semaphore 9, value 10
Semaphore 0, value 1
Semaphore 1, value 2
Semaphore 2, value 3
Semaphore 3, value 4
Semaphore 4, value 5
Semaphore 5, value 6
Semaphore 6, value 7
Semaphore 7, value 8
Semaphore 8, value 9
Semaphore 9, value 10
#
```

The `IPC_STAT` command is used to retrieve the current information about a semaphore or semaphore array. The data is retrieved into a structure called `semid_ds` and contains a variety of parameters. The application that reads this information is shown in Listing 16.10. We read the semaphore information at line 18 using the **semctl** API function and the `IPC_STAT` command. The information captures is then emitted at lines 27–49.

ON THE CD

Listing 16.10 Reading Semaphore Information Using `IPC_STAT` (on the CD-ROM at `./source/ch16/semstat.c`)

```
1:   #include <stdio.h>
2:   #include <sys/sem.h>
3:   #include <time.h>
4:   #include "common.h"
5:
6:   int main()
7:   {
8:     int semid, ret;
9:     struct semid_ds sembuf;
10:
```

```
    union semun {
      int val;
      struct semid_ds *buf;
      unsigned short *array;
    } arg;

    /* Get the semaphore with the id MY_SEM_ID */
    semid = semget( MY_SEM_ID, 1, 0 );

    if (semid >= 0) {

      arg.buf = &sembuf;
      ret = semctl( semid, 0, IPC_STAT, arg );

      if (ret != -1) {

        if (sembuf.sem_otime) {
          printf( "Last semop time %s",
                   ctime( &sembuf.sem_otime ) );
        }

        printf( "Last change time %s",
                 ctime( &sembuf.sem_ctime ) );

        printf( "Number of semaphores %ld\n",
                 sembuf.sem_nsems );

        printf( "Owner's user id %d\n",
                 sembuf.sem_perm.uid );
        printf( "Owner's group id %d\n",
                 sembuf.sem_perm.gid );

        printf( "Creator's user id %d\n",
                 sembuf.sem_perm.cuid );
        printf( "Creator's group id %d\n",
                 sembuf.sem_perm.cgid );

        printf( "Permissions 0%o\n",
                 sembuf.sem_perm.mode );

      }

    }
```

```
    return 0;
}
```

Three of the fields shown can be updated through another call to **semctl** using the IPC_SET call. The three updateable parameters are the effective user ID (sem_perm.uid), the effective group ID (sem_perm.gid), and the permissions (sem_perm.mode). The following code snippet illustrates modifying the permissions:

```
/* First, read the semaphore information */
arg.buf = &sembuf;
ret = semctl( semid, 0, IPC_STAT, arg );
/* Next, update the permissions */
sembuf.sem_perm.mode = 0644;
/* Finally, update the semaphore information */
ret = semctl( semid, 0, IPC_SET, arg );
```

Once the IPC_SET **semctl** has completed, the last change time (sem_ctime) is updated to the current time.

Finally, the IPC_RMID command permits us to remove a semaphore or semaphore array. A code snippet demonstrating this process is shown below:

```
int semid;
...
semid = semget( the_key, NUM_SEMAPHORES, 0 );
ret = semctl( semid, 0, IPC_RMID );
```

Note that if any processes were currently blocked on the semaphore, they would be immediately unblocked with an error return and errno would be set to EIDRM.

semop

The **semop** API function provides the means to acquire and release a semaphore or semaphore array. The basic operations provided by **semop** are to decrement a semaphore (acquire one or more semaphores) or to increment a semaphore (release one or more semaphores). The API for the **semop** function is defined as:

```
int semop( int semid, struct sembuf *sops, unsigned int nsops );
```

The semop takes three parameters: a semaphore identifier (semid), a sembuf structure, and the number of semaphore operations to be performed (nsops). The semaphore structure defines the semaphore number of interest, the operation to perform, and a flags word that can be used to alter the behavior of the operation. The `sembuf` structure is shown below:

```
struct sembuf {
    unsigned short sem_num;
    short sem_op;
    short sem_flg;
};
```

As you can imagine, the `sembuf` array can produce very complex semaphore interactions. We can acquire one semaphore and release another in a single **`semop`** operation.

Let's look at a simple application that acquires 10 semaphores in one operation. This application is shown in Listing 16.11.

An important difference to notice here is that rather than specify a single `sembuf` structure (as we did in single semaphore operations), we specify an array of `sembufs` (line 9). We identify our semaphore array at line 12; note again that we specify the number of semaphores (`nsems`, or number of semaphores, as argument 2). We build out our `sembuf` array as acquires (with a `sem_op` of −1), and also initialize the `sem_num` field with the semaphore number. This specifies that we want to acquire each of the semaphores in the array. If one or more aren't available, the operation blocks until all semaphores can be acquired.

At line 26, we perform the **`semop`** API function to acquire the semaphores. Upon acquisition (or error), the `semop` function returns to the application. As long as the return value is not -1, we've successfully acquired the semaphore array. Note that we could specify -2 for each `sem_op`, which would require that two counts of the semaphore would be required for successful acquisition.

ON THE CD

Listing 16.11 Acquiring an Array of Semaphores Using **`semop`** (on the CD-ROM at `./source/ch16/semaacq.c`)

```
 1:    #include <stdio.h>
 2:    #include <sys/sem.h>
 3:    #include <stdlib.h>
 4:    #include "common.h"
 5:
 6:    int main()
 7:    {
 8:      int semid, i;
 9:      struct sembuf sb[10];
10:
11:      /* Get the semaphore with the id MY_SEM_ID */
```

```
    semid = semget( MY_SEMARRAY_ID, 10, 0 );

    if (semid >= 0) {

      for (i = 0 ; i < 10 ; i++) {
        sb[i].sem_num = i;
        sb[i].sem_op = -1;
        sb[i].sem_flg = 0;
      }

      printf( "semaacq: Attempting to acquire semaphore %d\n",
               semid );

      /* Acquire the semaphores */
      if (semop( semid, &sb[0], 10 ) == -1) {

        printf("semaacq: semop failed.\n");
        exit(-1);

      }

      printf( "semaacq: Semaphore acquired %d\n", semid );

    }

    return 0;
}
```

Next, let's look at the semaphore release operation. We'll include only the changes from Listing 16.11, as they're very similar (on the CD-ROM at `./source/ch16/semarel.c`). In fact, the only difference is the `sembuf` initialization:

```
for ( i = 0 ; i < 10 ; i++ ) {
  sb[i].sem_num = i;
  sb[i].sem_op = 1;
  sb[i].sem_flg = 0;
}
```

In this example, we increment the semaphore (release) instead of decrementing it (as was done in Listing 16.11).

The `sem_flg` within the `sembuf` structure permits us to alter the behavior of the **`semop`** API function. Two flags are possible, as shown in Table 16.5.

TABLE 16.5 Semaphore Flag Options (`sembuf.sem_flg`)

Flag	Purpose
SEM_UNDO	Undo the semaphore operation if the process exits.
IPC_NOWAIT	Return immediately if the semaphore operation cannot be performed (if the process would block) and return an errno of EAGAIN.

Another useful operation that can be performed on semaphores is the wait-for-zero operation. In this case, the process is blocked until the semaphore value becomes zero. This is performed by simply setting the `sem_op` field to zero, as:

```
struct sembuf sb;
...
sb.sem_num = 0;         // semaphore 0
sb.sem_op = 0;          // wait for zero
sb.sem_flg = 0;         // no flags
...
```

As with previous **semop**s, setting `sem_flg` with `IPC_NOWAIT` causes **semop** to return immediately if the operation would block with an `errno` of `EAGAIN`.

Finally, if the semaphore is removed while a process is blocked on it (via a **semop** operation), the process becomes immediately unblocked and an `errno` value is returned as `EIDRM`.

USER UTILITIES

GNU/Linux provides the `ipcs` command to explore semaphores from the command line. The `ipcs` utility provides information on a variety of resources; we'll explore its use below for investigating semaphores.

The general form of the `ipcs` utility for semaphores is:

```
# ipcs -s
```

This presents all the semaphores that are visible to the calling process. Let's now look at an example where we create a semaphore (as was done in Listing 16.1):

```
# ./semcreate
semcreate: Created a semaphore 1769475
# ipcs -s

—— Semaphore Arrays ——
key        semid      owner      perms      nsems
0x0000006f 1769475    mtj        666        1

#
```

Here, we see our newly created semaphore (`key 0x6f`). We can get extended information about the semaphore using the `-i` option. This allows us to specify a specific semaphore ID, for example:

```
# ipcs -s -i 1769475

Semaphore Array semid=1769475
uid=500  gid=500        cuid=500        cgid=500
mode=0666, access_perms=0666
nsems = 1
otime = Not set
ctime = Fri Apr  9 17:50:01 2004
semnum     value      ncount     zcount     pid
0          0          0          0          0

#
```

Here we see our semaphore in greater detail. We see the owner and creator process and group IDs, permissions, number of semaphores (`nsems`), last **semop** time, last change time, and the details of the semaphore itself (`semnum` through `pid`). The value represents the actual value of the semaphore (zero after creation). If we were to perform the release operation (see Listing 15.3) and then perform this command again, we'd see:

```
# ./semrel
semrel: Releasing semaphore 1769475
semrel: Semaphore released 1769475
# ipcs -s -i 1769475

Semaphore Array semid=1769475
uid=500  gid=500        cuid=500        cgid=500
mode=0666, access_perms=0666
nsems = 1
otime = Fri Apr  9 17:54:44 2004
```

```
ctime = Fri Apr  9 17:50:01 2004
semnum      value       ncount      zcount      pid
0           1           0           0           20494

#
```

Note here that our value has increased (based upon the semaphore release), and other information (such as `otime` and `pid`) has been updated given a semaphore operation having been performed.

We can also delete semaphores from the command line using the `ipcrm` command. To delete our previously created semaphore, we'd simply use the `ipcrm` command as follows:

```
# ipcrm -s 1769475
[mtj@camus ch16]$ ipcs -s

—— Semaphore Arrays ——
key         semid       owner       perms       nsems

#
```

As with the `semop` and `semctl` API functions, the `ipcrm` command uses the semaphore identifier to specify the semaphore to be removed.

SUMMARY

In this chapter, we introduced the semaphore API and its application of interprocess coordination and synchronization. We began with a whirlwind tour of the API and then followed with a detailed description of each command including examples of each. Finally, we reviewed the `ipcs` and `ipcrm` commands and demonstrated their debugging and semaphore management capabilities.

SEMAPHORE APIS

```
#include <sys/types.h>
#include <sys/ipc.h>
#include <sys/sem.h>
int semget( key_t key, int nsems, int semflg );
int semop( int semid, struct sembuf *sops, unsigned int nsops );
int semctl( int semid, int semnum, int cmd, ... );
```

17 Shared Memory Programming

In This Chapter

- Introduction to Shared Memory
- Creating and Configuring Shared Memory Segments
- Using and Protecting Shared Memory Segments
- Locking and Unlocking Shared Segments
- Using the `ipcs` and `ipcrm` Utilities

INTRODUCTION

Shared memory APIs is the final topic of inter-process communication that we'll detail in this book. Shared memory allows two or more processes to share a chunk of memory (mapped to each of the process's individual address spaces) so that each can communicate with all others. Shared memory goes even further, as we'll see in this chapter.

Recall from Chapter 12, "Introduction to Sockets Programming," that the address spaces for parent and child processes are independent. The parent process could create a chunk of memory (such as declaring an array), but once the fork completes, the parent and child see different memory. On GNU/Linux, all processes have unique virtual address spaces, but the shared memory API permits a process to attach to a common (shared) address segment.

With all this power comes some complexity. For example, when processes share memory segments, they must also provide a means to coordinate access to them.

This is commonly provided via a semaphore (by the developer), which can be contained within the shared memory segment itself. We'll look at this specific technique in this chapter.

If shared memory segments have this disadvantage, why not use an existing IPC mechanism that has built in coordination, such as message queues? The answer also lies in the simplicity of shared memory. Using a message queue, one process writes a message to a queue, which involves a copy from the user's address space to the kernel space. When another user reads from the message queue, another copy is performed from the kernel's address space to the new user's address space. The benefit of shared memory is that we minimize copying in its entirety. The segment is shared between the two processes in their own address spaces, so bulk copies of data are not necessary.

Because processes share the memory segment in each of their address spaces, copies are minimized in sharing data. For this reason, shared memory can be the fastest form of IPC available within GNU/Linux.

QUICK OVERVIEW OF SHARED MEMORY

For the impatient reader, we'll now take a quick look at the shared memory APIs. In the next section, we'll dig into the API further. Here we'll look at code snippets to create a shared memory segment, get an identifier for an existing one, configure a segment, attach and detach, and also some examples of processes using them.

Using the shared memory API requires the function prototypes and symbols to be available to the application. This is done by including (at the top of the C source file):

```
#include <sys/ipc.h>
#include <sys/shm.h>
```

Create a Shared Memory Segment

To create a shared memory segment, we use the **`shmget`** API function. Using **`shmget`**, we specify a unique shared memory ID, the size of the segment, and finally a set of flags (see Listing 17.1). The `flags` argument, as we saw with message queues and semaphores, includes both access permissions and a command to create the segment (`IPC_CREAT`).

Listing 17.1 Creating a Shared Memory Segment with **shmget** (on the CD-ROM at `./source/ch17/shmcreate.c`)

```
1:    #include <stdio.h>
2:    #include <sys/shm.h>
3:    #include "common.h"
4:
5:    int main()
6:    {
7:      int shmid;
8:
9:      /* Create the shared memory segment using MY_SHM_ID */
10:     shmid = shmget( MY_SHM_ID, 4096, 0666 | IPC_CREAT );
11:
12:     if ( shmid >= 0 ) {
13:
14:       printf( "Created a shared memory segment %d\n", shmid );
15:
16:     }
17:
18:     return 0;
19:   }
```

At line 10 in Listing 17.1, we create a new shared memory segment that's 4KB in size. The size specified must be evenly divisible by the page size of the architecture in question (typically 4KB). The return value of **`shmget`** (stored in `shmid`) can be used in subsequent calls to configure or attach to the segment.

To identify the page size on a given system, simply call the **`getpagesize`** *function. This returns the number of bytes contained within a system page.*

```
#include <unistd.h>
int getpagesize( void );
```

Getting Information on a Shared Memory Segment

We can also get information about a shared memory segment and even set some parameters. The **`shmctl`** API function provides a number of capabilities. Here, we'll look at retrieving information about a shared memory segment.

In Listing 17.2, we first get the shared memory identifier for the segment using the **`shmget`** API function. Once we have this, we can call **`shmctl`** to grab the current stats. To **`shmctl`** we pass the identifier for the shared memory segment, the command to grab stats (`IPC_STAT`), and finally a buffer in which the data will be written.

This buffer is a structure of type `shmid_ds`. We'll look more at this structure in the "`shmctl`" section, later in this chapter. Upon successful return of **`shmctl`**, identified by a zero return, we emit our data of interest. Here, we emit the size of the shared memory segment (`shm_segsz`) and the number of attaches that have been performed on the segment (`shm_nattch`).

Listing 17.2 Retrieving Information about a Shared Memory Segment (on the CD-ROM at `./source/ch17/shmszget.c`)

```
#include <stdio.h>
#include <sys/shm.h>
#include <errno.h>
#include "common.h"

int main()
{
  int shmid, ret;
  struct shmid_ds shmds;

  /* Get the shared memory segment using MY_SHM_ID */
  shmid = shmget( MY_SHM_ID, 0, 0 );

  if ( shmid >= 0 ) {

    ret = shmctl( shmid, IPC_STAT, &shmds );

    if (ret == 0) {

      printf( "Size of memory segment is %d\n", shmds.shm_segsz );
      printf( "Number of attaches %d\n", (int)shmds.shm_nattch );

    } else {

      printf( "shmctl failed (%d)\n", errno );

    }

  } else {

    printf( "Shared memory segment not found.\n" );

  }

  return 0;
}
```

Attaching and Detaching a Shared Memory Segment

In order to use our shared memory, we must attach to it. Attaching to a shared memory segment maps the shared memory into our process's memory space. To attach to the segment, we use the **shmat** API function. This returns a pointer to the segment in the process's address space. This address can then be used by the process like any other memory reference. We detach from the memory segment using the **shmdt** API function.

In Listing 17.3, a simple application is shown to attach to and detach from a shared memory segment. We first get the shared memory identifier using **shmget** (at line 12). At line 16, we attach to the segment using **shmat**. We specify our identifier and an address (where we'd like to place it in our address space) and an options word (0). After checking, if this was successful (the return of a nonzero address from **shmat**), we detach from the segment at line 23 using **shmdt**.

ON THE CD

Listing 17.3 Attaching to and Detaching from a Shared Memory Segment (on the CD-ROM at `./source/ch17/shmattch.c`)

```
 1:   #include <stdio.h>
 2:   #include <sys/shm.h>
 3:   #include <errno.h>
 4:   #include "common.h"
 5:
 6:   int main()
 7:   {
 8:     int shmid, ret;
 9:     void *mem;
10:
11:     /* Get the shared memory segment using MY_SHM_ID */
12:     shmid = shmget( MY_SHM_ID, 0, 0 );
13:
14:     if ( shmid >= 0 ) {
15:
16:       mem = shmat( shmid, (const void *)0, 0 );
17:
18:       if ( (int)mem != -1 ) {
19:
20:         printf( "Shared memory was attached in our "
21:                 "address space at %p\n", mem );
22:
23:         ret = shmdt( mem );
24:
25:         if (ret == 0) {
26:
```

```
            printf("Successfully detached memory\n");

          } else {

            printf("Memory detached Failed (%d)\n", errno);

          }

        } else {

          printf( "shmat failed (%d)\n", errno );

        }

      } else {

        printf( "Shared memory segment not found.\n" );

      }

      return 0;
    }
```

Using a Shared Memory Segment

Now let's look at two processes that use a shared memory segment. For brevity, we'll pass on the error checking for this example. We'll first look at the write example. In Listing 17.4, we see a short example that uses the `strcpy` standard library function to write to the shared memory segment. Since the segment is just a block of memory, we cast it from a `void` pointer to a character pointer in order to write to it (avoiding compiler warnings) at line 16. It's important to note that a shared memory segment is nothing more than a block of memory, and anything you would expect to do with a memory reference is possible with the shared memory block.

ON THE CD

Listing 17.4 Writing to a Shared Memory Segment (on the CD-ROM at `./source/ch17/shmwrite.c`)

```
1:    #include <stdio.h>
2:    #include <sys/shm.h>
3:    #include <string.h>
4:    #include "common.h"
5:
6:    int main()
```

```
{
  int shmid, ret;
  void *mem;

  /* Get the shared memory segment using MY_SHM_ID */
  shmid = shmget( MY_SHM_ID, 0, 0 );

  mem = shmat( shmid, (const void *)0, 0 );

  strcpy( (char *)mem, "This is a test string.\n" );

  ret = shmdt( mem );

  return 0;
}
```

Now let's look at a read example. In Listing 17.5, we see a similar application to Listing 17.4. In this particular case, we read from the block of memory by using the `printf` call. In Listing 17.4 (the write application), we copied a string into the block with the `strcpy` function. Now in Listing 17.5, we emit that same string using `printf`. Note that the first process attaches to the memory, writes the string, and then detaches and exits. The next process attaches and reads from the memory. Any number of processes could read or write to this memory, which is one of the basic problems. We'll investigate some solutions in the section "Using a Shared Memory Segment," later in this chapter.

Listing 17.5 Reading from a Shared Memory Segment (on the CD-ROM at `./source/ch17/shmread.c`)

```
#include <stdio.h>
#include <sys/shm.h>
#include <string.h>
#include "common.h"

int main()
{
  int shmid, ret;
  void *mem;

  /* Get the shared memory segment using MY_SHM_ID */
  shmid = shmget( MY_SHM_ID, 0, 0 );

  mem = shmat( shmid, (const void *)0, 0 );

```

```
  printf( "%s", (char *)mem );

  ret = shmdt( mem );

  return 0;
}
```

Removing a Shared Memory Segment

To permanently remove a shared memory segment, we use the **shmctl** API function. We use a special command with `shmctl` called `IPC_RMID` to remove the segment (much as is done with message queues and semaphores). Listing 17.6 illustrates the segment removal.

Listing 17.6 Removing a Shared Memory Segment (on the CD-ROM at `./source/ch17/shmdel.c`)

```
#include <stdio.h>
#include <sys/shm.h>
#include <errno.h>
#include "common.h"

int main()
{
  int shmid, ret;

  /* Create the shared memory segment using MY_SHM_ID */
  shmid = shmget( MY_SHM_ID, 0, 0 );

  if ( shmid >= 0 ) {

    ret = shmctl( shmid, IPC_RMID, 0 );

    if (ret == 0) {

      printf( "Shared memory segment removed\n" );

    } else {

      printf( "shmctl failed (%d)\n", errno );

    }

  } else {

```

```
29:         printf( "Shared memory segment not found.\n" );
30:
31:       }
32:
33:       return 0;
34:     }
```

Once the shared memory segment identifier is found (line 11), we call **`shmctl`** with the `IPC_RMID` argument at line 15.

That completes our quick tour of the shared memory API. In the next section we'll dig deeper into the APIs and look at some of the more detailed aspects.

Shared Memory APIs

Now that we have a quick review behind us, let's dig down further into the APIs. Table 17.1 provides the shared memory API functions, along with their basic uses.

TABLE 17.1 Shared Memory API Functions and Uses

API Function	Uses
`shmget`	Create a new shared memory segment
	Get the identifier for an existing shared memory segment
`shmctl`	Get info on a shared memory segment
	Set certain info on a shared memory segment
	Remove a shared memory segment
`shmat`	Attach to a shared memory segment
`shmdt`	Detach from a shared memory segment

We'll address these API functions now in detail, identifying each of their uses with example source.

`shmget`

The **`shmget`** API function (like **`semget`** and **`msqget`**) is a multirole function. First, it can be used to create a new shared memory segment, and second, it can be used to get the ID of an existing shared memory segment. The result of the **`shmget`** API function (in either role) is a shared memory segment identifier that is to be used in all other shared memory functions. The prototype for the **`shmget`** function is defined as:

```
#include <sys/ipc.h>
#include <sys/shm.h>
int shmget( key_t key, size_t size, int shmflag );
```

The `key` argument specifies a system-wide identifier that uniquely identifies the shared memory segment. The `key` must be a nonzero value or the special symbol `IPC_PRIVATE`. The `IPC_PRIVATE` argument defines that we're creating a private segment (one that has no system-wide name). No other processes will find this segment, but it can be useful to create segments that are used only within a process or process group (where the return key can be communicated).

The `size` argument identifies the size of the shared memory segment to create. When we're interested in an existing shared memory segment, we leave this argument as zero (as it's not used by the function in the create case). As a minimum, the size must be `PAGE_SIZE` (or 4Kb). The size should also be evenly divisible by `PAGE_SIZE`, as the segment allocated will be in `PAGE_SIZE` chunks. The maximum size is implementation dependent, but typically is 4MB.

The `shmflag` argument permits the specification of two separate parameters. These are a command and an optional set of access permissions. The command portion can take one of three forms. The first is to create a new shared memory segment, where the `shmflag` is equal to the `IPC_CREAT` symbol. This returns the identifier for a new segment or an identifier for an existing segment (if it already exists). If we want to create the segment and fail if the segment already exists, then we can use the `IPC_CREAT` with the `IPC_EXCL` symbol (second form). If (`IPC_CREAT | IPC_EXCL`) is used and the shared segment already exists, then an error status is returned and `errno` is set to `EEXIST`. The final form simply requests an existing shared memory segment. In this case, we specify a value of zero for the command argument.

When creating a new shared memory segment, each of the read/write access permissions can be used except for the execute permissions. These permissions are shown in Table 17.2.

TABLE 17.2 Shared Memory Segment Permissions for **`shmget`** `msgflag` Argument

Symbol	Value	Meaning
S_IRUSR	0400	User read permission
S_IWUSR	0200	User write permission
S_IRGRP	0040	Group read permission
S_IWGRP	0020	Group read permission
S_IROTH	0004	Other read permission
S_IWOTH	0002	Other read permission

Let's now look at a few examples of the **shmget** function to create new shared memory segments or to access existing ones.

In this first example, we'll create a new private shared memory segment of size 4KB. Note that since we're using IPC_PRIVATE, we're assured of creating a new segment as no unique key is provided. We'll also specify full read and write permission to all (system, group, and user).

```
shmid = shmget( IPC_PRIVATE, 4096, IPC_CREAT | 0666 );
```

If the **shmget** API function fails, a -1 is returned (as shmid) with the actual error specified in the special errno variable (for this particular process).

Now let's look at the creation of a memory segment, with an error return if the segment already exists. In this example, our system-wide identifier (key) is 0x123, and we'll request a 64KB segment.

```
shmid = shmget( 0x123, (64 * 1024), (IPC_CREAT | IPC_EXCL | 0666)
);
if (shmid == -1) {
  printf( "shmget failed (%d)\n", errno );
}
```

Here we use the IPC_CREAT with IPC_EXCL to ensure that the segment doesn't exist. If we get a -1 return from **shmget**, an error occurred (such as the segment already existed).

Creating system-wide keys with **ftok** was discussed in Chapter 15, "IPC with Message Queues," and Chapter 16, "Synchronization with Semaphores." Please refer to those chapters for a detailed discussion of file-based key creation.

Finally, let's look at a simple example of finding the shared memory identifier for an existing segment.

```
shmid = shmget( 0x123, 0, 0 );
if ( shmid != -1 ) {
  // Found our shared memory segment
}
```

Here we specify only the system-wide key; segment size and flags are both zero (as we're getting the segment, not creating it).

A final point to discuss with shared memory segments creation is the initialization of the shared memory data structure that is contained within the kernel. The shared memory structure is shown in Listing 17.7.

Listing 17.7 The Shared Memory Structure (`shmid_ds`)

```
struct shmid_ds {
  struct ipc_perm shm_perm    /* Access permissions        */
  int     shm_segsz;        /* Segment size (in bytes)    */
  time_t shm_atime;         /* Last attach time (shmat)   */
  time_t shm_dtime;         /* Last detach time (shmdt)   */
  time_t shm_ctime;         /* Last change time (shmctl)  */
  unsigned short shm_cpid;    /* Pid of segment creator     */
  unsigned short shm_lpid;    /* Pid of last segment user   */
  short  shm_nattch;         /* Number of current attaches */
};
struct ipc_perm {
  key_t __key;
  unsigned short uid;
  unsigned short gid;
  unsigned short cuid;
  unsigned short cgid;
  unsigned short mode;
  unsigned short pad1;
  unsigned short __seq;
  unsigned short pad2;
  unsigned long int __unused1;
  unsigned long int __unused2;
};
```

Upon creation of a new shared memory segment, the `shm_perm` structure is initialized with the key and creator's user ID and group ID. Other initializations are shown in Table 17.3.

TABLE 17.3 Shared Memory Data Structure `init` on Creation

Field	Initialization
`shm_segsz`	Segment size provided to `shmget`
`shm_atime`	0
`shm_dtime`	0
`shm_ctime`	Current time
`shm_cpid`	Calling process's PID
`shm_lpid`	0
`shm_nattch`	0
`shm_perm.cuid`	Creator's process user ID
`shm_perm.gid`	Creator's process group ID

We'll return to these elements shortly when we discuss the control aspects of shared memory segments.

shmctl

The **shmctl** API function provides three separate functions. The first is to read the current shared memory structure (as defined at Listing 17.7) using the IPC_STAT command. The second is to write the shared memory structure using the IPC_SET command. Finally, a shared memory segment can be removed using the IPC_RMID command. The **shmctl** function prototype is shown below:

```
#include <sys/ipc.h>
#include <sys/shm.h>
int shmctl( int shmid, int cmd, struct shmid_ds *buf );
```

Let's begin by removing a shared memory segment. To remove a segment, we first must have the shared memory identifier. We then pass this identifier, along with the command IPC_RMID, to **shmctl**, as:

```
int shmid, ret;
...
shmid = shmget( SHM_KEY, 0, 0 );
if ( shmid != -1 ) {
  ret = shmctl( shmid, IPC_RMID, 0 );
  if (ret == 0) {
    // shared memory segment successfully removed.
  }
}
```

If no processes are currently attached to the shared memory segment, then the segment is removed. If there are currently attaches to the shared memory segment, then the segment is marked for deletion but not yet deleted. This means that only after the last process detaches from the segment will it be removed. Once the segment is marked for deletion, no processes may attach to the segment. Any attempt to attach results in an error return with errno set to EIDRM.

Internally, once the shared memory segment is removed, its key is changed to IPC_PRIVATE. *This disallows any new process from finding the segment.*

Next, let's look at the IPC_STAT command that can be used within **shmctl** to gather information about a shared memory segment. In Listing 17.7, we listed a number of parameters that define a shared memory segment. This structure can be read via the IPC_STAT command as shown in Listing 17.8.

ON THE CD

Listing 17.8 Shared Memory Data Structure Elements Accessible Through **shmctl** (on the CD-ROM at ./source/ch17/shmstat.c)

```
#include <stdio.h>
#include <sys/shm.h>
#include <errno.h>
#include <time.h>
#include "common.h"

int main()
{
  int shmid, ret;
  struct shmid_ds shmds;

  /* Create the shared memory segment using MY_SHM_ID */
  shmid = shmget( MY_SHM_ID, 0, 0 );

  if ( shmid >= 0 ) {

    ret = shmctl( shmid, IPC_STAT, &shmds );

    if (ret == 0) {

      printf( "Size of memory segment is %d\n",
                shmds.shm_segsz );
      printf( "Number of attaches %d\n",
                (int)shmds.shm_nattch );
      printf( "Create time %s",
                ctime( &shmds.shm_ctime ) );
      if (shmds.shm_atime) {
        printf( "Last attach time %s",
                ctime( &shmds.shm_atime ) );
      }
      if (shmds.shm_dtime) {
        printf( "Last detach time %s",
                ctime( &shmds.shm_dtime ) );
      }
      printf( "Segment creation user %d\n",
                shmds.shm_cpid );
      if (shmds.shm_lpid) {
        printf( "Last segment user %d\n",
                shmds.shm_lpid );
      }
      printf( "Access permissions 0%o\n",
                shmds.shm_perm.mode );
```

```
43:
44:         } else {
45:
46:           printf( "shmctl failed (%d)\n", errno );
47:
48:         }
49:
50:       } else {
51:
52:         printf( "Shared memory segment not found.\n" );
53:
54:       }
55:
56:       return 0;
57:     }
```

Listing 17.8 is rather self-explanatory. After getting our shared memory identifier at line 12, we use **shmctl** to grab the shared memory structure at line 17. Upon success of **shmctl**, we emit the various accessible data elements using `printf`. Note that at lines 27 and 31, we check that the time values are nonzero. If there have been no attaches or detaches, the value will be zero, and therefore there's no reason to convert it to a string time.

The final command available with **shmctl** is `IPC_SET`. This permits the caller to update certain elements of the shared memory segment data structure. These elements are shown in Table 17.4.

TABLE 17.4 Shared Memory Data Structure Writeable Elements

Field	Description
`shm_perm.uid`	Owner Process effective user ID
`shm_perm.gid`	Owner Process effective group ID
`shm_flags`	Access permissions
`shm_ctime`	Takes the current time of `shmctl.IPC_SET` action

The following code snippet illustrates setting new permission (see Listing 17.9). It's important that the shared memory data structure be read first to get the current set of parameters.

Listing 17.9 Changing Access Permissions in a Shared Memory Segment (on the CD-ROM at `./source/ch17/shmset.c`)

```
#include <stdio.h>
#include <sys/shm.h>
#include <errno.h>
#include <time.h>
#include "common.h"

int main()
{
  int shmid, ret;
  struct shmid_ds shmds;

  /* Create the shared memory segment using MY_SHM_ID */
  shmid = shmget( MY_SHM_ID, 0, 0 );

  if ( shmid >= 0 ) {

    ret = shmctl( shmid, IPC_STAT, &shmds );

    if (ret == 0) {

      printf("old permissions were 0%o\n", shmds.shm_perm.mode );

      shmds.shm_perm.mode = 0444;

      ret = shmctl( shmid, IPC_SET, &shmds );

      ret = shmctl( shmid, IPC_STAT, &shmds );

      printf("new permissions are 0%o\n", shmds.shm_perm.mode );

    } else {

      printf( "shmctl failed (%d)\n", errno );

    }

  } else {

    printf( "Shared memory segment not found.\n" );

  }
```

```
42:
43:        return 0;
44:    }
```

In Listing 17.9, we grab the current data structure for the memory segment at line 17 and then change the mode at line 23. We write this back to the segment's data structure at line 25 using the `IPC_SET` command, and then we read it back out at line 27. Not very exciting, but the key to remember is to read the structure first. Otherwise, the effective user and group IDs will be incorrect, leading to anomalous behavior.

One final topic for shared memory control that differs from message queues and semaphores is the ability to lock down segments so that they're not candidates for swapping out of memory. This can be a performance benefit, because rather than the segment being swapped out to the file system, it stays in memory and is therefore available to applications without having to swap it back in. This mechanism is therefore very useful from a performance standpoint. The **`shmctl`** API function provides the means both to lock down a segment and also to unlock.

The following examples illustrate the lock and unlock of a shared memory segment:

```
int shmid;
...
shmid = shmget( MY_SHM_ID, 0, 0 );
ret = shmctl( shmid, SHM_LOCK, 0 );
if ( ret == 0 ) {
  printf( "Shared Memory Segment Locked down.\n" );
}
```

Unlocking the segment is very similar. Rather than specify `SHM_LOCK`, we instead use the `SHM_UNLOCK` symbolic, as:

```
ret = shmctl( shmid, SHM_UNLOCK, 0 );
```

As before, a zero return indicates success of the **`shmctl`** call. Only a super-user may perform this particular command via **`shmctl`**.

shmat

Once the shared memory segment has been created, a process must attach to it to make it available within its address space. This is provided by the **`shmat`** API function. Its prototype is defined as:

```
#include <sys/types.h>
#include <sys/shm.h>
void *shmat( int shmid, const void *shmaddr, int shmflag );
```

The **shmat** function takes the shared memory segment identifier (returned by **shmget**), an address where the process would like to insert this segment in the process's address space (a desired address), and a set of flags. The desired address (`shmaddr`) is rounded down if the `SHM_RND` flag is set within `shmflags`. This option is rarely used because the process would need to have explicit knowledge of the available address regions within the process's address space. This method also is not entirely portable. To have the **shmat** API function automatically place the region within the process's address space, a `(const void *)NULL` argument is passed.

The caller can also specify the `SHM_READONLY` within `shmflags` to enforce a read-only policy on the segment for this particular process. This process must first have read permission on the segment. If `SHM_READONLY` is not specified, it is assumed that the segment is being mapped for both read and write. There is no write-only flag.

The return value from **shmat** is the address in which the shared memory segment is mapped into this process. A quick example of **shmat** is shown here:

```
int shmid;
void *myAddr;
/* Get the id for an existing shared memory segment */
shmid = shmget( MY_SHM_SEGMENT, 0, 0 );
/* Map the segment into our space */
myAddr = shmat( shmid, 0, 0 );
if ((int)myAddr != -1) {
  // Attach failed.
} else {
  // Attach succeeded.
}
```

Upon completion, `myAddr` will contain an address in which the segment is attached or `-1`, indicating that the segment failed to be attached. The return address can then be utilized by the process just like any other address.

The local address into which the shared memory segment is mapped may be different for every process that attaches to it. Therefore, no process should assume that since another mapped at a given address, it will be available to it at the same local address.

Upon successful completion of the **shmat** call, the shared memory data structure is updated as follows. The `shm_atime` field is updated with the current time (last attach time), shm_lpid is updated with the effective process ID for the calling process, and the shm_nattch field is incremented by 1 (the number of processes currently attached to the segment).

When a processes exits, its shared memory segments are automatically detached. Despite this, developers should detach from their segments using **shmdt** rather than relying on GNU/Linux to do this for them. Also, when a process forks into a parent and child, the child inherits any shared memory segments that were created previously by the parent.

shmdt

The **shmdt** API function detaches an attached shared memory segment from a process. When a process no longer needs access to the memory, this function frees it and also unmaps the memory mapped into the process's local address space that was occupied by this segment. The function prototype for the **shmdt** function is:

```
#include <sys/types.h>
#include <sys/shm.h>
int shmdt( const void *shmaddr );
```

The caller provides the address that was provided by **shmat** (as its return value). A return value of zero indicates a successful detach of the segment. Consider the following code snippet as a demonstration of the **shmdt** call:

```
int shmid;
void *myAddr;
/* Get the id for an existing shared memory segment */
shmid = shmget( MY_SHM_SEGMENT, 0, 0 );
/* Map the segment into our space */
myAddr = shmat( shmid, 0, 0 );
...
/* Detach (unmap) the segment */
ret = shmdt( myAddr );
if (ret == 0) {
  /* Segment detached */
}
```

Upon successful detach, the shared memory structure is updated as follows. The shm_dtime field is updated with the current time (of the **shmdt** call), the shm_lpid is updated with the process ID of the process calling **shmdt**, and finally the shm_nattach field is decremented by 1.

The address region mapped by the shared memory segment will be unavailable to the process and may result in a segment violation if an access is attempted.

If the segment had been previously marked for deletion (via a prior call to **shmctl** with the command of `IPC_RMID`) and the number of current attaches is zero, then the segment is removed.

Using a Shared Memory Segment

Shared memory can be a powerful mechanism for communication and coordination between processes. With this power comes some complexity. Since shared memory is a resource that's available to all processes that attach to it, we must coordinate access to it. One mechanism is to simply add a semaphore to the shared memory segment. If the segment represents multiple contexts, multiple semaphores can be created, each coordinating its respective access to a portion of the segment.

Let's look at a simple example of coordinating access to a shared memory segment. Listing 17.10 illustrates a simple application that provides for creating, using, and removing a shared memory segment. As we've already covered the creation and removal aspects in detail (lines 31–58 for create and lines 137–158 for remove), the use scenario (lines 59–111) is what we'll focus on here.

Our `block` (which represents our shared memory block) is `typdef`'d at lines 11–15. This contains our shared structure (`string`), a `counter` (as the index to our string), and our semaphore to coordinate access. Note that this was loaded into our shared structure at line 48.

Our use scenario begins by grabbing the user character passed as the second argument from the command line. This is the character we'll place into the buffer on each pass. We'll invoke this process twice with different characters to see each access the shared structure in a synchronized way. After getting the shared memory key (via **shmget** at line 69), we attach to the segment at line 72. The return value is the address of our shared block, which we cast to our block type (`MY_BLOCK_TYPE`). We then loop through a count of 2500, each iteration acquiring the semaphore, loading our character into the string array of the shared memory segment (our critical section), and then releasing the semaphore.

ON THE CD

Listing 17.10 Shared Memory Example Using Semaphore Coordination (on the CD-ROM at `./source/ch17/shmexpl.c`)

```
1:     #include <stdio.h>
2:     #include <sys/shm.h>
3:     #include <sys/sem.h>
4:     #include <string.h>
5:     #include <stdlib.h>
6:     #include <unistd.h>
7:     #include "common.h"
8:
9:     #define MAX_STRING      5000
```

```

typedef struct {
  int semID;
  int counter;
  char string[MAX_STRING+1];
} MY_BLOCK_T;


int main( int argc, char *argv[] )
{
  int shmid, ret, i;
  MY_BLOCK_T *block;
  struct sembuf sb;
  char user;

  /* Make sure there's a command */
  if (argc >= 2) {

    /* Create the shared memory segment and init it
     * with the semaphore
     */
    if (!strncmp( argv[1], "create", 6 )) {

      /* Create the shared memory segment and semaphore */

      printf("Creating the shared memory segment\n");

      /* Create the shared memory segment */
      shmid = shmget( MY_SHM_ID,
                       sizeof(MY_BLOCK_T), (IPC_CREAT | 0666) );

      /* Attach to the segment */
      block = (MY_BLOCK_T *)shmat( shmid, (const void *)0, 0 );

      /* Initialize our write pointer */
      block->counter = 0;

      /* Create the semaphore */
      block->semID = semget( MY_SEM_ID, 1, (IPC_CREAT | 0666) );

      /* Increment the semaphore */
      sb.sem_num = 0;
      sb.sem_op = 1;
      sb.sem_flg = 0;
      semop( block->semID, &sb, 1 );
```

```

        /* Now, detach from the segment */
        shmdt( (void *)block );

      } else if (!strncmp( argv[1], "use", 3 )) {

        /* Use the segment */

        /* Must specify also a letter (to write to the buffer) */
        if (argc < 3) exit(-1);

        user = (char)argv[2][0];

        /* Grab the shared memory segment */
        shmid = shmget( MY_SHM_ID, 0, 0 );

        /* Attach to it */
        block = (MY_BLOCK_T *)shmat( shmid, (const void *)0, 0 );

        for (i = 0 ; i < 2500 ; i++) {

          /* Give up the CPU temporarily */
          sleep(0);

          /* Grab the semaphore */
          sb.sem_num = 0;
          sb.sem_op = -1;
          sb.sem_flg = 0;
          if ( semop( block->semID, &sb, 1 ) != -1 ) {

            /* Write our letter to the segment buffer
             * (only if we have the semaphore).  This
             * is our critical section.
             */
            block->string[block->counter++] = user;

            /* Release the semaphore */
            sb.sem_num = 0;
            sb.sem_op = 1;
            sb.sem_flg = 0;
            if ( semop( block->semID, &sb, 1 ) == -1 ) {

              printf("Failed to release the semaphore\n");

            }
```

```
        } else {
          printf("Failed to acquire the semaphore\n");
        }
      }
      /* We're done, unmap the shared memory segment. */
      ret = shmdt( (void *)block );
    } else if (!strncmp( argv[1], "read", 6 )) {
      /* Here, we'll read the buffer in the shared segment */
      shmid = shmget( MY_SHM_ID, 0, 0 );
      if (shmid != -1) {
        block = (MY_BLOCK_T *)shmat( shmid, (const void *)0, 0 );
        /* Terminate the buffer */
        block->string[block->counter+1] = 0;
        printf( "%s\n", block->string );
        printf("length %d\n", block->counter);
        ret = shmdt( (void *)block );
      } else {
        printf("Unable to read segment.\n");
      }
    } else if (!strncmp( argv[1], "remove", 6 )) {
      shmid = shmget( MY_SHM_ID, 0, 0 );
      if (shmid != -1) {
        block = (MY_BLOCK_T *)shmat( shmid, (const void *)0, 0 );
```

```
        /* Remove the semaphore */
        ret = semctl( block->semID, 0, IPC_RMID );

        /* Remove the shared segment */
        ret = shmctl( shmid, IPC_RMID, 0 );

        if (ret == 0) {

          printf("Successfully removed the segment.\n");

        }

      }

    } else {

      printf( "Unknown command %s\n", argv[1] );

    }

  }

  return 0;
}
```

The key point of Listing 17.10 is that reading or writing from memory in a shared memory segment must be protected by a semaphore. Other structures can be represented in a shared segment, such as a message queue. The queue doesn't require any protection because it's protected internally.

Now let's look at an example run of our application shown in Listing 17.10. We create our share memory segment and then execute our use scenarios one after another (quickly). Note that we specify two different characters to differentiate which process had control for that position in the string. Once complete, we use the `read` command to emit the string (a snippet is shown here).

```
$ ./shmexpl create
Creating the shared memory segment
$ ./shmexpl use a &
$ ./shmexpl use b &
[1] 18254
[2] 18255
[1]+ Done
[2]+ Done
```

```
$ ./shmexpl read
aaaaaaaaaaaaaaaaaaaaaabbbbbbbbbbbbbbbbbbbaaabbb...
length 5000
$ ./shmexpl remove
Successfully removed the segment.
$
```

Note that in some cases, you can see an entire string of `a`'s and then all the `b`'s. It all comes down to executing the use cases quickly enough so that they each compete for the shared resource.

USER UTILITIES

GNU/Linux provides the `ipcs` command to explore IPC assets from the command line (including shared memory segments that are visible to the user). The `ipcs` utility provides information on shared memory segments as well as message queues and semaphores. We'll investigate its use for shared memory segments here.

The general form of the `ipcs` utility for shared memory segments is:

```
$ ipcs -m
```

This presents all of the shared memory segments that are visible to the process. Let's start by creating a shared memory segment (as shown in Listing 17.1):

```
$ ./shmcreate
Created a shared memory segment 163840
[mtj@camus ch17]$ ipcs -m
—— Shared Memory Segments ——
key        shmid      owner      perms      bytes      nattch     status
0x000003e7 163840     mtj        666        4096       0
$
```

We see here a new shared memory segment being available (`0x3e7 = 999`). Its size is 4,096 bytes, and we see that there are currently no attaches to this segment (`nattch = 0`). If we wanted to dig into this segment deeper, we could specify this shared memory segment specifically to `ipcs`. This is done with `ipcs` using the `-i` option:

```
$ ipcs -m -i 163840
Shared memory Segment shmid=163840
uid=500 gid=500 cuid=500        cgid=500
mode=0666       access_perms=0666
bytes=4096      lpid=0  cpid=15558      nattch=0
```

```
att_time=Not set
det_time=Not set
change_time=Thu May 20 11:44:44 2004
$
```

We now see some more detailed information, including the attach, detach and change times, last pid, created pid, and so on.

Finally, we can remove the shared memory segment using the `ipcrm` command. To remove our previously created shared memory segment, we simply provide the shared memory identifier, as:

```
$ ipcrm -m 163840
$
```

SUMMARY

In this chapter, we introduced shared memory in GNU/Linux and the APIs that control its use. We first introduced the shared memory APIs as a quick review and then provided a more detailed view of the APIs. As shared memory segments can be shared by multiple asynchronous processes, we illustrated the protection of a shared memory segment with a semaphore. Finally, we reviewed the `ipcs` utility and demonstrated its use as a debugging tool, as well as the `ipcrm` utility for removing shared memory segments from the command line.

REFERENCES

GNU/Linux `shmget`, `shmop` main pages

SHARED MEMORY APIS

```
#include <sys/types.h>
#include <sys/ipc.h>
#include <sys/shm.h>
int shmget( key_t key, size_t size, int shmflag );
int shmctl( int shmid, int cmd, struct shmid_ds *buf );
void *shmat( int shmid, const void *shmaddr, int shmflag );
int shmdt( const void *shmaddr );
```

18 Other Application Development Topics

In This Chapter

- Parsing Command-line Options with `getopt` and `getopt_long`
- Time Conversion Functions
- Gathering System Level Information with `sysinfo`
- Mapping Physical Memory with `mmap`
- Locking and Unlocking Memory Pages for Performance

INTRODUCTION

So far, we've discussed a large number of topics relating to some of the more useful GNU/Linux service APIs. We'll now look at a number of miscellaneous core APIs that will complete our exploration of GNU/Linux application development. This will include the `getopt` function to parse command-line options, time and time conversion functions, physical memory mapping functions, and memory locking for high-performance applications.

The C language provides the means to pass command-line arguments into a program as it begins execution. The C `main` *function may accept two arguments,* `argv` *and* `argc`*. The* `argc` *argument defines the number of arguments that were passed in, while* `argv` *is a character pointer array (vector), containing an element per argument. For example,* `argv[0]` *is a character pointer to the first argument (the program name), and* `argv[argc-1]` *points to the last argument.*

Parsing Command-Line Options with getopt and getopt_long

The **getopt** function provides a simplified API for extracting command-line arguments and their options from the command line. Most arguments take the form

```
<application> -f <f-arg>
```

where `-f` is a command-line option and `<f-arg>` is the option for `-f`. Function `getopt` can also handle much more complex argument arrangements, as we'll see in this section.

The function prototype for the `getopt` function is provided as:

```
#include <unistd.h>
int getopt( int argc, char * const argv[], const char *optstring );
extern char *optarg;
extern int optopt, optind;
```

The **getopt** function takes three arguments; the first two are the `argc` and `argv` arguments that are received through `main`. The third argument, `optstring`, represents our options specification. This consists of the options that we'll accept for our application. The option string has a special form. We define the characters that we'll accept as our options, and for each option that has an argument, we'll follow it with a :. Consider the following example option string:

```
"abc:d"
```

This will parse options such as `-a`, -b, -d, and also `-c <arg>`. We could also provide a double-colon, such as `"abc::d"`, which tells `getopt` that `c` uses an optional argument.

The **getopt** function returns an `int` that represents the character option. With this, three external variables are also provided as part of the `getopt` API. These are **optarg**, **optopt**, and **optind**. The `optarg` variable points to an option argument and is used to extract the option when one is expected. The `optopt` variable specifies the option that is currently being processed. The return value of **getopt** represents the variable. When **getopt** is finished parsing (returns `-1`), the `optind` variable represents the index of those arguments on the command line that were not parsed. For example, if a set of arguments are provided on the command line without any - options, then these arguments can be retrieved via the `optind` argument (we'll see an example of this shortly).

The application developer must ensure that all options are specified that are required for the application. The **getopt** *function will provide the parsing aspect of command-line arguments, but the application must determine whether the options specified are accurate.*

Let's now look at an example of **getopt** that demonstrates the features that we've touched upon (see Listing 18.1). At line 8 we call **getopt** to get the next option. Note that we call it until we get a `-1` return, and it iterates through all of the options that are available. If **getopt** returns `-1`, we exit our loop (to line 36).

At line 10, we start at `switch` construct to test the returns. If the `'h'` character was returned (line 12–14) we handled the `Help` option. At line 16, we handle the verbose option, which has an argument. Since we expect an integer argument after `-v`, we grab it using the `optarg` variable, passing it to `atoi` to convert it to an integer (line 17).

At line 20, we grab our `-f` argument (representing the filename). Since we're looking for a string argument, we can use `optarg` directly (line 21). At line 29, we test for any unrecognized options for which **getopt** will return `'?'`. We emit the actual option that was found with `optopt` (line 29).

Finally, at line 38–42, we emit any options found that were not parsed using the `optind` variable. The **getopt** internally moves the nonoption arguments to the end of the `argv` argument list. Therefore, we can walk from `optind` to `argc` to find these.

ON THE CD

Listing 18.1 Example Use of **getopt** (on the CD-ROM at `./source/ch18/opttest.c`)

```
1:    #include <unistd.h>
2:    #include <stdio.h>
3:
4:    int main( int argc, char *argv[] )
5:    {
6:      int c;
7:
8:      while ( (c = getopt( argc, argv, "hv:f:d" ) ) != -1 ) {
9:
10:       switch( c ) {
11:
12:         case 'h':
13:           printf( "Help menu.\n" );
14:           break;
15:
16:         case 'v':
17:           printf( "Verbose level = %d\n", atoi(optarg) );
18:           break;
19:
20:         case 'f':
21:           printf( "Filename is = %s\n", optarg );
22:           break;
23:
24:         case 'd':
```

```
            printf( "Debug mode\n" );
            break;

          case '?':
            printf( "Unrecognized option encountered -%c\n", optopt );

          default:
            exit(-1);

      }

    }

    for ( c = optind ; c < argc ; c++ ) {

      printf( "Non option %s\n", argv[c] );

    }


    /*
     *  Option parsing complete...
     */

    return 0;
  }
```

Many new applications support not only short option arguments (such as `-a`) but also longer options (such as `–command=start`). The **getopt_long** function provides the application developer with the ability to parse both types of option arguments. The **getopt_long** function has the prototype:

```
#include <getopt.h<
int getopt_long( int argc, char * const argv[],
                     const char *optsring,
                     const struct option *longopts, int *longindex
);
```

The first three arguments (`argc`, `argv`, and `opstring`) mirror the **getopt** function. What differs for **getopt_long** are the final two arguments: `longopts` and `longindex`. The `longopts` argument is a structure that defines the set of long arguments that are desired to be parsed. This structure is defined as

```
struct option {
  const char *name;
  int has_arg;
  int *flag;
  int val;
};
```

where `name` is the name of the long option (such as `command`) and `has_arg` represents the argument that may follow the option (`0` for no argument, `1` for a required option, and `2` for an optional argument). The `flag` reference determines how the return value is provided. If `flag` is not `NULL`, the return value is provided by the fourth argument, `val`; otherwise the return value is returned by **`getopt_long`**.

Let's now look at an example of **`getopt_long`**. In this example, our application will accept the following arguments:

```
–start
–stop
–command <command<
```

As we'll see, the **`getopt_long`** function is a perfect example of the use of a data structure to simplify the job of coding (see Listing 18.2).

The first item to note is that for the **`getopt_long`** function, we must include `getopt.h` (rather than `unistd.h`, as was done for `getopt`). Our option data structure is defined at lines 4–9. At line 5, we define the element for the `–start` option. We define the name `start`, specify that it has no options, and then define the character that **`getopt_long`** will return once this option is found `('s')`. The `–stop` option is defined similarly (but return `'t'` on recognition). The `–command` option identifies a required argument to follow (as defined by `required_argument`) and will return `'c'` when found on the command line.

At line 16 and 17, we see the call to **`getopt_long`**, which specifies our option string (`'stc:'`) and our options structure (`longopts`). The return value, like function **`getopt`**, is `-1` for no further options or a single character (as defined in the options structure).

When `–start` is encountered, an `'s'` is returned and handled at lines 21–23. The `–stop` option is found, a `'t'` is returned and handled at lines 25–27. When `–command` is parsed, **`getopt_long`** returns `'c'`, and we emit the command option at line 30 using the `optarg` variable. Finally, we identify unrecognized options at lines 33–36 when `'?'` is returned from **`getopt_long`** (or an unknown option).

Listing 18.2 Simple Example of **getopt_long** to Parse Command-line Options (on the CD-ROM at `./source/ch18/optlong.c`)

```
#include <stdio.h<
#include <getopt.h<

static struct option longopts[] = {
  { "start",   no_argument,       NULL,  's' },
  { "stop",    no_argument,       NULL,  't' },
  { "command", required_argument, NULL,  'c' },
  { NULL,      0,                 NULL,   0 }
};


int main( int argc, char *argv[] )
{
  int c;

  while ( (c = getopt_long( argc, argv, "stc:",
               longopts, NULL)) != -1 ) {

    switch( c ) {

      case 's':
        printf( "Start!\n" );
        break;

      case 't':
        printf( "Stop!\n" );
        break;

      case 'c':
        printf( "Command %s!\n", optarg );
        break;

      case '?':
      default:
        printf( "Unknown option\n");
        break;

    }

  }

  return 0;
}
```

Any application that requires command-line configurability can benefit from **getopt** or **getopt_long**.

Time API

GNU/Linux provides a wide variety of functions to deal with time (as in time of day). Time is commonly represented by the tm structure, which is identified as:

```
struct tm {
    int  tm_sec;   /* seconds         (0..59)           */
    int  tm_min;   /* minutes         (0..59)           */
    int  tm_hour;  /* hours           (0..23)           */
    int  tm_mday;  /* day of the month (1..31)          */
    int  tm_mon;   /* month           (1..12)           */
    int  tm_year;  /* year            (200x)            */
    int  tm_wday;  /* day of the week (0..6, 0 = Monday */
    int  tm_yday;  /* day in the year (1..366)          */
    int  tm_isdst; /* daylight savings time (0, 1, -1)  */
};
```

A simplified representation is defined as the time_t structure, which simply represents the time in seconds (since the epoch 00:00:00 UTC, January 1, 1970). The time functions that we'll review in this section are these:

```
#include <time.h<
time_t time( time_t *t );
struct tm *localtime( const time_t *timep );
struct tm *gmtime( const time_t *timep );
char *asctime( const struct tm *tm );
char *ctime( const time_t *timepDay  );
time_t mktime( struct tm *tm );
```

Grabbing the current time can be done with the **time** function, as:

```
time_t currentTime;
currentTime = time( NULL );
```

Where NULL is passed to return the local time from the **time** function. The time can also be loaded into a variable by passing a time_t reference to the function:

```
time_t currentTime;
(void)time( &currentTime );
```

With this time stored, we can now convert it into the tm structure using the **localtime** function. Putting it together with **time**, we get:

```
time_t currentTime;
struct tm *tm_time;
currentTime = time( NULL );
tm_time = localtime( &currentTime );
printf( "%02d:%02d:%02d\n",
 tm_time-<tm_hour, tm_time-<tm_min, tm_time-<tm_sec );
```

Converting **time** to an ASCII string is easily provided using the **asctime** or **ctime** function. The **ctime** function takes a `time_t` reference, while **asctime** takes a `tm` structure, as:

```
time_t currentTime;
struct tm *tm_time;
currentTime = time( NULL );
printf("%s\n", ctime( &currentTime ) );
tm_time = localtime( &currentTime );
printf("%s\n", asctime( tm_time ) );
```

The **gmtime** function breaks down a `time_t` variable into a `tm` structure, but in Coordinated Universal Time (CUT). This is the same as GMT (Greenwich Mean Time). The **gmtime** function is illustrated as:

```
tm_time = gmtime( &currentTime );
```

Finally, the **mktime** function converts the `tm` structure into the `time_t` format. It is demonstrated as:

```
tm_time = gmtime( &currentTime );
```

The entire set of functions is illustrated in the simple application shown in Listing 18.3.

ON THE CD

Listing 18.3 Demonstration of Time Conversion Functions (on the CD-ROM at `./source/ch18/time.c`)

```
#include <time.h<
#include <stdio.h<

int main()
{
  time_t currentTime;
  struct tm *tm_time;

  currentTime = time( NULL );
```

```
  tm_time = localtime( &currentTime );

  printf( "from localtime %02d:%02d:%02d\n",
          tm_time-<tm_hour, tm_time-<tm_min, tm_time-<tm_sec );

  printf( "from ctime %s\n", ctime( &currentTime ) );

  printf( "from asctime/localtime %s\n", asctime( tm_time ) );

  tm_time = gmtime( &currentTime );

  printf( "from gmtime %02d:%02d:%02d\n",
          tm_time-<tm_hour, tm_time-<tm_min, tm_time-<tm_sec );

  printf( "from asctime/gmtime %s\n", asctime( tm_time ) );

  currentTime = mktime( tm_time );

  printf( "from ctime/mktime %s\n", ctime( &currentTime ) );

  return 0;
}
```

Executing this application yields the following result:

```
$ ./time
from localtime 22:53:02
from ctime Tue Jun  1 22:53:02 2004

from asctime/localtime Tue Jun  1 22:53:02 2004

from gmtime 04:53:02
from asctime/gmtime Wed Jun  2 04:53:02 2004

from ctime/mktime Wed Jun  2 05:53:02 2004

$
```

Gathering System Information with `sysinfo`

The **`sysinfo`** command allows an application to gather high-level information about a system, some of it very useful. The API for the **`sysinfo`** command is:

```
int sysinfo( struct sysinfo *info );
```

The **`sysinfo`** command returns zero on success and fills the `info` structure as defined by Table 18.1. Note that all sizes are provided in the units defined by `mem_unit`.

TABLE 18.1 Elements and Meaning for `struct sysinfo`

Element	Description
`uptime`	The current uptime of this system in seconds
`loads[0]`	System load average for 1 minute
`loads[1]`	System load average for 5 minutes
`loads[2]`	System load average for 15 minutes
`totalram`	Total usable main memory
`freeram`	Available main memory
`sharedram`	Amount of memory that's shared
`bufferram`	Amount of memory used by buffers
`totalswap`	Total swap space
`freeswap`	Free swap space
`procs`	Number of currently active processes
`totalhigh`	Total amount of high memory
`freehigh`	Free amount of high memory
`mem_unit`	Memory unit size in bytes

Gathering the system information is very simple, as illustrated in Listing 18.4. Note that **`uptime`** has been further decomposed to provide a more meaningful representation (line 16–25).

ON THE CD

Listing 18.4 Sample Use of **`sysinfo`** Function (on the CD-ROM at `./source/ch18/sysinfo.c`)

```
1:    #include <sys/sysinfo.h<
2:    #include <stdio.h<
3:
4:    int main()
5:    {
6:      struct sysinfo info;
7:      int ret;
8:      int days, hours, minutes, seconds;
9:
```

```
    ret = sysinfo( &info );

    if (ret == 0) {

      printf( "Uptime is %ld\n", info.uptime );

      days = info.uptime / (24 * 60 * 60);
      info.uptime -= (days * (24 * 60 * 60));
      hours = info.uptime / (60 * 60);
      info.uptime -= (hours * (60 * 60));
      minutes = info.uptime / 60;
      info.uptime -= (minutes * 60);
      seconds = info.uptime;

      printf( "Uptime %d Days %d Hours %d Minutes %d Seconds\n",
              days, hours, minutes, seconds );
      printf( "One minute load average %ld\n", info.loads[0] );
      printf( "Five minute load average %ld\n", info.loads[1] );
      printf( "Fifteen minute load average %ld\n", info.loads[2] );
      printf( "Total Ram Available %ld\n", info.totalram );
      printf( "Free Ram Available %ld\n", info.freeram );
      printf( "Shared Ram Available %ld\n", info.sharedram );
      printf( "Buffer Ram Available %ld\n", info.bufferram );
      printf( "Total Swap Size %ld\n", info.totalswap );
      printf( "Available Swap Size %ld\n", info.freeswap );
      printf( "Processes running: %d\n", info.procs );
      printf( "Total high memory size %ld\n", info.totalhigh );
      printf( "Available high memory %ld\n", info.freehigh );
      printf( "Memory Unit size %d\n", info.mem_unit );
    }

    return 0;
  }
```

Much of this information can also be gathered from the `/proc` filesystem, but that is more difficult due to the parsing that's necessary. Some information is provided in `/proc/uptime`, `/proc/meminfo`, and `/proc/loadavg`.

The `proc` *filesystem is a virtual filesystem that contains runtime information about the state of the operating system. The* `proc` *filesystem can be used to inquire about various features by '*`cat`*'ing files in the* `proc` *filesystem. For example, we could identify all processes in the system (*`/proc/#`*), information about the CPU (*`/proc/cpuinfo`*), the devices found on the PCI buses (*`/proc/pci`*), the kernel modules currently loaded (*`/proc/modules`*), and much more runtime information.*

Mapping Memory with mmap

While not completely related to shared memory, the **mmap** API function provides the means to map file contents into user program space. The prototype function for **mmap** (and **munmap** to unmap the memory) is defined as:

```
#include <sys/mman.h>
void *mmap( void *start, size_t length, int prot, int flags,
                int fd, off_t offset );
int munmap( void *start, size_t length );
```

The **mmap** function takes in a file byte offset (`offset`) and tries to map it to the address defined by the caller (`start`) from the file descriptor `fd` (we'll look at what this means shortly). Commonly, the `start` address is defined as `NULL`, allowing **mmap** to map it to whatever local address it chooses. This address (local mapping) is returned by the **mmap** function. The length of the region is defined by `length`. The caller defines the desired memory protection through the `prot` argument, which can be `PROT_EXEC` (region may be executed), `PROT_READ` (region may be read), `PROT_WRITE` (region may be written), or `PROT_NONE` (pages can't be accessed). Combinations of protections can be defined. Finally, the type of mapped object is defined by the `flags` argument. This can be `MAP_FIXED` (fail if the `start` address can't be used for the local mapping), `MAP_SHARED` (share this mapping with other processes), or `MAP_PRIVATE` (create a private copy-on-write mapping). The caller must specify `MAP_SHARED` or `MAP_PRIVATE`.

The `offset` and `start` arguments must be on page boundaries. The `length` argument should be a multiple of the page size.

GNU/Linux also provides some other nonstandard `flags` that are defined in Table 18.2.

TABLE 18.2 Nonstandard Flags for **mmap**

Flag	Use
MAP_NORESERVE	Used with `MAP_PRIVATE` to instruct the kernel not to reserve swap space for this region.
MAP_GROWSDOWN	Used by the kernel for stack allocation.
MAP_ANONYMOUS	This region is not backed by a file (`fd` and offset arguments of `mmap` are therefore ignored).

The **munmap** function simply unmaps the memory mapped by **mmap**. To **munmap**, the address returned by **mmap** is provided along with the length (which was also specified to **mmap**).

Let's look at an example of **mmap** and **munmap**. In Listing 18.5, we find an application that maps physical memory and makes it available for read. This application first creates a file descriptor of the file `/dev/mem`, which represents the physical memory available.

It's important to note that `/dev/mem` should be used only for read operations. Writing can be dangerous to the point of crashing your system. Finally, it's also a large security hole, and therefore its use should be avoided.

Our sample application allows the base address and length to be defined on the command line (lines 15–32). At line 34, we open `/dev/mem` for read. The file `/dev/mem` permits access to physical memory space (for which we'll use **mmap** to map physical memory into the process's memory space). At lines 38–40, we use **mmap** to map the requested region into the process's address space. The return value is the address from which the memory can be accessed. We then perform a loop to read and `printf` the addresses and their contents (lines 45–57). We finally clean up by unmapping the memory with **munmap** (line 59) and then closing the `/dev/mem` file descriptor at line 67.

Listing 18.5 Mapping Physical Memory with **mmap** (on the CD-ROM at `./source/ch18/phymap.c`)

```
 1:     #include <stdio.h>
 2:     #include <unistd.h>
 3:     #include <fcntl.h>
 4:     #include <stdlib.h>
 5:     #include <errno.h>
 6:     #include <sys/mman.h>
 7:
 8:     int main( int argc, char *argv[] )
 9:     {
10:       int fd;
11:       unsigned char *addr, *waddr, count;
12:       int length;
13:       off_t offset;
14:
15:       if (argc < 3) {
16:
17:         printf( "Usage is phymap <address< <length<\n");
18:         exit(-1);
```

```
    }

    if (argv[1][1] == 'x') {
      sscanf( argv[1], "0x%x", &offset );
    } else {
      sscanf( argv[1], "%d", &offset );
    }

    if (argv[2][1] == 'x') {
      sscanf( argv[2], "0x%x", &length );
    } else {
      sscanf( argv[2], "%d", &length );
    }

    fd = open( "/dev/mem", O_RDONLY );

    if (fd != -1) {

      addr = (unsigned char *)mmap( NULL, length,
                                     PROT_READ, MAP_SHARED,
                                     fd, offset );


      if (addr != NULL) {

        waddr = addr;

        for ( count = 0 ; count < length ; count++ ) {

          if ( (count % 16) == 0 ) {

            printf( "\n%8p : ", waddr );

          }

          printf( "%02x ", *waddr++ );

        }

        munmap( addr, length );

      } else {

```

```
                printf("Unable to map memory.\n");

            }

            close(fd);
            printf("\n");

        }

        return 0;
    }
```

Using this application (let's call it `phymap`) is illustrated below. Here we peek at the address `0x000c1000` for 64 bytes.

```
# ./phymap 0x000c1000 64
addr = 0x40017000 (Success)

0x40017000 : 56 57 a0 49 04 e8 c0 fe 32 ed 41 c1 e9 03 32 e4
0x40017010 : e8 0d 67 2e 8b b5 e6 08 2e 8a 44 0c d1 e8 f7 e1
0x40017020 : 5f 5e 07 5a 59 c3 33 c9 a4 fe c1 8a 2c 0a ed 75
0x40017030 : f7 fe c1 a4 32 ed c3 50 53 32 e4 b0 11 b3 80 e8
#
```

The address shown here is the local address that's been mapped in our address space. While it's different than our requested `0x000c1000`, the `0x40017000` represents the same region for our process.

Locking and Unlocking Memory

Let's look at an additional set of functions that are very useful to high-performance applications. The memory-locking functions permit a process to lock some or all of its storage that it is never swapped out of memory. The result is greater performance for the application, since it never has to suffer paging penalties in a busy system, but this does require the proper permissions.

The functions of interest for locking and unlocking memory are:

```
#include <sys/mman.h>
int mlock( const void *addr, size_t len );
int munlock( const void *addr, size_t len );
int mlockall( int flags );
int munlockall( void );
```

The `addr` and `len` arguments must be on page boundaries. Let's now look at the lock/unlock pairs of functions in detail.

The **`mlock`** function takes an address (`addr`) for which the memory pages that represent the region are to be locked. The `len` argument defines the size of the region to lock (meaning that one or more pages may be locked by the operation). A return value of zero means that the pages are locked. When the application is finished with the memory, a call to **`munlock`** makes the pages available for swapping. A return of zero represents success of the unlock system call.

The following code example illustrates locking a page of memory and then unlocking it (see Listing 18.6). Our buffer is a locally created array of characters.

Listing 18.6 Locking and Unlocking a Memory Page (on the CD-ROM at `./source/ch18/lock.c`)

```
 1:     #include <stdio.h>
 2:     #include <sys/mman.h>
 3:
 4:     char data[4096];
 5:
 6:     int main()
 7:     {
 8:       int ret;
 9:
10:       ret = mlock( &data, 1024 );
11:
12:       printf("mlock ret = %d\n", ret);
13:
14:       ret = munlock( &data, 1024 );
15:
16:       printf("munlock ret = %d\n", ret);
17:
18:       return 0;
19:     }
```

At line 10, we call **`mlock`** with a reference to our global buffer (`data`) and its length. We unlock this page by calling **`mlock`** with our buffer and size (identically to the call to **`mlock`**). The entire page containing the buffer is locked, in the event it falls under the bounds of a page.

Child processes do not inherit memory locks (created by **`mlock`** *or* **`mlockall`***). Therefore, if a region of memory is created and subprocesses also created to operate upon it, each child process should perform its own* **`mlock`**.

The **mlockall** API function locks all memory (disable paging) for a process's entire memory space. This includes not only the code and data for the process, but also its stack, shared libraries, shared memory, and other memory mapped files.

The **mlockall** function takes a single argument, which represents the scope of the lock to be provided. The user can specify MCL_CURRENT, which locks all pages that are currently mapped into the address space of the process, or MCL_FUTURE, which locks all pages that will be mapped into the address space of the process in the future. For example:

```
/* Lock currently mapped pages */
mlockall( MCL_CURRENT);

/* Lock all future pages */
mlockall( MCL_FUTURE );
```

We can also define that all current and future pages are locked into memory by performing a bitwise OR the flags together, such as:

```
mlockall( MCL_CURRENT | MCL_FUTURE );
```

If insufficient memory is available to lock the current set of pages, an ENOMEM error is returned. If MCL_FUTURE is used, and insufficient memory exists to lock a growing process stack, the kernel will throw a SIGSEGV (segment violation) signal for the process, causing it to terminate.

The **munlockall** system call reenables paging for the pages mapped for the calling process. It takes no arguments and returns zero on success:

```
/* Unlock the pages for this process */
munlockall();
```

The **munlockall** system call should be called once the process has completed its real-time processing.

SUMMARY

In this chapter, we covered a number of system calls that are useful in the development of application and tools software. These included the **getopt** and **getopt_long** calls for parsing command-line arguments, a variety of time and time conversion functions, the **sysinfo** call to gather high-level system information, the **mmap** system call to map files, and finally two pairs of page-locking functions to help build high-performance applications (by avoiding page swapping penalties).

API SUMMARY

```
#include <unistd.h>
int getopt( int argc, char * const argv[], const char *optstring );
extern char *optarg;
extern int optopt, optind;

#include <getopt.h>
int getopt_long( int argc, char * const argv[],
                     const char *optsring,
                     const struct option *longopts, int *longindex );
extern char *optarg
extern int optopt, optind;

#include <time.h>
time_t time( time_t *t );
struct tm *localtime( const time_t *timep );
struct tm *gmtime( const time_t *timep );
char *asctime( const struct tm *tm );
char *ctime( const time_t *timepDay  );
time_t mktime( struct tm *tm );


#include <sys/mman.h>
void *mmap( void *start, size_t length, int prot, int flags,
                int fd, off_t offset );
int munmap( void *start, size_t length );
int mlock( const void *addr, size_t len );
int munlock( const void *addr, size_t len );
int mlockall( int flags );
int munlockall( void );
```

Part IV GNU/Linux Shells and Scripting

In this part of the book, we'll look at the topic of scripting languages. While scripting languages are historically tied to shells, scripting is a much larger topic. We'll cover bash scripting, domain-specific languages such as Sed and Awk, and finally building interpreters with flex and bison.

Chapter 19, "GNU/Linux Commands"

GNU/Linux includes a large set of commands that aid in software development. We'll look at a variety of the most commonly used commands and cover them in a tutorial fashion, including numerous examples.

Chapter 20, "The Bourne-Again Shell (bash)"

The bash shell is the de facto standard shell for GNU/Linux. Shell programming is very beneficial to understand because it permits developers to code repetitive tasks quickly. Shell programming can be slower than compiling to the native instruction set, but efficiency isn't always the most important aspect of development. Compiling source to a native image takes time. The great advantage to scripting is that you can execute your script immediately. This makes scripting languages perfect for prototyping. Many scripting languages also allow the script to be performed interactively, which makes it much easier for them to be understood.

Chapter 21, "Editing with sed"

The sed utility (stream editor) is a noninteractive text editing utility that is quite useful in performing global editing of files. sed can be used on a single file or many files, and although it is cryptic, it is generally a useful tool that solves a number of text editing problems. This chapter introduces sed and describes its use in a number of simple and complex examples.

Chapter 22, "Text Processing with Awk"

While some consider the Awk utility an advancement of Sed, it is in its own right a high-level procedural programming language. Awk was designed for text pattern processing on files, but it has evolved into a compact language that is useful in a number of areas. Awk is very convenient in prototyping because it is interpreted, and programs can be developed and tested quickly. The Awk programming language is useful and can be very beneficial to GNU/Linux developers.

Chapter 23, "Parser Generation with flex and bison"

The development of parsers was traditionally limited to those with extensive experience in parsing and compiler theory. With the introduction of lex and yacc (and the GNU replacements, flex and bison), building parsers is much simplified as the difficult work is done for you. The flex tool is a lexical analyzer generator that creates programs for tokenization of input files. The bison tool, working in concert with the lexer, generates a grammar parser to validate the correctness of input tokens. Each of these tools takes an input file defining the token structure and grammar and produces C source to perform tasks. This makes the development of parsers simple and very maintainable. In this chapter, flex and bison are introduced through multiple examples of tokenization and grammar parsing.

19 GNU/Linux Commands

In This Chapter

- Standard In, Out, and Error
- Invoking Shell Scripts
- Redirection
- Discussion of Important GNU/Linux Commands

INTRODUCTION

In this chapter, we'll look at some of the basics of standard GNU/Linux shells in addition to the important commands that are used frequently.

REDIRECTION

Let's first look at a basic topic in GNU/Linux shell use, that of input or output redirection.

The concept of redirection is simply that of redirecting our input or output to something other than the default. For example, the standard output of most commands will be redirected to the shell window. We can redirect the output of a command to a file using the > output redirection symbol. For example:

```
ls -la
```

will generate a file listing and emit the results to the shell window. We could instead redirect this to a file as:

```
ls -la > ls-out.txt
```

What would have been emitted to the shell window will now be present in the output file `ls-out.txt`.

Rather than accept input from the keyboard, we can accept input from another source. For example, the command `cat` will simply emit its standard input (or files named on its command line) to its standard out. We can redirect the contents of a file to `cat` using the < input redirection symbol.

```
cat < ls-out.txt
```

We can also build more complicated redirection structures. For example, using the `|` pipe symbol, we can chain a number of generators and filters together. For example, consider the following command sequence:

```
find . -name '*.[ch]' -print | xargs grep "mtj" | more
```

This is actually three different commands that stream their output from left to right. The first command searches the subdirectory tree (from the current directory) looking for all files that fit a certain pattern. The pattern defined is '`*.[ch]`', which means all files that end in `.c` or `.h`. These are passed to the next command, `xargs`, which is a special command to read from standard input and pass to the embedded command. In this case, it's `grep`. The `grep` command is a text search utility that will search all files passed to it from the previous stage for the string term `mtj`. For files that pass the search criteria, the lines that contain the search term are emitted to the next stage, the `more` command. The `more` command simply ensures that the user is able to see all output before it scrolls by. When a screen full of output is present, the user must type `return` for `more` to continue and potentially present a new screen's worth of data.

Standard In/Out/Error

For each application, three special file descriptors are automatically created. These are called standard-input, standard-output, and standard-error (see Figure 19.1).

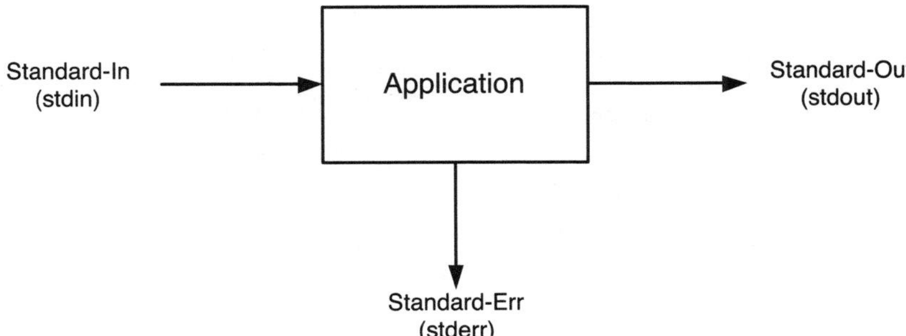

FIGURE 19.1 Program input and output.

Herein, we'll refer to these by their shortened names for brevity. The `stdin` descriptor is commonly the keyboard. Descriptors `stdout` and `stderr` are the terminal or window attached to the shell (`stdout` for program results, `stderr` for program errors, warnings, and status). The output descriptors are split to provide greater flexibility for emitting information to the user. While `stdout` and `stderr` share the same default output device, they can be split as desired by the developer.

Recall from our previous discussion that we can redirect `stdout` to a file as:

```
prog > out.txt
```

where the output of `prog` would be redirected to the file `out.txt`. Note that if we wanted to append our output to `out.txt` rather than replace the file altogether, we would use the double redirect, as:

```
prog >> out.txt
```

We could redirect only the error output as:

```
prog 2> error-out.txt
```

Note that we're using a constant number here to represent `stderr`. The file descriptors that are defined for our three standard I/O descriptors are shown in Table 19.1.

If instead we wanted to redirect both the `stdout` and `stderr` to a file (`out.txt`), we could do the following:

```
prog 1> out.txt 2>&1
```

TABLE 19.1 File Descriptors for Standard I/O Descriptors

Descriptor	Description
0	Standard input (`stdin`)
1	Standard output (`stdout`)
2	Standard error (`stderr`)

For the opposite scenario, we could redirect the `stdout` output to the `stderr` descriptor as:

```
prog 1>&2
```

We could also redirect output to unique files. For example, if we wanted our `stdout` to go to `out.txt` and `stderr` to go to `err.txt`, we could do the following:

```
prog 1>out.txt 2>err.txt
```

To verify that descriptor routing is working the way you expect, the script in Listing 19.1 can be used to test.

Listing 19.1 Descriptor Routing Test Script (on CD-ROM at `./source/ch19/redirtest.sh`)

```
#!/bin/bash
echo "stdout test" >&1
echo "stderr test" >&2
```

Finally, consider another example that demonstrates ordering of redirection.

```
prog 2>&1 1>out.txt
```

In this example, we redirect `stderr` to `stdout` and then redirect `stdout` (not including `stderr`) to the file `out.txt`. This has the effect of consolidating both the `stdout` and `stderr` to the file `out.txt`.

Environment Variables

An environment variable is a named object that contains information for use by the shell and other applications. A number of standard environment variables exist, such as the `PWD` variable, but the user can create his own for his applications (or change existing variables). We can inspect the `PWD` variable by echoing its contents with the `echo` command:

A process can be viewed as an environment and inherits the environment variables from its parent (such as the shell, which is also a process). A process or script may also have local variables.

```
$ echo $PWD
PWD=/home/mtj
$
```

The `PWD` environment (or shell) variable identifies our current (present) working directory. We could create our own using the `declare` or `export` bash built-in command.

A script can make environment variables available to child processes, but only by exporting them. A script cannot export back to the parent process. Let's look at a couple of examples. The `declare` built-in command can be used to declare variables with specific attributes. The `export` built-in command is used to create a variable and mark it to be passed to child processes in the environment. These commands are illustrated as:

```
$ declare -x myvar="Hello"
$ echo $myvar
Hello
$ export myothervar="Hi"
$ echo $myothervar
Hi
$
```

A number of other useful environment variable commands exist. For example, if no argument is provided to `export`, then it will emit all of the variables available to the environment (which can also be done with the `declare` and `set` commands).

Script Invocation

When we invoke a command or script, the command or script must be in our binaries path (`PATH` environment variable) in order for it to be found. We can view our path by echoing the `PATH` environment variable as:

```
$ echo $PATH
/usr/local/bin:/bin:/usr/bin:/usr/X11R6/bin:/home/mtj/bin
```

If we've created a script that does not exist in the path defined (for example, in our current working directory), then we will have to invoke it as:

```
./script.sh
```

The `./` tells the interpreter that the shell script we're invoking is located in the current directory. Otherwise, we'd get an error message telling us that the script could not be found.

Two special directory files exist that are important in Linux development. The . file represents the current directory, while the .. file represents the parent directory. For example, if we provide the command `cd .`*, there's no visible change because we've changed our current directory to the current directory. More interesting, the* `cd ..` *command changes the current directory to the parent directory. These special files can be seen by viewing the current subdirectory with* `ls -la`.

If this is undesirable, we could easily update our path environment variable to add our current working subdirectory as:

```
$ export PATH=$PATH:./
$ echo $PATH
/usr/local/bin:/bin:/usr/bin:/usr/X11R6/bin:/home/mtj/bin:./
```

The script `script.sh` can now be invoked directly without needing to prefix ./.

File links are special files that are used to provide a reference to another file. Two types of links exist: hard *links and* soft *links (otherwise known as symbolic links). A hard link is a new entry in the directory file that points to an existing file. The hard link is indistinguishable from the original file. The problem with hard links is that they must reference files within the same filesystem. Soft links are regular files themselves and simply contain a pointer to the actual file. Soft links can be absolute (point to a file with a full path) or relative (the path being relative from the current location). Soft links can be moved, while retaining their linkage to the original file. Note that the special files . and .. are in fact hard links to the absolute directories.*

The `ln` *command can be used to construct symbolic links.*

Basic GNU/Linux Commands

Now that we have some basic understanding of redirection and the standard I/O descriptors (`stdin`, `stdout`, and `stderr`), let's explore the more useful of the GNU/Linux commands. We'll take an interactive approach for investigating these commands, compared to simply telling you what the command does and all the available options for it. The commands are in no particular order, and therefore you can explore each command independently of any other if desired.

Every command in GNU/Linux is itself a process. The shell manipulates the `stdin`, `stdout`, *and* `stderr` *for commands that are executed in it.*

tar

The GNU `tar` command (named for Tape ARchive) is a useful and versatile archiving and compressing utility. Targets for `tar` can be files or directories, where the directories are recursed to gather the full contents of the directory tree. Let's look at uses of the `tar` utility, investigating a variety of the options as we go.

We can create a new archive using `tar` as follows:

```
tar cf mytar.tar mydir/
```

We specify two options for creating an archive of directory `mydir`. The `c` option instructs `tar` to create a new archive with the name (identified via `f` option) `mytar.tar`. The final arguments are the list of files and/or directories to archive (for which all files in `mydir` will be included in the archive).

Now let's say we want to take our `tar` file (also called a tarball) and re-create the subdirectory and its contents (otherwise known as the extract option). We could do this (in another subdirectory) by simply typing:

```
tar xf mytar.tar
```

If we'd like to know the contents of our tarball without having to unarchive the contents, we could do this:

```
tar tf mytar.tar
```

which lists all files with their directory paths intact.

By adding the `v`, or verbose option, we can view the operation of the `tar` utility as it works. For both creating and extracting archives, we use the verbose option as:

```
tar cvf mytar.tar mydir/
tar xvf mytar.tar
```

One of the most important aspects of the tar utility is the automatic compression of the tarball. This is performed using the `z` option and works symmetrically for both creation and extraction of archives, as:

```
tar czf mytar.tgz mydir/
tar xzf mytar.tgz
```

A compressed tarball can take more time to create and extract, but this can still be beneficial, especially if the set of files is intended to be transferred over the Internet.

cut

The GNU/Linux cut utility can quickly cut elements of each line in a file using one of two types of specification. The user can define the desired data in terms of fields in the file or based upon numbered sequences of characters.

Let's first look at the basic format of the cut utility and then look at some examples of how it can be used. As we discussed, the cut utility can operate in two modes. In the first mode, cut extracts based upon field specifications using a delimiting character:

```
cut -f[spec] -d[delimiter] file
```

In the second mode, cut extracts based upon character position specifications:

```
cut -c[spec] file
```

The spec argument is a list of comma-separated ranges. A range can be represented as follows (base 1) as shown in Table 19.2.

TABLE 19.2 Range Specs for the cut Utility

Range	Meaning
n	Nth character (-c) or field (-f)
n-	Nth character (-c) or field (-f) from the end of the line
n-m	From nth to mth character (-c) or field (-f)
-m	From first to Mth character (-c) or field (-f)

Let's now look at some examples of the cut utility. We'll explore the field-based cut first, using the sample file in Listing 19.2.

ON THE CD

Listing 19.2 Sample File (passwd) for Field-based Cutting (on the CD-ROM at ./source/ch19/passwd)

```
bob:x:500:500::/home/bob:/bin/bash
sally:x:501:501::/home/sally:/bin/sh
jim:x:502:502::/home/jim:/bin/tcsh
dirk:x:503:503::/home/dirk:/bin/ash
```

We'll experiment with cut interactively, looking now only at the command but also exactly what cut will produce given the field specification. First, let's say we want to cut and emit the first field (the name). This is a simple case for `cut`, demonstrated as:

```
$ cut -f1 -d: passwd
bob
sally
jim
dirk
$
```

In this example, we specify to cut field 1 (`-f1`) using the colon character as the delimiter (`-d:`). If we wanted to know the home directory rather than the user name, we simply update the field to point to this element of `passwd` (field 6), as:

```
$ cut -f6 -d: passwd
/home/bob
/home/sally
/home/jim
/home/dirk
$
```

We could also extract multiple fields, such as the user name (field 1) and the preferred shell (field 7):

```
$ cut -f1,7 -d: passwd
bob:/bin/bash
sally:/bin/bash
jim:/bin/tcsh
dirk:/bin/ash
$
```

Ordering is important to the `cut` utility. For example, if we had specified `-f7,1` instead in the previous example, the result would have been the same (ordering of fields in the original file is retained).

Let's now look at some examples of character position specifications. In these examples, we'll pipe our input from another command. The `ls` command lists the contents of a directory, for example:

```
$ ls -la
total 20
drwxrwxr-x    2 mtj      mtj          4096 Feb 17 20:08 .
drwxr-xr-x    6 mtj      mtj          4096 Feb 15 20:47 ..
-rw-rw-r—     1 mtj      mtj          6229 Feb 16 17:59 ch12.txt
-rw-r—r—      1 mtj      mtj           145 Feb 17 20:02 passwd
```

If we were interested only in the file size (includes the date and name of the file), we could do the following:

```
$ ls -la | cut -c38-
 4096 Feb 17 20:08 .
 4096 Feb 15 20:47 ..
 6229 Feb 16 17:59 ch12.txt
  145 Feb 17 20:02 passwd
$
```

If we're interested only in the size of the file and the file's name, we could use the following:

```
$ ls -la | cut -c38-42,57-
 4096.
 4096..
 6229ch12.txt
  145passwd
$
```

Note in this example that the two `cut` regions include no space between them. This is because `cut` simply segments the regions of the file and emits no spaces between those regions. If we were interested in a space, we could have updated the command as follows (taking a space from the data itself):

```
$ ls -la | cut -c38-43,57-
 4096 .
 4096 ..
 6229 ch12.txt
  145 passwd
$
```

The `cut` utility is very simple, but it's also a very useful utility with quite a bit of flexibility. The `cut` utility isn't the only game in town; later on we'll look at `sed` and `awk` and their capabilities for text filtering and processing.

paste

The `paste` command takes data from one or more files and binds them together into a new stream (with default emission to `stdout`). Consider the files shown in Listing 19.3 and 19.4.

ON THE CD

Listing 19.3 The Fruits File (on the CD-ROM at `./source/ch19/fruits.txt`)

```
Apple
Orange
Banana
Papaya
```

ON THE CD

Listing 19.4 The Tools File (on the CD-ROM at `./source/ch19/tools.txt`)

```
Hammer
Pencil
Drill
Level
```

Using the `paste` utility, we can bind these files together as demonstrated below:

```
$ paste fruits.txt tools.txt
Apple   Hammer
Orange  Pencil
Banana  Drill
Papaya  Level
$
```

If we wanted some delimiter other than tabs between our consecutive elements, we could specify a new one using the `-d` option. For example, we could use a `:` character instead, using:

```
$ paste -d: fruits.txt tools.txt
Apple:Hammer
Orange:Pencil
Banana:Drill
Papaya:Level
$
```

Rather than pair consecutive elements in a vertical fashion, we could instead pair them horizontally using the `-s` option:

```
$   paste -s fruits.txt tools.txt
Apple   Orange  Banana  Papaya
Hammer  Pencil  Drill   Level
$
```

Note that we could specify more than two files if desired.

Let's now look at one final example to illustrate the `paste` utility. Recall from our discussion of `cut` that it wasn't possible to alter the order of fields pulled from a file. The following short script provides the utility of listing the filename and then the size of the file (see Listing 19.5).

ON THE CD

Listing 19.5 Simple Reversed `ls` Utility Using `cut` and `paste` (on the CD-ROM at `./source/ch19/newls.sh`)

```
#!/bin/bash
ls -l | cut -c38-42 > /tmp/filesize.txt
ls -l | cut -c57- > /tmp/filename.txt
paste /tmp/filename.txt /tmp/filesize.txt
```

In this example, we first cut the file sizes from the `ls -l` command and store the result to `/tmp/filesize.txt`. We grab the filenames next and store them to `/tmp/filename.txt`. Next, using the `paste` command, we merge the results back together, reversing the order. Executing this script on a small directory results in the following:

```
$ ./newls.sh
fruits.txt        27
newls.sh         133
tools.txt         26
$
```

NOTE

Note that the use of the `/tmp` directory in Listing 19.5 is useful for temporary files. Temporary files can be written to `/tmp` because they are not persistent across system boots. In some cases, files are removed from `/tmp` as part of a daily or weekly cleanup process.

sort

The `sort` utility is useful for sorting a data file in some defined order. In the simplest case, where the key for `sort` is the beginning of the file, we specify the file. Take for example our `tools.txt` file shown in Listing 19.4. We can easily sort this as:

```
$ sort tools.txt
Drill
Hammer
Level
Pencil
$
```

We could reverse sort this file by simply adding the `-r` (reverse) option to the command line.

We can also sort based upon a key defined by the user. Consider the sample text file shown in Listing 19.6. This file contains five lines with three columns, with none of the columns being presorted.

Listing 19.6 Sample File for Sorting (on the CD-ROM at `./source/ch19/table.txt`)

```
5 11 eee
4  9 ddd
3 21 aaa
2 24 bbb
1  7 ccc
```

To specify a column to sort, we use the `-k` (or key) option. The key represents the column for which we desire the file to be sorted. We can specify more than one key by simply separating them with commas. To sort based upon the first column, we can specify the key as column one (or not, as it will be the default):

```
$ sort -k 1 table.txt
1  7 ccc
2 24 bbb
3 21 aaa
4  9 ddd
5 11 eee
$
```

To sort the second column, another option is required to perform a numeric sort. The space character that comes before the single-digit numbers is significant and therefore will preclude a numeric sort. Therefore, we'll use the `-n` option to force a numeric sort. For example:

```
$ sort -k 2 -n table.txt
1  7 ccc
4  9 ddd
5 11 eee
3 21 aaa
2 24 bbb
$
```

One other useful option allows the specification of a new delimiter, in the event spaces or tabs are not used. The `-t` option allows us to use a different delimiter, such as `-t:` to specify the colon character as the field separator.

find

The find utility is a powerful but complex utility that permits searching the filesystem for files based upon given criteria. Rather than walk through the plethora of options available for find, we'll demonstrate some of the more useful patterns.

To find all files that end in .c and .h in the current subdirectory, the following command can be used:

```
find . -name '*.[ch]' -print
```

The . specifies that we wish to start at the current subdirectory. The -name argument refers to what we're searching for, in this case any file ('*') that ends in either c or h. Finally, we specify -print to emit the search results to standard-out.

For each of the search results, we can execute a command using -exec. This permits us to invoke a command on each file that was found based upon the search. For example, if we wanted to change the file permissions of all of the files found to read-only, we could do the following:

```
find . -name '*.[ch]' -exec chmod 444 {} \;
```

We can also restrict the type of files that we'll look at using the type modifier. For example, if we wanted to look only at regular files (a common occurrence when searching for source files), we could do the following:

```
find . -name '*.[ch]' -type f -print
```

The f argument to -type represents regular files. We could also look specifically for directories, symbolic links, or special devices. Table 19.3 provides the type modifiers that are available.

TABLE 19.3 Type Modifiers Available to find (-type)

Modifier	Description
b	Block Device
c	Character Device
d	Directory
p	Pipe (named FIFO)
f	Regular File
l	Symbolic Link
s	Socket

One final useful `find` use is identifying files within a directory that have been changed within a certain period of time. The following command (using `mtime`) identifies the files that have been modified in a given time range (multiples of 24 hours). The following command identifies files that have been modified in the last day.

```
find  -name '*' -type f -mtime -1 -print
```

To find files modified in the last week, we could update the `mtime` argument to `-7`. The `atime` argument can be used to identify recently accessed files. The `ctime` argument identifies files whose status was changed recently.

wc

The `wc` utility is very useful to count the number of characters, words, or lines within a file.

The following samples illustrate the possibilities of the `wc` utility.

```
wc -m file.txt        # Count characters in file.txt
wc -w file.txt        # Count words in file.txt
wc -l file.txt        # Count lines in file.txt
```

All three counts can be emitted by accumulating the arguments, as:

```
wc -l -w -m file.txt
```

Regardless of the order of the flags, the order of count emission is always lines, then words, and then characters.

grep

The `grep` command permits searching one or more files for a given pattern. The format of the `grep` command can be complex, but the simpler examples are quite useful. Let's look at a few simple examples and discuss what they achieve.

In its simplest form, we can search a single file for a given search string, as:

```
grep "the" file.txt
```

The result of this example is each line emitted to `stdout` that contains the word `the`. Rather than specify a single file, we can use the wildcard to check multiple files, as:

```
grep "the" *.txt
```

In this case, all files in the current subdirectory that end in `.txt` are checked for the search string.

When the wildcard is used, the filename from which the search string is found is emitted before each line that's emitted to `stdout`.

If the line location of the file where the search string is found is important, the `-n` option can be used.

```
grep -n "the" *.txt
```

Each time the search string is found, the filename and line number where the string was found are emitted before the line. If we're interested only in whether the particular string was found in a file, we can use the `-l` option to simply emit filenames, rather than the lines from which the string was found:

```
grep -l "the" *.txt
```

When we're searching specifically for words within a file, the `-w` option can be helpful in restricting the search to whole words only:

```
grep -w "he" *.txt
```

This option is very useful as when searching for `he`; we'll also find occurrences of the word `the`. The `-w` option restricts the search only to words that match, so `the` and `there` will not result in a match to `he`.

SUMMARY

In this chapter, we explored some of the basics of commanding in the GNU/Linux shell. The standard input and output descriptors were investigated (`stdin`, `stdout`, and `stderr`), along with redirection and command pipelining. We also looked at ways of invoking shell scripts, including the addition of the current working directory to the default search path. Finally, we demonstrated some of the more useful GNU/Linux commands, including `tar` (file archives), `find` (file search), `grep` (string search), and some other useful text processing commands (`cut`, `paste`, and `sort`).

20 Bourne-Again Shell (bash)

In This Chapter

- An Introduction to Bash Scripting
- Scripting versus Interactive Shell Use
- User Variables and Environmental Variables
- Arithmetic in Bash
- Tests, Conditionals, and Loops in Bash
- Script Input and Output
- Dissecting of Useful Scripts

INTRODUCTION

In this chapter, we'll explore script programming in the Bourne-Again SHell, otherwise known as bash. Bash is the de facto standard command shell and shell scripting language on Linux and other UNIX systems. We'll investigate some of the major language features of bash, including variables, arithmetic, control structures, input and output, and function creation. This follows the flow of all other scripting chapters in this book, allowing the reader to easily understand the similarities and differences of each covered language.

Preliminaries

Before jumping in to scripting with bash, let's look at some preliminaries that are necessary to run bash scripts. You can tell which shell you're currently using by interrogating the `SHELL` environment variable:

```
$ echo $SHELL
/bin/bash
$
```

The echo command is used to print to the screen. We print the contents of the SHELL variable, accessing the variable by preceding it with the $ symbol. The result is the shell that we're currently operating on, in this case, bash. Technically, it printed out the location of the shell we're using (the bash interpreter is found within the /bin subdirectory).

If we happened not to be using bash, we could simply invoke bash to start that interpreter:

```
$ echo $SHELL
/bin/csh
$ bash
$
```

In this case, we were using another command shell (C-shell here). We invoke bash to start this interpreter for further commanding.

Sample Script

Let's now write a simple script as a first step to bash scripting. The source for the script is shown in Listing 20.1.

Listing 20.1 First Bash Script (on the CD-ROM at ./source/ch20/first.sh)

```
#!/bin/bash
echo "Welcome to $SHELL scripting."
exit
```

When invoked, this script simply emits the line "Welcome to /bin/bash scripting." and then exits. If you tried to enter this script (named first.sh) and execute it as ./first.sh, you'd notice that it didn't work. You probably saw something like this:

```
$ ./first.sh
-bash: ./first.sh: Permission denied.
$
```

The problem here is that the script is not executable. We must first change the attributes of the file to tell GNU/Linux that it can be executed. This is done using the chmod command, as illustrated below:

```
$ chmod +x first.sh
$ ./first.sh
Welcome to /bin/bash scripting.
$
```

We use `chmod` to set the `execute` attribute of the file `first.sh`, telling GNU/Linux that it can be executed. After trying to execute again, we see the expected results.

The question we could ask ourselves now is, even though we've told GNU/Linux that the file can be executed, how does it know that the file is a bash script? The answer is the "shebang" line in our script. You'll notice that the first line of the script starts with `#!` (also called *shabang*) followed by the path and interpreter (`/bin/bash`). This defines that bash is the shell to be used to interpret this file. If the file had contained a Perl script, it would have begun `#! /bin/perl`. We can also add comments to our script simply by preceding them with a `#` character.

Since our interpreter is also our shell, we can perform this command interactively, such as:

```
$ echo "Welcome to $SHELL scripting."
Welcome to /bin/bash scripting.
$
```

Now that we have some basics under our belt, let's dig into bash and work through some more useful scripts.

BASH SCRIPTING

In this chapter, we'll take a look at the bash scripting language. The following sections will identify the necessary language constructs for application development, including variables, operations on variables, conditionals, looping, and functions. We'll also demonstrate a number of sample applications to illustrate scripting principles.

Variables

Any worthwhile language permits the creation of variables. In bash, variables are untyped, which means that all variables are in essence strings. This doesn't mean we can't do arithmetic on bash variables, which we'll explore shortly. We can create a variable and then inspect it very easily as:

```
$ x=12
$ echo $x
12
$
```

In this example, we create a variable `x` and bind the value `12` to it. We then echo the variable `x` and find our value. Note that the lack of space between the variable name, the equals, and the value is relevant. There can be no spaces, otherwise an error will occur. Note also that to reference a variable, we precede it with the dollar sign. This variable is scoped (exists) for the life of the shell from which this sequence was performed. Had this sequence of commands been performed within a script (such as `./test.sh`, the variable `x` would not exist once the script was completed.

As bash doesn't type its variables, we can create a string variable similarly:

```
$ b="my string"
$ echo $b
my string
$
```

Note that single quotes also would have worked here. An interesting exception is the use of the *backtick*. Consider the following:

```
$ c='echo $b'
$ echo $c
my string
$
```

The backticks have the effect of evaluating the contents within the backticks, and in this case the result is assigned to the variable `c`. When we emit variable `c`, we find the value of the original variable `b`.

The bash interpreter defines a set of environment variables that effectively define the environment. These variables exist when the bash shell is started (though others can be created using the `export` command). Consider the script in Listing 20.2, which makes use of special environment variables to identify the environment of the script.

ON THE CD

Listing 20.2 Sample Script Illustrating Standard Environmental Variables (on the CD-ROM at `./source/ch20/env.sh`)

```
#!/bin/bash

echo "Welcome to host $HOSTNAME running $OSTYPE."
echo "You are user $USER and your home directory is $HOME."
echo "The version of bash running on this system is $BASH_VERSION."
sleep 1
echo "This script has been running for $SECONDS second(s)."
exit
```

Upon executing this script, we see the following on a sample GNU/Linux system:

```
$ ./env.sh
Welcome to host camus running linux-gnu.
You are user mtj and your home directory is /home/mtj.
The version of bash running on this system is 2.05b.0(1)-release.
This script has been running for 1 second(s).
$
```

Note that we added a useless sleep call to the script (to stall it for one second) so that we could see that the SECONDS variable was working. Also note that the SECONDS variable can be emitted from the command line, but this value represents the number of seconds that the shell has been running.

Bash provides a variety of other special variables. Some of the more important ones are shown in Table 20.1. These can be echoed to understand their formats.

TABLE 20.1 Useful Environment Variables

Variable	Description
$PATH	Default path to binaries
$PWD	Current working directory
$OLDPWD	Last working directory
$PPID	Process ID of the interpreter (or script)
$#	Number of arguments
$0, $1, $2, ...	Arguments
$*	All arguments as a single word

One final word about variables in bash and then we'll move on to some real programming. We can declare variables in bash, providing some form of typing. For example, we can declare a constant variable (cannot be changed after definition) or declare an integer or even a variable whose scope will extend beyond the script. Examples of these variables are shown below interactively:

```
$ x=1
$ declare -r x
$ x=2
-bash: x: readonly variable
$ echo $x
1
$ y=2
$ declare -i y
```

```
$ echo $y
2
$ persist='$PWD'
$ declare -x persist
$ export | grep persist
declare -x persist="/home/mtj"
```

The last item may require a little more discussion. In this example, we create a variable persist and assign the current working subdirectory to it. We declare for exporting outside of the environment, which means if it had been done in a script, the variable would remain once the script had completed. This can be useful to allow scripts to alter the environment or to return variables.

Simple Arithmetic

We can perform simple arithmetic on variables, but there are differences from normal assignments that we've just reviewed. Consider the source in Listing 20.3.

Listing 20.3 Simple Script Illustrating Arithmetic with Variables (on the CD-ROM at ./source/ch20/arith.sh)

```
1:    #!/bin/bash
2:
3:    x=10
4:    y=5
5:
6:    let sum=$x+$y
7:    diff=$(($x - $y))
8:    let mul=$x*$y
9:    let div=$x/$y
10:   let mod=$x%$y
11:   let exp=$x**$y
12:
13:   echo "$x + $y = $sum"
14:   echo "$x - $y = $diff"
15:   echo "$x * $y = $mul"
16:   echo "$x / $y = $div"
17:   echo "$x ** $y = $exp"
18:   exit
```

At lines 3 and 4, we create two local variables called x and y and assign values to them. We then illustrate simple math evaluations using two different forms. The first form uses the let command to assign the evaluated expression to a new variable. No spaces are permitted in this form. The second example uses the $((<expr>))

form. Note in this case that spaces are permitted, potentially making the expression much easier to read.

Bitwise Operators

Standard bitwise operators are also available in bash. These include bitwise left shift (<<), bitwise right shift (>>), bitwise AND (&), bitwise OR (|), bitwise negate (~), bitwise NOT (!), and bitwise XOR (^). The following interactive session illustrates these operators:

```
$ a=4
$ let b="$a<<1"
$ echo $b
8
$ b=8
$ c=4
$ echo $(($c|$d))
12
$ echo $((0xc^0x3))
15
$
```

Logical Operators

Traditional logical operators can also be found within bash. These include the logical AND (&&) and logical OR (||). The following interactive session illustrates these operators:

```
$ echo $((2 && 0))
0
$ echo $((4 && 1))
1
$ echo $((3 || 0))
1
$ echo $((0 || 0))
0
```

In the next section, we'll investigate how these can be used in conditionals for decision points.

Conditional Structures

Bash provides the typical set of conditional constructs. In this section, we'll explore each of these constructs and also investigate some of the other available conditional expressions that can be used.

Conditionals

In this section, we'll look at conditionals. The `if/then` construct provides a decision point after evaluating a test construct. The test construct returns a value as its result. The result of the test construct is zero for true (test succeeds and subsequent commands are executed) or nonzero for false (test fails else section, if available, is executed). Let's look at a simple example to illustrate (see Listing 20.4).

Listing 20.4 Simple Script Illustrating Basic `if/then/else` Construct (on the CD-ROM at `./source/ch20/cond.sh`)

```
#!/bin/bash
a=1
b=2
if [[ $a -eq $b ]]
then
  echo "equal"
else
  echo "unequal"
fi
```

Note that the result of the test construct is the inverse of what you would expect. The is because the exit status of a command is 0 for success/normal and != 0 to indicate an error.

After creating two variables, we test them for equality using the `-eq` comparison operator. If the test construct is true, we perform the commands contained in the `then` block. Otherwise, if an `else` block is present, this is executed (the test construct was false). `else-if` chains can also be constructed, as shown in Listing 20.5.

Listing 20.5 Simple Script Illustrating the `if/then/elif/then/fi` Construct (on the CD-ROM at `./source/ch20/cond2.sh`)

```
#!/bin/bash
x=5
y=8
if [[ $x -lt $y ]]
then
  echo "$x < $y"
elif [[ $x -gt $y ]]
then
  echo "$x > $y"
elif [[ $x -eq $y ]]
```

```
    then
      echo "$x == $y"
    fi
```

In this example, we test the integers for using the `-lt` operator (less-than), `-gt` (greater-than), and finally `-eq` (equality). Other operators are shown in Table 20.2.

Test constructs can also utilize strings such as is illustrated in Listing 20.6. In this example, we'll also look at two forms of the `if/then/fi` construct that provide identical functionality.

TABLE 20.2 Integer Comparison Operators

Operator	Description
`-eq`	is equal to
`-ne`	is not equal to
`-gt`	is greater than
`-ge`	is greater than or equal to
`-lt`	is less than
`-le`	is less than or equal to

Listing 20.6 Simple Script Illustrating the `if/then/fi` Construct (on the CD-ROM at `./source/ch20/cond3.sh`)

```
1:   #!/bin/bash
2:   str="ernie"
3:   if [[ $str = "Ernie" ]]
4:   then
5:     echo "It's Ernie"
6:   fi
7:
8:
9:   if [[ "$str" == "Ernie" ]]; then echo "It's Ernie"; fi
```

After creating a string variable at line 2, we test it against a constant string at line 3. The = operator tests for string equality, as does the == operator. At line 9, we look at a visibly different form of the `if/then/fi` construct. As it's represented on one line, semicolons are used to separate the individual commands.

In Listing 20.5, we saw the use of the string equality operators. In Table 20.3, we see some of the other string comparison operators.

TABLE 20.3 String Comparison Operators

Operator	Description
=	is equal to
==	is equal to
!=	is not equal to
<	is alphabetically less than
>	is alphabetically greater than
-z	is null
-n	is not null

As a final look at test constructs, let's look some of the more useful file test operators. Consider the script shown in Listing 20.7. In this script, we emit some information about a file (based upon its attributes). We use the file test operators to determine the attributes of the file.

ON THE CD

Listing 20.7 Determine File Attributes Using File Test Operators (on the CD-ROM at ./source/ch20/fileatt.sh)

```
#!/bin/sh
thefile="test.sh"

if [ -e $thefile ]
then
  echo "File Exists"

  if [ -f $thefile ]
  then
    echo "regular file"
  elif [ -d $thefile ]
  then
    echo "directory"
  elif [ -h $thefile ]
  then
    echo "symbolic link"
  fi

else
  echo "File not present"
fi
```

```
22:
23:     exit
```

The first thing to notice in Listing 20.7 is the embedded `if/then/fi` construct. Once we identify that the file exists at line 4 using the `-e` operator (returns true of the file exists), we continue to test the attributes of the file. At line 8, we check to see whether we're dealing with a regular file (`-f`), in other words a real file as compared to a directory, a symbolic link, and so forth. The file tests continue with a directory test at line 11 and finally a symbolic link test at line 14. Lines 19–21 close out the initial existence test by emitting whether the file was actually present.

A large number of file test operators are provided by bash. Some of the more useful operators are shown in Table 20.4.

TABLE 20.4 File Test Operators

Operator	Description
`-e`	Test for file existence
`-f`	Test for regular file
`-s`	Test for file with nonzero size
`-d`	Test for directory
`-h`	Test for symbolic link
`-r`	Test for file read permission
`-w`	Test for file write permission
`-x`	Test for file execute permission

For the file test operators shown in Table 20.4, a single file argument is provided for each of the tests. Two other useful file test operators compare the dates of two files, illustrated as:

```
if [ $file1 -nt $file2 ]
then
  echo "$file is newer than $file2"
elif [ $file1 -ot $file2 ]
then
  echo "$file1 is older than $file2"
fi
```

The file test operator `-nt` tests whether the first file is *newer than* the second file, while `-ot` tests whether the first file is *older than* the second file.

If we're more interested on the reverse of a test, for example, whether a file is not a directory, then the `!` operator can be used. The following code snippet illustrates this use:

```
if [ ! -d $file1 ]
then
  echo "File is not a directory"
fi
```

One special case to note is when you have a single command to perform based upon the success of a given test construct. Consider the following:

```
[ -r myfile.txt ] && echo "the file is readable."
```

If the test succeeds (the file `myfile.txt` is readable), then the command that follows is executed. The logical AND operator between the test and command ensures that only if the initial test construct is true will the command that follows be performed. If the test construct is false, the rest of the line is ignored.

This has been a quick introduction to some of the bash test operators. The "Resources" section at the end of this chapter provides more information to investigate further.

`case` Construct

Let's look at another conditional structure that provides some advantages over standard `if` conditionals when testing a large number of items. The `case` command permits a sequence of test constructs utilizing integers or strings. Consider the example shown in Listing 20.8.

Listing 20.8 Simple Example of the `case/esac` Construct (on the CD-ROM at `./source/ch20/case.sh`)

```
#!/bin/bash
var=2

case "$var" in
  0) echo "The value is 0" ;;
  1) echo The value is 1 ;;
  2) echo The value is 2 ;;
  *) echo The value is not 0, 1, or 2
esac

exit
```

The case construct shown in Listing 20.8 illustrates testing an integer among 3 values. At line 4, we set up the case construct using $var. At line 5, the test against 0 is performed, and if it succeeds, the commands that follow are executed. Lines 6 and 7 test against values 1 and 2. Finally at line 8, the default * simply says that if all previous tests failed, this line will be executed. At line 9, the case structure is closed. Note that the command list within the tests ends with ;;. This indicates to the interpreter that the commands are finished, and either another case test or the closure of the case construct is coming. Note that the ordering of case tests is important. Consider if line 8 had been the first test instead of the last. In this case, the default would always succeed, which isn't what is desired.

We can also test ranges within the test construct. Consider the script shown in Listing 20.9 that tests against the ranges 0-5 and 6-9. The special form [0-5] is used to define a range of values between 0 and 5 inclusive.

ON THE CD

Listing 20.9 Simple Example of the case/esac Construct (on the CD-ROM at ./source/ch20/case2.sh)

```
#!/bin/bash
var=2

case $var in
  [0-5] ) echo The value is between 0 and 5 ;;
  [6-9] ) echo The value is between 6 and 9 ;;
  *) echo It's something else...
esac

exit
```

The case construct can be used to test characters as well. The script shown in Listing 20.10 illustrates character tests. Also shown is the concatenation of ranges, here [a-zA-z] tests for all alphabetic characters, both lower- and uppercase.

ON THE CD

Listing 20.10 Another Example of the case/esac Construct Illustrating Ranges (on the CD-ROM at ./source/ch20/case3.sh)

```
#!/bin/bash

char=f

case $char in
  [a-zA-z] ) echo An upper or lower case character ;;
  [0-9]    ) echo A number ;;
  *        ) echo Something else ;;
```

```
esac

exit
```

Finally, strings can also be tested with the `case` construct. A simple example is shown in Listing 20.11. In this example, a string is checked against four possibilities. Note that at line 7, the test construct is made up of two different tests. If the name is `Marc` or `Tim`, then the test is satisfied. We use the logical OR operator in this case, which is legal within the `case` test construct.

ON THE CD

Listing 20.11 Simple String Example of the `case/esac` Construct (on the CD-ROM at `./source/ch20/case4.sh`)

```
1:   #!/bin/bash
2:
3:   name=Tim
4:
5:   case $name in
6:     Dan        ) echo It's Dan. ;;
7:     Marc | Tim ) echo It's me. ;;
8:     Ronald     ) echo It's Ronald. ;;
9:     *          ) echo I don't know you. ;;
10:  esac
11:
12:  exit
```

This has been the tip of the iceberg as far as `case` test constructs go—many other types of conditionals are possible. The "Resources" section at the end of this chapter provides other sources that dig deeper into this area.

Looping Structures

Let's now look at how looping constructs are performed within bash. We'll look at the two most commonly used constructs; the `while` loop and the `for` loop.

`while` Loops

The `while` loop simply performs the commands within the `while` loop as long as the conditional expression is true. Let's first look at a simple example that counts from 1 to 5 (shown in Listing 20.12).

ON THE CD

Listing 20.12 Simple `while` Loop Example (on the CD-ROM at `./source/ch20/loop.sh`)

```
#!/bin/bash

```

```
 3:    var=1
 4:
 5:    while [ $var -le 5 ]
 6:    do
 7:      echo var is $var
 8:      let var=$var+1
 9:    done
10:
11:    exit
```

In this example, we define our looping conditional at line 5 (`var <= 5`). While this condition is true, we print out the value and increment `var`. Once the condition is false, we fall through the loop to `done` (at line 9) and ultimately exit the script.

Loops may also be nested. The sample script in Listing 20.13 illustrates this. In this example, we generate a multiplication table of sorts using two variables. Lines 4 to 18 define the `outer` loop, while lines 8 to 14 define the `inner`. The only difference, as in other high-level languages, is that the `inner` loop is indented to show the structure of the code.

ON THE CD

Listing 20.13 Nested `while` Loop Example (on the CD-ROM at `./source/ch20/loop2.sh`)

```
 1:    #!/bin/bash
 2:    outer=0
 3:
 4:    while [ $outer -lt 5 ] ; do
 5:
 6:      inner=0
 7:
 8:      while [ $inner -lt 3 ] ; do
 9:
10:        echo $outer * $inner = $(($outer * $inner))
11:
12:        inner=$(expr $inner + 1)
13:
14:      done
15:
16:      let outer=$outer+1
17:
18:    done
19:
20:    exit
```

Another interesting item to note about the script in Listing 20.13 is the arithmetic expressions used. In the `outer` loop we find the use of `let` to assign `outer` to itself plus one. The `inner` loop uses the `expr` command, which is an expression evaluator.

`for` Loops

The `for/in/do/done` construct in bash allows us to loop through a range of variables. This differs from the classical `for` loop that is available in high-level languages such as C, but bash's perspective is very useful and offers some capabilities not found in C. Listing 20.14 provides a very simple `for` loop.

Listing 20.14 Simple `for` Loop Example (on the CD-ROM at `./source/ch20/forloop.sh`)

```
#!/bin/bash

echo Counting from 1 to 5

for val in 1 2 3 4 5
do
  echo -n $val
done
echo

exit
```

The result of this script is:

```
# ./test.sh
Counting from 1 to 5
1 2 3 4 5
#
```

Of course, we can emulate the C `for` loop mechanism very simply as shown in Listing 20.15.

Listing 20.15 Simple `for` Loop Example Using C-like Construct (on the CD-ROM at `./source/ch20/forloop2.sh`)

```
#!/bin/bash

for ((var=1 ; var <= 5 ; var++))
do
  echo -n $var
done
echo

exit
```

This code in Listing 20.15 is identical to the original `for-in` loop shown in Listing 20.14.

We can also use strings within our looping range, as illustrated in Listing 20.16. This script simply iterates through the string's provided range.

Listing 20.16 Another `for` Loop Illustrating String Ranges (on the CD-ROM at `./source/ch20/forloop3.sh`)

```
#!/bin/bash

echo -n The first four planets are
for planet in mercury venus earth mars ; do
  echo -n  $planet
done
echo .

exit
```

Where bash shines over traditional high-level languages is in the capability to replace ranges with results of commands. Let's look at a more complicated example of the `for-in` loop that uses the replacement symbol `*`, which means the files in the current subdirectory (see Listing 20.17).

Listing 20.17 Listing the Users on the System Using Wildcard Replacement (on the CD-ROM at `./source/ch20/forloop4.sh`)

```
#!/bin/bash

# Save the current directory
curwd=$PWD

# Change the current directory to /home
cd /home

echo -n Users on the system are:

# Loop through each file (via the wildcard)
for user in *; do
  echo -n  $user
done
echo

# Return to the previous directory
cd $curwd

exit
```

Input and Output

We've seen some examples already of output using the `echo` command. This command simply emits the provided string to the display. We also saw suppression of the `newline` character using the `-n` option. The `echo` command also provides the means to emit data through a binary interface. For example, to emit horizontal tabs, the `\t` option can be used, as:

```
echo -e \t\t\t\tIndented text.
```

Some of the other options that exist are shown in Table 20.5. To enable interpretation of these strings, the `-e` option must be specified before the string.

TABLE 20.5 Special Sequences in Echoed Strings

Sequence	Interpretation
`\b`	Backspace
`\f`	Form feed
`\n`	Newline
`\r`	Carriage return
`\t`	Horizontal tab
`\v`	Vertical tab
`\\`	Backslash
`\NNN`	ASCII code of octal value

We can accept input from the user using the `read` command. The `read` command provides a number of options, a few of which we'll investigate. First, let's look at the basic form of a `read` command via the interactive bash shell:

```
# read var
test string
# echo $var
test string
# read -s var
# echo $var
silent input
```

In the first form, we read a string from the user and store it into variable `var`. The second form of read we specify the `-s` flag. The `-s` flag represents silent input, which means that characters that are entered in response to a `read` are not echoed back to the screen. In this case, we typed `silent input`, and we see this after the variable is echoed back.

Some of the other options that exist for the `read` command are shown in Table 20.6

TABLE 20.6 Other Options for the `read` Command

Option	Description
`-a`	Input is assigned into an array, starting with index 0.
`-d`	Character to use to terminate input (rather than newline).
`-n`	Maximum number of characters to read.
`-p`	Prompt string displayed to prompt user for input.
`-s`	Silent mode (don't echo input characters).
`-t`	Timeout in seconds for read.
`-u`	File descriptor to read from rather than terminal.

Functions

Bash allows us to break scripts up into more manageable pieces by creating functions. Functions can be very simple, such as:

```
function <name> () {
    sequence of command
}
```

As in C, the function must be declared before it can be called. Let's now look at a simple example function that sums together two numbers that are passed in from the caller (see Listing 20.18).

ON THE CD

Listing 20.18 Creating a Function That Utilizes Parameters (on the CD-ROM at `./source/ch20/func.sh`)

```
#!/bin/bash

function sum ()
{
```

```
 5:
 6:      echo $(( $1 + $2 ))
 7:
 8:    }
 9:
10:    sum 5 10
```

In this example, we declare a new function called `sum` (lines 3–8), which emits the sum of the two parameters passed to it. Recall that `$1` represents the first parameter and `$2` represents the second. So what does `$0` represent? Just as in C, the first argument from the perspective of a main program is the name of the program itself that was called. In this case, the name is the script file itself. What happens if the caller doesn't provide all of the necessary parameters (passes only one parameter instead of two)? This would be a good time for some error checking, so we can update the script as shown in Listing 20.19.

ON THE CD

Listing 20.19 Adding Error Checking to Our Previous Function (on the CD-ROM at `./source/ch20/func2.sh`)

```
 1:    #!/bin/bash
 2:
 3:    function sum ()
 4:    {
 5:
 6:      if [ $# -ne 2 ] ; then
 7:        echo usage is sum <param1> <param2>
 8:        exit
 9:      fi
10:
11:      echo $(( $1 + $2 ))
12:
13:    }
14:
15:    sum 5 10
```

Note in this version (updated from Listing 20.18) that error checking is now performed in lines 6–9. We test the special variable `$#`, which represents the number of parameters passed to the constant 2. Since we're expecting two arguments to be passed to us, we `echo` the use and exit if two parameters are not present.

NOTE

The parameter variables are dependent upon context. So, in Listing 20.19, the `$1` *parameter at line 14 will be different from the* `$1` *present at line 6.*

We can also return values from functions. We use the `return` command to actually return the value from the function, and then we use the special variable `$?` to access this value from the caller. See the example shown in Listing 20.20.

Listing 20.20 Adding Function `return` to Our Previous `sum` Function (on the CD-ROM at `./source/ch20/func3.sh`)

```
1:      #!/bin/bash
2:
3:      function sum ()
4:      {
5:
6:        if [ $# -ne 2 ] ; then
7:          echo usage is sum <param1> <param2>
8:          exit
9:        fi
10:
11:       return $(( $1 + $2 ))
12:
13:     }
14:
15:     sum 5 10
16:     ret=$?
17:
18:     echo $ret
```

In this version, rather than `echo` the result of the summation, we return it to the caller at line 11. At line 16, we grab the result of the function using the special `$?` variable. This variable represents the exit status of the last function called.

Sample Scripts

Now that we've covered some of the basic elements of scripting, from variables to conditional and looping structures, let's look at some sample scripts that actually provide some useful functionality.

Simple Directory Archive Script

The goal of the first script is to provide a subdirectory archive tool. The single parameter for the tool is a subdirectory that will be archived using the `tar` utility, with the resulting archive file stored in the current working subdirectory. The script source can be found in Listing 20.21.

Listing 20.21 Directory Archive Script (on the CD-ROM at `./source/ch20/archive.sh`)

```
1:    #!/bin/bash
2:
3:    # First, do some error checking
4:    if [ $# -ne 1 ] ; then
5:      echo Usage is ./archive.sh <directory-name>
6:      exit -1
7:    fi
8:
9:    if [ ! -e $1 ] ; then
10:     echo Directory does not exist
11:     exit -1
12:   fi
13:
14:   if [ ! -d $1 ] ; then
15:     echo Target must be a directory.
16:     exit -1
17:   fi
18:
19:   # Remove the existing archive
20:   archive=$1.tgz
21:
22:   if [ -f $archive ] ; then
23:     rm -f $archive
24:   fi
25:
26:   # Archive the directory
27:   tar czf $archive $1
28:
29:   exit
```

As is probably apparent right away, the script in Listing 20.21 is mostly error checking. We first ensure that there's a single argument to the script at lines 4–7. We then check that the target provided actually exists at lines 9–12, and that it's a directory at lines 14–17. At line 20, we create the archive file by appending the extension `.tgz` to the end. If the archive exists, we remove it at lines 22–24. Finally, we call the `tar` utility to create the archive at line 27. We specify three arguments to `tar`: `c` to create the archive, `z` to filter using `gzip`, and `f` to specify the filename for the archive.

Files Updated/Created Today Script

The goal of this script is to recursively search a directory to print any files that have been updated today. This is a relatively simple task that is also simple to express in bash.

The following sample script illustrates some other concepts not yet covered. See Listing 20.22 for the full script. This script is made up of three parts. The first part (lines 61–63) invokes the script based upon the user's call. The second part (lines 48–59) does some basic error checking and then starts the recursive process by interrogating the current subdirectory. Finally, the last part is the recursive function that looks at all files within a given subdirectory. Upon finding a new directory, the `recurse()` function is called again to dig down further into the tree.

When we call the `fut.sh` script, two functions are declared (`recurse()` and `main()`). The script ultimately ends up at line 61, where the main function is called using the first argument passed to the script as the argument passed to main().

In function `main()` (line 48), we begin by storing the current data in the format `YYYY-MM-DD`. This is performed using the `date` command, specifying the desired format in double quotes. We store this result into a variable called `today`. Note that `today` isn't local to `main()`; it can be used in other functions afterward.

It is possible to declare a variable as local to a function. This is useful if we wish to store information in a function for recursive uses. To declare a local variable, we simply insert the `local` *keyword before the variable.*

The `main()` function continues by storing the argument (the directory to `recurse`) in variable `checkdir` (line 52). We then test `checkdir` to see if it's empty (has zero length) at line 54. If it is empty, we store . to `checkdir`, which represents the current subdirectory. This entire test was done so that if the user passed no arguments, we'd simply use the current subdirectory as the argument default. Finally, we call the `recurse()` function with the `checkdir` variable (line 58).

The bulk of the script is found in function `recurse()`. This function recursively digs into the directory to find any files that have changed today. The first thing we do is `cd` into the subdirectory passed to us (line 15). Note that when . is passed, we `cd` into the current directory (in other words, no change takes place). We then iterate through the files in the subdirectory (line 18).

The first thing to check for a given file is to see if it's another directory (line 21). If it is, we simply call the `recurse()` function (recursively) to dig into this subdirectory (line 22). Otherwise, we check to see if the file is a regular file (line 25). If it is, we perform an `ls` command on the file, gathering a long time format (line 27). The time style format of this `ls` command (`long-iso`) happens to match the format that we gathered in `main()` to represent the date for today.

At line 29, we search the `ls` line (`longfile`) using our date stored in `today`. This is done through the `grep` command. We pipe the contents of `longfile` to `grep` to search for the `today` string. If the string is found in `longfile`, the line will simply result, otherwise a blank line will result. At line 31, we check to see if the `check` variable (the result of the `grep`) is a nonzero length string. If so, we emit the current file and continue the process at line 18 to get the next file in the directory.

Once we've exhausted the file list from line 18, we exit our loop at line 39. We check to see if the directory passed to us was not .. If not, we `cd` up one directory (because we `cd`'d down one directory at line 15). If our directory was identified as `. (the current directory),` we avoid `cd`'ing up one level.

ON THE CD

Listing 20.22 Files Updated/Created Today Script (on the CD-ROM at `./source/ch20/fut.sh`)

```
1:  #!/bin/bash
2:  #
3:  # fut.sh
4:  #
5:  #  Find files created/updated today.
6:  #
7:  #  Usage is:
8:  #
9:  #    fut.sh <dir>
10: #
11:
12: function recurse()
13: {
14:   # ‘cd’ down into the named directory
15:   cd $1
16:
17:   # Iterate through all of the files
18:   for file in * ; do
19:
20:     # If the file is a directory, recurse
21:     if [ -d $file ] ; then
22:       recurse $file
23:     fi
24:
25:     if [ -f $file ] ; then
26:
27:       longfile='ls -l —time-style=long-iso $file'
28:
29:       check='echo $longfile | grep $today'
30:
31:       if [ -n $check ] ; then
32:
33:         echo $PWD/$file
34:
35:       fi
```

```
        fi
      done
      if [ $1 != . ] ; then
        cd ..
      fi
    }
    function main()
    {
      today='date +%Y-%m-%d'
      checkdir=$1
      if [ -z $checkdir ] ; then
        checkdir=.
      fi
      recurse $checkdir
    }
    main $1
    exit
```

SCRIPTING LANGUAGE ALTERNATIVES

Bash is one option of scripting, but GNU/Linux offers a number of shells and scripting language alternatives. In addition to bash, GNU/Linux offers csh/tcsh (Berkeley C-shell), ksh (the Korn shell), zsh (Z-shell), and others. For higher-level scripting, there are numerous languages, such as Python, Ruby, Perl, Tcl, Scsh, and others. So many languages, so little time. . .

SUMMARY

In this chapter, we took a quick tour of the bash scripting shell. We explored variables in bash, including special variables that describe the environment. The basics of scripting were introduced, along with a demonstration of simple numerical methods in bash. Fundamental concepts in bash were also reviewed, including tests, conditionals, and a number of looping constructs. Methods for input and output in scripts were also discussed, in addition to bash's function specification constructs. Finally, a number of useful scripts were discussed and dissected to illustrate bash scripting.

RESOURCES

[Cooper04] "Advanced Bash-Scripting Guide" at *http://www.tldp.org/LDP/abs/html/*

[GNU02] "Bash Reference Manual" at *http://www.gnu.org/software/bash/manual/bashref.html*

[Ramey04] "Bash FAQ" at *ftp://ftp.cwru.edu/pub/bash/FAQ*

21 Editing with sed

In This Chapter

- Quick Introduction to sed
- Discussion of sed Spaces
- Typical sed Command-line Options
- sed and Regular Expressions
- sed Numerical and Pattern Ranges
- Most Useful sed Commands

INTRODUCTION

The sed utility is a very useful utility, but it can also be one of the most cryptic. sed is a stream editor that alters text that flows through it using a number of types of transformations. sed does not alter the original file provided to it but instead provides the transformed text to stdout. sed is one of the oldest tools in UNIX, written in the early 1970s by Lee McMahon. sed operations are stream operations (as the name implies) where simple scripts provide filtering and transformation of text. Consider Figure 21.1, which shows a simple example of sed use with a graphical illustration.

Let's pick apart this sed command and look at it a little further to understand how sed works for this case (the substitute command, s).

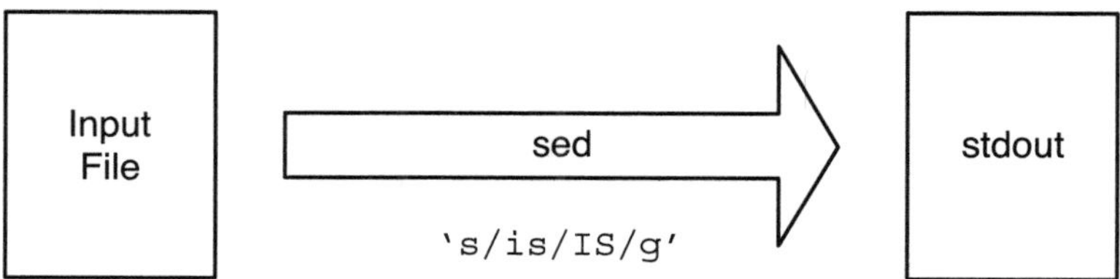

FIGURE 21.1 The `sed` model as a text filter.

The full `sed` command (for the case shown in Figure 21.1) is illustrated in Figure 21.2.

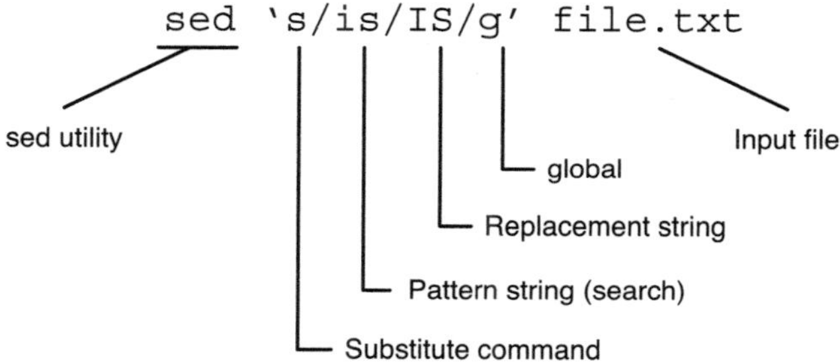

FIGURE 21.2 Anatomy of a simple `sed` invocation.

We invoke `sed` using the `sed` command and then provide a script to use on the input stream (defined by the third parameter, `file.txt`). The `sed` script represents a substitute transformation where a pattern is searched in the input stream (pattern string) and, when found, is replaced by the replacement string. Note that the command (`s`), pattern search string, and replacement string are all delimited by the `/` character. This character can be used for search and replacement; we'll look at how this is done shortly.

We end our `sed` script with a `g` to indicate that we wish to perform the search and replace over the entire stream. If `g` were not provided, only the first occurrence would be replaced. We could instead replace `g` with a number, which would indicate the particular occurrence to change. For example, ending with `/2` would indicate to change only the second occurrence.

Anatomy of a Simple Script

To complete our introduction, let's look at this script in action. We'll use the file shown in Listing 21.1 to illustrate our first simple script.

ON THE CD

Listing 21.1 Sample File for sed Illustration (on the CD-ROM at `./source/ch21/file.txt`)

```
This is a sample text string which is going to be used.

This is one more string that can be used.

Finally, this is the last string in the test set.
```

Now using our previous sed script on Listing 21.1, we see the following:

```
# sed 's/is/IS/g' file.txt
ThIS IS a sample text string which IS going to be used.

ThIS IS one more string that can be used.

Finally, thIS IS the last string in the test set.

#
```

Each occurrence of `is` is replaced with `IS` over the entire file. Consider now what happens if we omit the final `g`:

```
# sed 's/is/IS/' file.txt
ThIS is a sample text string which is going to be used.

ThIS is one more string that can be used.

Finally, thIS is the last string in the test set.

#
```

Note that instead of replacing only the first occurrence in the file, it replaced only the first occurrence in each line.

We can create two transforms by using the `-e` option (which is implied in our earlier example). Consider this example:

```
# sed -e 's/is/IS/g' -e 's/IS/is/g' file.txt
This is a sample text string which is going to be used.

This is one more string that can be used.

Finally, this is the last string in the test set.

#
```

In this case, we convert `is` to `IS`, but then we follow this with the reverse transform. The result is the same as the original file. The key here is the use of the `-e` (script or expression) to provide multiple transformations on the same input text.

`sed` *makes only one pass over the input file and is therefore very efficient.* `sed` *operates by reading a single line from* `stdin`, *executes the series of editing commands on it (potentially a series of commands), and then writes the output to* `stdout`.

Now that we have a quick introduction to `sed`, let's look at some of `sed`'s other capabilities. The `sed` utility can be quite broad and complex. Therefore, we'll focus in this chapter on the more useful aspects of `sed`.

sed SPACES (BUFFERS)

Within `sed`, there are a number of spaces (or buffers) on which `sed` commands operate. Each input line is first copied to the pattern space. The pattern space is the temporary holding space on which `sed` commands are performed. Once the provided `sed` scripts have all been performed on the pattern space, the line is copied to the output. Also available in `sed` is the hold space. Within `sed`, we can copy the contents of the pattern space to the hold space and retrieve them at some later time. The hold space is just a temporary buffer, but it can be useful to remember certain lines.

TYPICAL sed COMMAND-LINE OPTIONS

Rather than specify scripts on the command line, `sed` scripts can also be specified in a file. This can be useful when complex `sed` scripts are needed. To invoke a scripted file with `sed`, simply use the `-f` option to specify the script file.

We can suppress automatic emission of `sed` output by using the `-n` flag. When we look at the print command, we'll see where this flag can be useful.

When dealing with very large files, we can achieve better intermediate performance with the `-u` flag. In this unbuffered mode, `sed` loads smaller amounts of data from the input files and flushes data to the output more often. This means better visual performance (we see results more often), but it may represent worse overall performance.

REGULAR EXPRESSIONS

A key aspect of `sed` is its use of regular expressions. A regular expression is a pattern that can match text strings. Regular expressions are a formal language from which very complex patterns can be expressed. Let's look at a few examples (using the sed delimiter for completeness):

/dog/	Matches any occurrence of dog
/[a-z]/	Matches a single character a through z
/[a-zA-Z]/	Matches single characters a through z and A through Z
/[0-9]/	Matches all single digits (0 through 9)
/0[ab]1/	Matches 0a1 and 0b1
/Z*/	Matches zero or more occurrences of Z ("Z, ZZ, ZZZ, and so on)
/Z?/	Match zero or one instance of Z (" Z)
/[^0-9]/	Matches any single character other than digits
/t.m/	Matches any occurrence of t separate by one character followed by m, such as tim, tom, and so on

These patterns illustrate a number of special symbols used within regular expressions. For example, the [] indicates a range of characters. The - symbol is used to define the range. The * character indicates that a character may repeat zero or more times. The ? specifies that one or zero instances of the character is used for the match. The ^ character indicates the characters that are NOT used for the match. Finally, the . character matches any character.

Two other regular expressions (called *anchors*) that we'll explore with sed in this chapter include ^ (different context than before) to match at the beginning of the line and $ to match at the end of the line. For example:

/^The/	Matches "The" at the beginning of a line
/end.$/	Matches "end." at the end of a line
/^T.*./	Matches lines that start with T and end with .

RANGES AND OCCURRENCES

The range of lines to be processed by sed can be restricted using ranges. A line range can consist of a single-line definition or a range of lines. For example (using the substitute command for illustration):

```
5s/this/that/
```

replaces occurrences of `this` with `that` on line 5. We could specify that only the first five lines are operated upon, such as:

```
1,5s/this/that/
```

where `1,5` represents the range of lines one through five. If instead we wanted the range of line five to the end of the file, this could be provided using the end of file symbol `$`:

```
5,$s/this/that/
```

When we provide no range, it applies to all lines. One final useful modifier is the `!` command. This tells `sed` not to apply to the given line range. Using this with our prior example:

```
5,$!s/this/that/
```

This tells `sed` to apply the substitution to all lines, excluding line five through the end of the file.

Rather than specify line ranges, we can also define ranges based upon patterns. Consider the following example:

```
'/^The/s/this/that/'
```

This substitutes `that` for `this` for any line that begins with `The`. We can also perform substitution between pattern ranges, as:

```
'/start/,/end/s/index/idx/g'
```

This replaces `index` with `idx` after a line containing `start` is found and ends the replacement after a line containing `end` is found.

ESSENTIAL sed COMMANDS

Now that we have some basic details of `sed` under our belts, let's look at some of the most useful commands. We'll dig in a little further to these commands to explore some of their other uses.

Substitute (s)

The substitute command, as we've explored already, provides a simple search and replacement over the input stream. As we've already investigated this command, we'll cover some of the other aspects not yet touched upon. The format of the substitute is as follows:

```
[address1[,address2]]s/pattern/replacement/[n|g]
```

An optional address or address range can be specified before the substitute command. A single address indicates that the substitution is restricted to the particular line or search expression. Two addresses or patterns (separated by a comma) indicate a range for the substitution. The flags (shown at the end of the command specification) are `n`, representing a number, and `g`, representing `global`. If a number is specified, the substitution is performed only on that occurrence. For example:

```
sed '/this/that/' file.txt
```

replaces the first occurrence (of each line) contained in the input file, `file.txt`. We could achieve the same thing with the script:

```
sed '/this/that/1' file.txt
```

If we wanted the second occurrence of `this` to be replaced with `that`, we'd simply specify `n` as `2`. The global flag specifies that all occurrences on all inputs lines will be replaced.

Delete (d)

The delete command simply deletes the lines that match our defined restriction. For example, if we wanted to remove the first five lines of a file, we could use the following `sed` command:

```
sed '1,5d' file.txt
```

We could tell `sed` to delete all but the first five lines using the delete command and reversing the restriction with `!`:

```
sed '1,5!d' file.txt
```

Print (p)

The print command can be thought of as the reverse of delete. One difference is that we must specify the `-n` flag to avoid double-printing the output. To reproduce our earlier delete example of emitting only the first five lines:

```
sed -n '1,5p' file.txt
```

We could also instruct `sed` to emit only those lines that contain a particular search string, such as those containing a `:` at the beginning of the line:

```
sed -n '/^:/p' file.txt
```

Appending (a), Inserting (i), and Changing (c) Lines

We can also append, insert, or change entire lines based upon range or pattern. To insert a blank line after a line is found containing `start`, we could perform the following append command:

```
sed '/start/a\ ' file.txt
```

We could insert a blank line at the end of the file using the insert command:

```
sed '$i\ ' file.txt
```

We can change a line entirely as follows. This script looks for the word `secret` and replaces the entire line with `DELETED`:

```
sed '/secret/c\DELETED' file4.txt
```

Finally, we can perform all three commands using the `{}` symbols to group the commands. The following script will emit a start line before the line matching the keyword, a stop line after the matching line, and then emit `DELETED` for the line itself (on the CD-ROM at `./source/ch21/multi.sed`).

ON THE CD

```
sed '/secret/{
i\
-start
a\
-end
c\
DELETED
}' file.txt
```

The grouping shown here can apply to other commands, but with some restrictions. For example, when the delete command is used, all commands after it in the grouping are ignored because the delete command terminates the editing session.

Quit (q)

The quit command, as the name implies, simply ends the `sed` editing session. This command can be quite useful. Consider the following example:

```
sed '10q' file.txt
```

This command emits the first 10 lines of the `file.txt`, and when the tenth line is reached, the `sed` session is ended. This particular command emulates another useful GNU/Linux command called `head` (which emits the head of a file).

Transformation (y)

We can transform text using the y command. This provides for replacing one character for another. We provide two sets of characters: the first refers to the search set and the second the replacement set. For example, if we encounter an A, we replace with an a, and so on.

To convert all uppercase letters to lowercase, we could use the following:

```
sed 'y/ABCDEFGHIJKLMNOPQRSTUVWXYZ/abcdefghijklmnopqrstuvwxyz/' file
```

The pattern and replacement strings represent a one-to-one correspondence for which the transformation will take place. For this reason, each string must be of equal length.

Line Numbering (=)

The = command is used to emit the current line number. This can be used to emit the line numbers that match a given search string or to simply emit the number of lines in the input file.

For example, if we're interested in which lines contain a given search pattern, we could accomplish this with the following sed script:

```
sed -n '/This/=' file.txt
```

This script results in line numbers emitted (one per line) for each line that contains the word This.

Holding the Pattern Space (h)

Using the hold buffer, we can store away the pattern space to the hold buffer and then operate on the pattern buffer directly. The following example illustrates emitting both the altered line and the unaltered line. Also illustrated here are comments within sed scripts (all characters after the # symbol).

```
sed '{
# store the pattern space to the hold buffer
h
# perform the substitution on the pattern space
s/is/IS/
# append the unaltered line to the pattern space
G
}' file.txt
```

This script will emit the altered line and follow it immediately with the unaltered line (the line not having been altered by the substitute command).

SUMMARY

While we've only scratched the surface of text processing with `sed` in this short introduction, we've seen some of the power provided by this little utility. We took a quick tour of regular expressions and their use in `sed`, as well as the use of numerical and pattern-based ranges to restrict `sed`'s processing. We then reviewed some of the most-used `sed` commands, including substitution, printing, deleting, and transforming. We ended with a quick discussion of the use of the hold buffer, which gives `sed` the capability of memory.

SOME USEFUL sed ONE-LINERS

```
# Emit the first 10 lines of a file (such as 'head')
sed 10q file.txt
# Emit a double-spaced version of the file
sed 'G' file.txt
# Emit number of lines in input file
sed -n '$=' file.txt
# Emit the last line of a file
sed '$!d' file.txt
# Emit all lines greater than 30 characters in length
sed -n '/^.\{30\}/p' file.txt
# Emit all non-blank lines
sed '/^$/d' file.txt
# Remove all blank lines at the top of a file
sed '/./,$!d' file.txt
```

REFERENCES

"On the Early History and Impact of Unix Tools to Build the Tools for a New Millennium," from *Netizens: An Anthology*, by Ronda and Michael Hauben, June 12, 1996. Found at *http://www.columbia.edu/~rh120/ch106.x09I*

"Sed . . . The Stream Editor" at *http://www.cornerstonemag.com/sed/*

22 Text Processing with awk

In This Chapter

- An Introduction to awk
- Simple awk Scripting
- Complex Applications in awk
- Conditional and Looping Constructs in awk
- Awk Built-in Functions

INTRODUCTION

The awk programming language is a scripting language that can be used for very simple single-line applications to large applications. Awk is a general-purpose language, but it excels at its original task of text processing. In this chapter, we'll take a tour of the awk programming language, illustrating by numerous examples how it can be used by the developer. Awk takes over where sed left off, but each has its own individual strengths.

Short History

The first version of awk was first released in 1977 by Alfred Aho, Peter Weinberger, and Brian Kernighan (its name is the combination of the first letter of the authors' last names). Its first integration was into version 7 AT&T UNIX (as all three worked for Bell Labs at the time). Awk has gone through a variety of changes in its life. Its syntax and notation borrow both from shell scripting languages and also C.

Awk Structure

The higher-level structure of awk programs is conceptually very simple (see Figure 22.1).

```
awk' BEGIN { }

     pattern { actions }
     ...

     END { }            '&lt;filenames&gt;
```

FIGURE 22.1 Structure of an awk program.

The structure of an awk program can be split into three sections. The BEGIN section is performed before the first line is read from the input file(s), and the END section is performed after the last line is processed from the input file(s). Between the optional BEGIN and END sections is the awk pattern/action section. For each input file specified, each of the patterns is compared in order, and if a pattern matches, then its associated action is performed.

We'll now split our discussion of awk into two sections. In the first, we'll look at simple awk programs that can be specified on the command line, and then we'll look at building more complex awk programs for scripting.

COMMAND-LINE awk

Let's start by looking at some simple awk scripts that demonstrate its behavior. We'll use the following data in the file missiles.txt (which contains data about Cold War nuclear delivery platforms). The data (Listing 22.1) is delimited with : and contains five fields (missile name, length, weight, range, and speed).

ON THE CD

Listing 22.1 Missile Data for awk Scripts. (This information can also be viewed at the Strategic Air Command Web site at *http://www.strategic-air-command.com* and on the CD-ROM at ./source/ch22/missiles.txt)

```
Thor:65:109330:1725:10250
Snark:67:48147:6325:650
```

```
Jupiter:55:110000:1976:9022
Atlas:75:260000:6300:17500
Titan:98:221500:6300:15000
Minuteman III:56:65000:6300:15000
Peacekeeper:71:195000:6000:15000
```

We can emit lines in much the same way that `sed` did, but including a search expression as the pattern and the print command as the action:

```
# awk '/Thor/{print}' missiles.txt
Thor:65:109330:1725:10250
#
```

Without a pattern, we simply emit the entire file:

```
# awk '{print}' missiles.txt
Thor:65:109330:1725:10250
Snark:67:48147:6325:650
Jupiter:55:110000:1976:9022
Atlas:75:260000:6300:17500
Titan:98:221500:6300:15000
Minuteman III:56:65000:6300:15000
Peacekeeper:71:195000:6000:15000
#
```

Rather than emit the entire line, we can emit selected fields instead. `Awk` automatically splits the line (otherwise known as a *record*) into the fields delimited by the colon. So if we wanted to emit the missile and range for the Thor missile, we could do this as:

```
# awk -F: '/Thor/{print $1 ":" $5}' missiles.txt
Thor:10250
#
```

Note that we specify the delimiter as : using the `-F` command-line option. Each field is parsed to a `$` variable. The first field is defined as `$1`, the second as `$2`, and so on. The entire record is defined as `$0`.

We can add additional text to make our output more reasonable by simply including more text for the print command (the command is actually one line):

```
    # awk -F: '/Thor/{print "Missile " $1 " has a range of " $5 "
miles"}' missiles.txt
    Missile Thor has a range of 10250 miles
    #
```

Arithmetic expressions are also possible on the data. Consider this example, which emits those missiles that have a range of 12,000 miles or more:

```
# awk -F: '$5 > 12000 { print $1 }' missiles.txt
Atlas
Titan
Minuteman III
Peacekeeper
#
```

In this example, our pattern is the test of `$5` (the range field) being greater than 12,000. When this test pattern is satisfied, our action is to emit the first field (the name of the missile).

`Awk` provides a number of built-in variables that can be useful. For example, if we wanted to know the number of records in the file, we could use the optional `END` section with the `NR` built-in variable (Number of the Record):

```
# awk 'END { print NR }' missiles.txt
7
#
```

Note here that we emit `NR` at the end, so it's the total number of records that were in the file. If we emit `NR` at each line, it's the number of that given line, such as:

```
# awk -F: '{ print NR, $0 }' missiles.txt
1 Thor:65:109330:1725:10250
2 Snark:67:48147:6325:650
3 Jupiter:55:110000:1976:9022
4 Atlas:75:260000:6300:17500
5 Titan:98:221500:6300:15000
6 Minuteman III:56:65000:5300:15000
7 Peacekeeper:71:195000:6000:15000
#
```

Given the range and speed data, we can calculate how long it takes to reach its maximum target (roughly calculated as range over speed). This is provided as:

```
# awk -F: '{ printf "%15s %3.2f\n", $1, $4/$5}' missiles.txt
           Thor 0.17
          Snark 9.73
        Jupiter 0.22
          Atlas 0.36
          Titan 0.42
  Minuteman III 0.35
```

```
    Peacekeeper 0.40
#
```

Here we demonstrate simple arithmetic (`$4/$5` to compute the time to target) but also the use of `printf` within `awk`. Rather than simply print the results (as we've done in previous examples), we use the `printf` command to provide a more structured output. We specify size and alignment for our string (missile name) and also the format of our time-to-target result. From this data, we can see that the Snark has the longest time of flight (it's also the slowest of the missiles shown here), and Thor the least.

Let's look at one final example in this command-line section that demonstrates a bit more of the arithmetic properties of `awk`. Let's say that we have one of each of these missiles, and we want to know their combined weight. This is easily calculated, using each of the three `awk` sections, as:

```
# awk 'BEGIN {FS=":"} {wt += $3} END {print wt}' missiles.txt
1008977
#
```

Let's look at each of the three sections to see what's going on. In the first section (`BEGIN`), we specify our field separator (using the built-in `FS` variable). We could also specify this on the command line (with the `-F` option), but this use makes it part of the script and is therefore less error prone. For each record that we find, we sum the weight field (field 3). Note that we did not initialize our `wt` variable, as `awk` will automatically initialize it to zero when it's created. After we've processed the last record, our `END` section is performed where we simply print the weight total (just a tad over 1 million pounds or 457,664.27 kilograms).

We've looked at some of `awk`'s built-in variables so far (such as `FS` and `NR`). These and other built-in variables are available for use. A list of some of the most useful is shown in Table 22.1.

TABLE 22.1 Awk's Built-in Variables

Variable	Description
`NR`	Input record number
`NF`	Number of fields in the current record
`FS`	Field separator (default space and tab)
`OFS`	Output field separator (default space)
`RS`	Input record separator (default newline)
`ORS`	Output record separator (default newline)
`FILENAME`	Current input filename

SCRIPTED awk

Now let's dig in further and explore some of awk's other capabilities. In this section, we'll move beyond the simple single-line scripts and look at how applications can be developed in awk.

Scripting applications in awk allows us to build bigger and more complex applications. Let's start with an update to our previous application (printing the total weight of all missiles). In this example, we'll sum all of the numeric data and add headers and trailers to the data (see Listing 22.2).

ON THE CD

Listing 22.2 Expanding Our Summing Application (on the CD-ROM at `./source/ch21/tabulate.awk`)

```
 1:    BEGIN {
 2:
 3:      FS = ":"
 4:      printf "\n           Name     Length      Range"
 5:      printf "     Speed     Weight\n"
 6:
 7:    }
 8:
 9:    {
10:      printf "%15s  %8d  %8d  %8d  %8d\n", $1, $2, $4, $5, $3
11:
12:      len += $2
13:      wt  += $3
14:      rng += $4
15:      spd += $5
16:
17:    }
18:
19:    END {
20:
21:      printf "\n         Totals    ——     ——"
22:      printf "  ——    ——\n"
23:      printf "                  %8d  %8d  %8d  %8d\n\n",
24:             len, rng, spd, wt
25:
26:    }
```

Listing 22.2 illustrates the three awk sections. We define our field separator and emit a header line at lines 3–5 within the BEGIN section. Next, for each record in the input file, we emit the fields of the record in an order different than that of the original itself (note line 10). Lines 10–15 are performed for each record (as there's no pattern here, only an action that defaults to each record). Lines 12–15 simply sum each of the fields that we desire an accumulation of. We keep track of the total lengths (`len`), total weight (`wt`), total range (`rng`), and finally total speed (`spd`). The `END` section (which is performed after the last record is processed) emits the totals. Note the use of `printf` here to better control the format of the output.

We invoke this script (called `tabulate.awk`) as follows (with sample results shown given our input file from Listing 22.1):

```
# awk -f tabulate.awk missiles.txt
              Name    Length     Range     Speed    Weight
              Thor        65      1725     10250    109330
             Snark        67      6325       650     48147
           Jupiter        55      1976      9022    110000
             Atlas        75      6300     17500    260000
             Titan        98      6300     15000    221500
     Minuteman III        56      5300     15000     65000
       Peacekeeper        71      6000     15000    195000
            Totals    ——         ——        ——        ——
                         487     33926     82422   1008977
#
```

Granted, the data is meaningless, but it illustrates how we can process the input data and format the output data.

Let's now update our application to find the extremes of the data. In this example, we'll store the missile that is the longest, heaviest, has the longest range, and is the fastest. This example illustrates a very interesting aspect of `awk` that is not provided in many other languages that we use every day (dynamic and associative arrays).

Listing 22.3 shows our new application, which is similar to our original in Listing 22.2. In our `BEGIN` section (lines 1–7) we set up our field separator and then emit our table header.

For each record in the file, we have a default action (lines 9–29). We check each of our test elements (length, weight, range, and speed), and if we find one that exceeds the current (default of zero), then we save it and the current record. Recall that numeric variables are automatically initialized to zero. Note here that saving the current record is done with an associative array that is also dynamic. We've not declared our array (`saved`) or its size. Our index is a string that identifies the particular record of interest. Note that we can remove an element from our associative array using the `delete` command, such as:

```
delete saved["longest"]
```

which will remove the entry identified by the `longest` index.

Once the last record is processed, we perform our `END` section (lines 31–52). Here we simply emit our data stored from the previous extreme's capture. Note the use of the `split` command, which provides the means to split a line into its individual fields (as is done automatically when the record is read). The `split` command takes a string (stored in our associative array) and another variable that will represent our array of split elements.

ON THE CD

Listing 22.3 Finding and Storing the Extremes (on the CD-ROM at `./source/ch21/order.awk`)

```
 1:    BEGIN {
 2:
 3:      FS = ":"
 4:      printf "\n           Name    Length     Range"
 5:      printf "     Speed    Weight\n"
 6:
 7:    }
 8:
 9:    {
10:      if ($2 > longest) {
11:        saved["longest"] = $0
12:        longest = $2
13:      }
14:
15:      if ($3 > heaviest) {
16:        saved["heaviest"] = $0
17:        heaviest = $3
18:      }
19:
20:      if ($4 > longest_range) {
21:        saved["longest_range"] = $0
22:        longest_range = $4
23:      }
24:
25:      if ($5 > fastest) {
26:        saved["fastest"] = $0
27:        fastest = $5
28:      }
29:    }
30:
31:    END {
32:
33:      printf "--------   ----   ----"
34:      printf "  ----   ----\n"
```

```

    split( saved["longest"], var, ":")
    printf "%15s  %8d  %8d  %8d  %8d (Longest)\n\n",
           var[1], var[2], var[4], var[5], var[3]

    split( saved["heaviest"], var, ":")
    printf "%15s  %8d  %8d  %8d  %8d (Heaviest)\n\n",
           var[1], var[2], var[4], var[5], var[3]

    split( saved["longest_range"], var, ":")
    printf "%15s  %8d  %8d  %8d  %8d (Longest Range)\n\n",
           var[1], var[2], var[4], var[5], var[3]

    split( saved["fastest"], var, ":")
    printf "%15s  %8d  %8d  %8d  %8d (Fastest)\n\n",
           var[1], var[2], var[4], var[5], var[3]

  }
```

The sample output, given our previous data file, is shown below:

```
# awk -f order.awk missiles.txt
           Name    Length      Range      Speed     Weight
           ____    ______      _____      _____     ______
          Titan        98       6300      15000     221500 (Longest)
          Atlas        75       6300      17500     260000 (Heaviest)
          Snark        67       6325        650      48147 (Longest Range)
          Atlas        75       6300      17500     260000 (Fastest)
#
```

Awk does provide some shortcuts to simplify our application. Consider the following replacement to the END section of Listing 22.3 (see Listing 22.4).

ON THE CD

Listing 22.4 Replacement of the END Section of Listing 22.3 (on the CD-ROM at ./source/ch21/order2.awk)

```
  END {

    printf "————————  ————  ————"
    printf "  ————  ————\n"

    for (name in saved) {

      split( saved[name], var, ":")
```

```
 9:        printf "%15s  %8d  %8d  %8d  %8d (%s)\n\n",
10:               var[1], var[2], var[4], var[5], var[3], name
11:
12:      }
13:
14:    }
```

In this example, we illustrate awk's for loop, but using an index other than an integer (what we commonly think of for iterating through a loop). At line 6, we walk through the indices of the saved array ("longest", "heaviest", "longest_range", and "fastest"). Using name at line 8, we split out the entry in the saved array for that index and emit the data as we did before.

OTHER awk PATTERNS

The awk programming language can be used for other tasks besides file processing. Consider this example that simply emits a table of various data (Listing 22.5).

Here we illustrate an awk program that processes no input file (as our code exists solely in the BEGIN section, no file is ever sought). We perform a for loop using an integer iterator and emit the index, the square root of the index (sqrt), the natural logarithm of the index (log), and finally a random number between 0 and 1.

ON THE CD

Listing 22.5 Generating a Data Table (on the CD-ROM at ./source/ch21/table.awk)

```
BEGIN {
  for (i = 1 ; i <= 10 ; i ++ ) {
    printf( "%2d %f %f %f\n", i, sqrt(i), log(i), rand() )
  }
}
```

We could use a while loop instead of a for loop as shown in Listing 22.6.

ON THE CD

Listing 22.6 Generating a Data Table Using a while Loop (on the CD-ROM at ./source/ch21/table2.awk)

```
BEGIN {
  i = 1
  while (i <= 10) {
    printf( "%2d %f %f %f\n"", i, sqrt(i), log(i), rand() )
    i++
  }
}
```

So awk gives us the basic looping and control constructs that we'd expect from a high-level language, but within a pattern-matching architecture.

This tour hopefully gives you a taste for the capabilities of the awk programming language, but there is much more. Awk provides a number of other built-in functions for reading a line from the input file (getline), searching for a substring within a string (index), returning the length of a string (length), and an sprintf command for string formatting.

SUMMARY

In this chapter, we took a quick tour of the awk programming language. We explored a number of text-processing applications using both one-line and multiline scripts. We also investigated the variety of control forms available in awk inherited from the C language, including loop constructs and conditionals.

USEFUL awk ONE-LINERS

Awk is a great language for useful one-line programs. Below are a number of useful awk scripts that can be coded in a single line.

```
# Emit every line that is not blank
awk 'NF > 0 {print}' file.txt
# Emit the number of lines in a file
awk '{num_lines++} END{print num_lines}' file.txt
# or (another number of lines example)
awk 'END { print NR }' file.txt
# Emit the first 10 lines of a file (like head)
awk 'NR < 11 { print $0 }' file.txt
# Print each line, preceded by its line number
awk '{print NR, $0}' missiles.txt
# Count the number of lines that contain 'PATTERN'
awk '/PATTERN/{num++} END{ print num }' file.txt
```

23 Parser Generation with `flex` and `bison`

In This Chapter

- An Introduction to Lexical Analysis
- An Introduction to Parser Generation
- The `flex` and `bison` Utilities
- The `flex` and `bison` Specification Files
- Lexer and Parser Integration

INTRODUCTION

The topic of parser construction using two very well-known tools will be the focus of this chapter. We'll take a quick tour of lexical analysis and grammar processing and then investigate the respective tools. We'll investigate a few examples, including a simple firewall configuration parser. The `flex` (fast lexical analyzer generator) and `bison` (GNU parser generator) will be the focus of this chapter.

Lexical Analysis and Grammar Parsing

Let's begin with a short introduction to parser construction. A parser gives us the ability to process a file that has a known structure and grammar. Rather than build this parser from scratch, we can use tools to specify the parser. This is both faster and a less error prone approach to parser construction and is therefore very useful.

Parsers are very useful, and we'll develop many of them in our careers. Though few of us will take on the task of building full-featured compilers, parsers are useful in a variety of areas. For example, configuration files for larger applications can require complex parsers to be built to describe how the application's behavior is specified. The techniques we'll discuss here make building these kinds of parsers a snap.

The first task in parsing is the tokenization of our input file. In this phase, we break our input file down into its representative chunks. For example, breaking down the C source fragment:

```
if (counter < 9) counter++;
```

would result in the tokens parsed:

```
'if', '(', 'counter', '<', '9', ')', 'counter', '++', ';'
```

That's quite a few tokens, but we'll see why this is necessary shortly. Identifying how the tokens are made up is the specification part of lexical analysis. Further, the lexical analyzer will return metadata that describes what was parsed. For example, rather than the tokens, additional data is returned, such as:

```
IF_TOKEN  OPEN_PAREN  VARIABLE  OP_LESSTHAN  NUMBER  CLOSE_PAREN
VARIABLE  UNARY_INC  SEMICOLON
```

This is useful because when we're trying to understand whether the tokens have meaning, it's not important which variable we're dealing with, but just that we have a variable (identified by the `VARIABLE` token metadata).

Once we've broken down our input file into tokens, these tokens can be passed to our grammar parser to understand if they are meaningful. For example, consider the following simple grammar to define an `if` statement:

```
IF_STMT:
    IF_TOKEN OPEN_PAREN TEST CLOSE_PAREN EXPRESSION SEMICOLON
```

This defines an `if` statement rule (`IF_STMT`), which specifies that we'll have an `IF_TOKEN` (`if`), a test expression surrounded by parents (`OPEN_PAREN TEST CLOSE_PAREN`), followed by an `EXPRESSION` terminated by a `SEMICOLON`. Further:

```
TEST:
    VARIABLE  OPERATOR  NUMBER
OPERATOR:
    OP_EQUALITY  |  OP_LESSTHAN |  OP_GREATERTHAN
```

defines that our test is represented by a variable, an operator (one of three), and a number. Finally, our simple EXPRESSION is:

```
EXPRESSION:
    VARIABLE UNARY_OP
UNARY_OP:
    UNARY_INC | UNARY_DEC
```

Now that we have our simple grammar rules specified, let's look at a couple of test fragments to see how it works. Consider our original code fragment. Figure 23.1 illustrates this if statement with its parse tree.

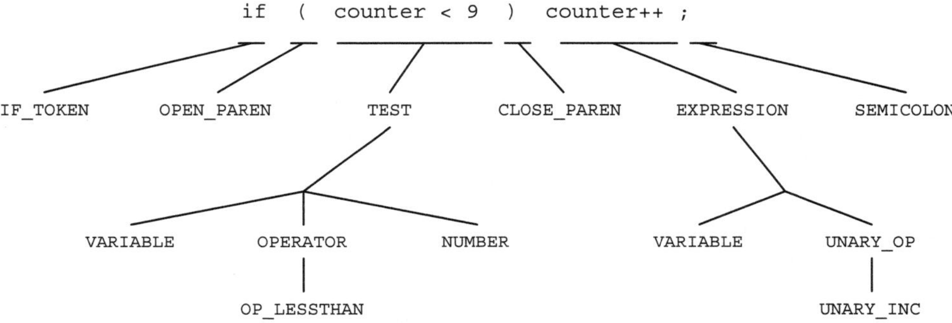

FIGURE 23.1 Parse tree for a sample C code fragment.

We can see in this diagram that each token from our C fragment is correctly recognized by our sample grammar. Each token is matched with our syntax and is therefore a proper line of C (in this simple example). Now let's look at another example, but in this case, an erroneous one (Figure 23.2).

In this case, two errors are found. First, we parse down the TEST subtree and find an issue. The TEST is made up of a VARIABLE followed by an OPERATOR and ends with a NUMBER. We find our VARIABLE (counter), but upon trying to recognize the OPERATOR, we find something that is not matched (!=). Since an invalid OPERATOR was defined, we'd signal an error and exit (or more intelligently identify the error, but then try to move on with the parse to see if any other errors are to be found). In the next case, we expect an EXPRESSION following the CLOSE_PAREN. We try to recognize the EXPRESSION as a VARIABLE and a UNARY_OP, but instead find a UNARY_OP and then a VARIABLE. Both elements are recognized, but syntactically, they represent an error. Instead of first finding a VARIABLE, we see a UNARY_OP and signal the error.

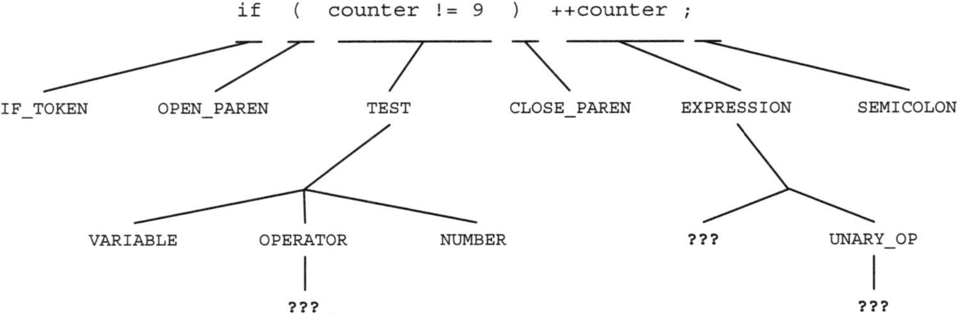

FIGURE 23.2 Incomplete parse tree for an erroneous C code fragment.

Note that the tree illustrates two concepts in parsing. The endpoints of the tree (leaves) represent the symbols of the grammar, whereas the edges represent the derivation of the grammar to the endpoints. When a sequence of tokens is properly matched in the tree to a set of nodes, our tokens are recognized by the grammar.

Lexer and Parser Communication

In a typical compiler, there will be numerous phases (commonly six). The lexical analyzer and grammar parser are two of the phases in compiler construction, but for our purposes here, they are all that are required. For the parsing of configuration files, we'll need to break our file down into tokens and then parse these into a parse tree. This is illustrated in Figure 23.3.

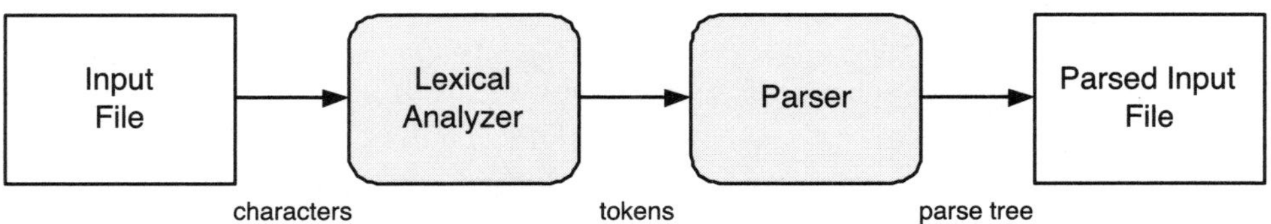

FIGURE 23.3 Typical phases in configuration parsing.

Given our input file, the lexical analyzer (or lexer) will take the file as characters and assemble them into tokens. The grammar parser takes the tokens and parses them into parse trees. Based upon the successful creation of the parse tree, we have syntactically correct code (or configuration). From here, we can utilize the parse tree to extract our configuration data.

The lexical analyzer (`flex`) takes a file of regular expressions and produces a finite automaton that recognizes the sequence of tokens for the target language. The parser generator (`bison`) takes a file defining a context free grammar and produces an LALR parser that recognizes the language.

While we focus here on `flex` and `bison`, some may recognize these tools by their early UNIX names `lex` and `yacc` (yet-another-compiler-compiler). `flex` and `bison` are GNU projects, therefore we'll utilize them instead. We'll look at the each of the tools separately and then bring them together in a single configuration example.

flex

The fast lexical analyzer generator (or `flex`) is a tool that uses a collection of regular expressions provided by the user to produce a lexical analyzer. The produced application can then be used to tokenize an input file from a sequence of characters to a sequence of tokens. The generated application is really nothing more than a finite state automaton derived from the set of regular expressions.

Let's now look at a simple example to understand what all of this means. First, the format of a `flex` input file has the format:

```
%{
<C declarations>
%}
<definitions>
%%
<rules>
%%
<user code>
```

The first section introduces a set of C declarations into the resulting lexer. Next are definitions, or simple name declarations that are used in the rules section (think of them as symbolic constant regular expressions). This section also contains start conditions for the lexical analysis, which can be used to conditionally activate rules in the rules section. Definitions have the form:

```
name        definition
```

The rules section defines the set of regular expressions used to tokenize the input. These rules define the finite automata to parse the tokens and take the form:

```
pattern        action
```

Finally, the last section is user code that will be integrated into the resulting lexical analyzer code. Now on to a real example. In Listing 23.1 we see a simple `flex` input file.

Listing 23.1 Simple `flex` Input File Example (on the CD-ROM at `./source/ch23/simple/example.fl`)

```
1:      %{
2:      #include <stdio.h>
3:      %}
4:
5:      NUM        [0-9]
6:      VAR        [a-zA-Z]
7:      WS         [ \t]
8:
9:      %%
10:     set        printf( "(STMT set) " );
11:     {NUM}+     printf( "(NUM %s) ", yytext );
12:     {VAR}+     printf( "(VAR %s) ", yytext );
13:     -          printf( "(OP minus) " );
14:     \+         printf( "(OP plus) " );
15:     =          printf( "(OP equal) " );
16:     ;          printf( "(END stmt) " );
17:     \n         printf( "\n\n" );
18:     {WS}+      /* Ignore whitespace */
19:     <<EOF>>    printf( "End of parse\n." ); yyterminate();
20:     %%
```

Note first the three sections in Listing 23.1. We see our definitions section (everything prior to the first `%%`). At line 2, we see some C code that will be integrated into the final lexer. Since we'll use some standard C library functions within our rules section, we tell `flex` to include the `stdio.h` header file so that these symbols (such as `printf`) can be resolved. We then provide three definitions that will be used later in our rules section (one can also think of these as aliases). First, the `NUM` definition defines a regular expression for a number. This defines that a number is a digit between 0 and 9. Next is the `VAR` definition, which specifies that variables are made up of lower- and uppercase letters. Finally, the `WS` definition indicates that whitespace is space or tab characters. Note that these are simply definitions; the rules section will utilize these to further specify tokens in our simple language.

The next section (between the two `%%` tags) identifies the rules. These represent patterns (regular expressions, herein called regex) and an action to take when the pattern is found. Consider the first rule at line 10. The rule defined here refers to the discovery of the token set. When the set is found, we perform the `printf` (to emit (stmt SET)). The next rule is a regular expression that's used to match numbers of arbitrary length. We use our definition of NUM (in braces, as required by flex) and follow it with a plus. The plus indicates that we're looking for one or more of these digits. So, if we find a number (0–9) and then a character that is not a number, the match is complete, and we perform the action. In this action, we emit the value that we've recognized, which flex stores in a character string called yytext. The next rule at line 12 is similar to our number rule but instead (per the definition at line 6) collects alpha characters.

At lines 13 and 15, we see some very simple recognizers. When we see a - symbol, we emit a minus string, and when we see a + (escaped with \ to delineate it from the one or more regex symbol), we emit a plus string. Line 17 looks for newlines and simply emits a newline to stdout.

One final point to note is line 19. This special symbol recognizes the end of the file, for which we simply end the session using the supplied yyterminate function.

Let's now look at how we can build and then test this lexer using stdin. To build our lexer, we simply provide the file (example.fl) to the flex utility, as:

```
# flex example.fl
```

The result is a file called lex.yy.c, which contains the lexer. We can now compile this file as follows:

```
# gcc -o example lex.yy.c -lfl
```

The trailing -lfl tells gcc to link in the flex library (libfl.a). Now we have an executable for which we can test. Let's look at a few examples typed from the shell (see Listing 23.2).

Listing 23.2 Test of Simple Lexer Specified in Listing 23.1

```
# ./example
set counter = 0;
(STMT set) (VAR counter) (OP equal) (NUM 0) (END stmt)

set counter=0;
(STMT set) (VAR counter) (OP equal) (NUM 0) (END stmt)

set set = set;
(STMT set) (STMT set) (OP equal) (STMT set) (END stmt)

set a = a + 1;
(STMT set) (VAR a) (OP equal) (VAR a) (OP plus) (NUM 1) (END stmt)
```

```
14:    abc123def
15:    (VAR abc) (NUM 123) (VAR def)
16:
17:    End of parse
18:    #
```

At line 1, we start the lexer and then begin providing input. Note at line 18 in Listing 23.1 that we ignore any whitespace that's encountered. This is why lines 2 and 5 of Listing 23.2 result in an identical set of tokens being generated. At line 8, we see the `set` reserved word used as a variable. The lexer simply parses this as a statement, which would presumably be caught by the grammar parser. The point here is that the lexer doesn't know and therefore simply parses the tokens as it sees them (even though they may not make sense).

One other item to note (to further clarify how the lexer works) is the input shown at line 14. In this case, no whitespace is provided to delimit the input. The lexer correctly notes that alpha characters are `VAR` tokens and numeric characters represent `NUM` tokens. In this case, the three tokens are correctly parsed from the input without any reference to delimiters.

Finally, at line 16 of Listing 23.2, we perform a Ctrl+D to end the parse. The lexer catches this (via the `<<EOF>>` symbol) and ends the parse session.

In these examples, the actions to the rule patterns have been to write to `stdout`. Later in this chapter, we'll look at how to connect the lexical analyzer to the grammar parser.

bison

Now that we have some understanding of the lexical analysis process, let's dig into grammar parsing. We'll evolve our last lexer to work with a new grammar parser built with `bison`.

The grammar parser that we'll build with `bison` works in concert with the lexical analyzer. It accepts the tokens recognized by the lexer and matches them with the grammar symbols to identify a properly structured input. The `bison` tool builds a bottom-up parser and, using a process known as shift-reduce, attempts to map all of the lexer data elements to grammar symbols.

A Simple Grammar

Let's first look at a simple example that models our input from Listing 23.2. What we want our grammar to represent is a set of `set` operations. These can take the form:

```
set counter = 1;
set counter = counter + 1;
set counter = lastcount;
set delta = counter - lastcount;
```

In this very simple grammar, we can reduce to the following rules. First each statement starts with a `set` command followed by a variable and a = symbol. After the assignment operator, we have what we'll call an expression and terminate with a `;`. So our first rule can take the form:

```
'set' VARIABLE '=' EXPRESSION ';'
```

Our `EXPRESSION` has one of four forms:

```
NUMBER
VARIABLE
VARIABLE OPERATOR NUMBER
VARIABLE OPERATOR VARIABLE
```

Finally, we'll support two operators within an expression. It can be either an addition or a subtraction:

```
'+'
'-'
```

This is reasonable so far for this very simple grammar. It should be clear now how the lexer and grammar parser work together. The lexer will break the input down into the necessary tokens, which the grammar parser takes, and using its rules, determines if the symbols can be recognized by the grammar.

Encoding the Grammar in bison

Let's now look at how the simple grammar can be represented for `bison`. The `bison` input file is similar to the flex input file. It starts with a section containing C code that is imported into the grammar parser, followed by a set of declarations:

```
%{
C Declarations
%}
Bison Declarations
%%
Grammar Rules
%%
C Code
```

The `bison` input file can be quite a bit more complex than the `flex` input file, so let's continue our illustration with a real example. In this example, we'll provide a grammar for our previous `set` example. This is shown in Listing 23.3.

Listing 23.3 bison Grammar File for Our Simple set Example (on the CD-ROM at `./source/ch23/setexample/grammar.y`)

```
%{
#include <stdio.h>

void yyerror( const char *str )
{
        fprintf( stderr, "error: %s\n", str );
}

int main()
{
        yyparse();

        return 0;
}
%}

%token SET NUMBER VARIABLE OP_MINUS OP_PLUS ASSIGN END

%%

statements:
        | statements statement
        ;


statement:
        SET VARIABLE ASSIGN expression END
        {
                        printf("properly formed\n");
        }
        ;


expression:
        NUMBER
        |
        VARIABLE
        |
        VARIABLE operator NUMBER
        |
        VARIABLE operator VARIABLE
        ;
```

```
43:
44:
45:     operator:
46:             OP_MINUS
47:             |
48:             OP_PLUS
49:             ;
```

In our first section of C declarations (lines 1–15), we see code that will be ported to the generated parser. We include any header files that are referenced in this C declarations section or in code in the rules section (such as the `printf` at line 29). Also of note are two functions that we'll provide here for the parser. The first is a special function called `yyerror` that is called by the generated parser if an error occurs. Next we see our `main` function. Normally, we would have our own `main` and call the parser, but for this simple example, we'll embed the `main` within the parser itself. The `main` calls the function `yyparse`, which is the grammar parser. The grammar parser invokes the lexical analyzer internally.

In the next section, `bison` declarations, we see a special symbol called `%token`. This identifies all the tokens that are used by the parser to recognize the grammar. The lexical analyzer must know about these as well and must be built to return them. We'll have another look at the lexer shortly and see how it's changed to support connectivity with the parser.

The bulk of our `bison` input file is the rules that make up the grammar. The rules section begins at the first `%%` symbol in the file (line 19). Our first rule is the statements rule (lines 21–23). A rule is made up of a name followed by a colon and then a series of tokens or nonterminals with an optional action sequence (within a set of braces), terminated by a ;. The `statements` rule has two possibilities. The blank after the colon means that there may be no more tokens to parse (end of file) or (`|`) or another possibility. If we had placed `statement` as the only possibility, then we could parse a single statement and no more. By specifying `statements` before `statement`, we allow the possibility of parsing more than one statement in our input.

The `statement` rule (lines 26–31) permits the parse of a single kind of statement (a variable assignment). Our statement must begin with a `set` command followed by a variable name and an assignment operator. We then see an expression rule, which will be used to cover a number of different possibilities. Our statement ends with a semicolon (represented here as an `END` token). Note here that we provide an optional action sequence for the statement. If our rule matches without encountering any errors, the optional command is performed within the parser. Here we simply emit that the input provided was properly formed within the grammar.

The `expression` rule (lines 34–42) defines the varieties of expressions that are possible (the right value, or `rvar`, of our statement). There are four possibilities that we consider legal. It can be a number, another variable, a sum of a variable and a variable or number, or a difference between a variable and a variable or number. The `operator` rule (lines 45–49) covers the two types of operators that are legal (a - as defined by the `OP_MINUS` token or a + as defined by `OP_PLUS`).

The goal of this parse, as defined by the code that we've inserted into the grammar, is simply to identify a properly formed statement.

`bison` *grammars are expressed in a machine-readable Backus-Naur Form (or BNF). A BNF is a formal syntax used to express context-free grammars and is a widely used notation to express grammars for computer programming languages.*

Hooking the Lexer to the Grammar Parser

Now that we have our grammar parser done, let's look at how to modify the lexer so that both know about the proper set of tokens and also how to connect the lexer to the parser. Let's start by looking at the upgraded flex input file, and we'll describe the changes to support the connectivity.

Listing 23.4 Upgraded `flex` Input File (on the CD-ROM at `./source/ch23/setexample/tokens.l`)

```
%{
#include <stdio.h>
#include "grammar.tab.h"
%}

NUM      [0-9]
VAR      [a-zA-Z]
WS       [ \t]

%%
set              return SET;
{NUM}+           return NUMBER;
{VAR}+           return VARIABLE;
-                return OP_MINUS;
\+               return OP_PLUS;
=                return ASSIGN;
;                return END;
\n               /* Ignore whitespace */
{WS}+            /* Ignore whitespace */
%%
```

The first point to note is that in order to know which tokens (constant names) the parser generator is using we must include them in the lexer. When the parser generator is built with bison, it also generates a header file that contains the tokens that were specified (via the %token symbol). These can now be included within the flex input file to connect them together. In the compile phase of the lexer, we can easily find if there are disconnects (due to missing symbol errors).

The second point to note is that we're no longer just printing the token that was recognized, but instead returning the token to the caller. The caller in this case will be the parser generator.

Now let's walk through the process of building a parser from both the flex and bison input files. In this example, our bison grammar input file is named grammar.y, and our flex input file is called tokens.l. We start by building the parser itself because the lexer is dependent upon the header file generated here:

```
# ls
grammar.y  tokens.l
# bison -d grammar.y
# ls
grammar.tab.c  grammar.tab.h  grammar.y  tokens.l
#
```

We tell bison (via the -d flag) to generate the extra output file (which contains our macro definitions) so that we can hook it up to the lexer. We see here that two files were created: grammar.tab.c and grammar.tab.c. The C source file is our parser source, and the .h file our macro definitions file.

Next, let's build our lexer using flex:

```
# flex tokens.l
# ls
# grammar.tab.c  grammar.tab.h  grammar.y  lex.yy.c  tokens.l
```

We invoke flex with our flex input file, which results in a new file called lex.yy.c. This is our lexical analyzer. Finally, we can build these together with gcc and generate our parser as:

```
# gcc grammar.tab.c lex.yy.c -o parser -lfl
```

Now we have our parser executable in a file called parser. Let's try it out and see how it works (see Listing 23.5).

Listing 23.5 Sample Use of the `set` Parser

```
# ./parser
set counter = 1;
properly formed
set counter = counter + 1;
properly formed
set counter   =   lastCount;
properly formed
set deltaCount = counter - lastCount;
properly formed
set set = set;
error: parse error
#
```

We can see from this use that it properly recognizes our grammar. When we encounter an error, it simply exits with a status. We can also exit the parser by pressing Ctrl+D.

Note that our last example, where an error was detected, presents an interesting case. The `set` token is valid as far as variables go (given our variable rule in the lexer at line 13 of Listing 23.4), but given the precedence of our regular expressions, it's detected as the `SET` token rather than a variable. Given our grammar, this is defined as an error.

So now we've built a parser that will identify whether the input is correct given the tokens that can be recognized and the grammar. The next step is actually doing something with the data, which we'll look at in the next section.

BUILDING A SIMPLE CONFIGURATION PARSER

Let's now continue with another example of parser construction with `flex` and `bison`, but in this case, we'll use the data that's parsed. In this example, we'll create a parser for an e-mail firewall configuration that defines from whom we'll accept mail and also from whom we'll reject e-mail. In this example, mail that that's not specified as allowed but also not explicitly specified as disallowed will simply be quarantined for later review by the e-mail recipient. The goal of such an e-mail firewall is to quickly identify e-mail that we expect, disallow e-mail that we don't want, and then cache the rest for later review.

Our configuration file is illustrated in Listing 23.6. We have two sections that define those e-mail addresses that we'll allow (those that we'll immediately allow to make it to the recipient) and also those that we'll disallow (reject or ignore).

ON THE CD

Listing 23.6 Sample E-Mail Firewall Configuration File (on the CD-ROM at ./source/ch23/config/config.file)

```
allow {
  mtj@mtjones.com
  dan_5422@yahoo.com
  albert@camus.com
}
disallow {
  spammer@spamcorp.com
  viagera@pharma.com
}
```

The structure of this file (Listing 23.6) is very simple. Those e-mail addresses in the `allow` section are permitted, while those in the `disallow` section are rejected.

Our parser will be constructed in the two phases that we've demonstrated in our first example. Our configuration file lexer will be defined in a file called `config.l`, and our grammar parser in `config.y`. Rather than taking our configuration file from `stdin`, in this example we'll allow the configuration file to be read from a file.

Configuration File Lexical Analyzer

The lexer is the first phase of our configuration file parser. It will break down the file into its basic elements and return them to the grammar parser. The tokens that we'll permit, from Listing 23.6, are minimal and consist of two reserved words (`allow` and `disallow`), section delimiters (`{` and `}`), and finally the e-mail addresses, which will consist of an aggregate of tokens. Whitespace and newlines will simply be ignored. Our `flex` input file for the configuration file lexer is shown in Listing 23.7.

ON THE CD

Listing 23.7 Configuration Parser `flex` Input File (on the CD-ROM at ./source/ch23/config/config.fl)

```
%{
#include <stdio.h>
#include "config.tab.h"
%}
```

```
7:   %%
8:   allow                      return ALLOW;
9:   disallow                   return DISALLOW;
10:  [a-zA-Z]+[a-zA-Z0-9_]*     yylval=strdup(yytext); return WORD;
11:  \{                         return OPEN_BRACE;
12:  \}                         return CLOSE_BRACE;
13:  \@                         return ATSYM;
14:  \.                         return PERIODSYM;
15:  \n                         /* Ignore end-of-line */
16:  [ \t]+                     /* Ignore whitespace  */
17:  %%
```

The first item to note is that we've specified that we'll need the parser generator header file to understand what tokens will be communicated (line 4). Next, at line 10, we see a new action for a string pattern. This particular pattern recognizes strings that begin with a letter (uppercase or lowercase) and then follow with optional characters of letters, numbers, or the _ character. When we recognize one of these, its value (what was tokenized) is copied into a special variable called `yylval`. When the lexer recognizes this token, it's stored in `yytext`, which allows the parser to see the actual token (in addition to the token type that's returned, in this case `WORD`).

Now we have a lexer that will tokenize our file and also return the special words that it finds (ignoring the reserved words, given their precedence in the table). Let's now explore the new parser (Listing 23.8).

ON THE CD

Listing 23.8 Configuration Parser `bison` Input File (on the CD-ROM at `./source/ch23/config/config.y`)

```
%{
#include <stdio.h>
#include <string.h>

void yyerror( const char *str )
{
        fprintf( stderr, "error: %s\n", str );
}


int main()
{
        FILE *infp;

        infp = fopen( "config.file", "r" );

        yyrestart( infp );
```

```

        yyparse();

        fclose( infp );

        return 0;
}


char address[10][80];
int  addrCount = 0;

%}

%token ALLOW OPEN_BRACE CLOSE_BRACE DISALLOW WORD ATSYM PERIODSYM

%%

configs:
        | configs config
        ;

config:
        allowed
        |
        disallowed
        ;

allowed: ALLOW OPEN_BRACE targets CLOSE_BRACE
        {
                int i;
                printf("Allow these addresses:\n");
                for ( i = 0 ; i < addrCount ; i++ ) {
                        printf( "\t%s\n", address[i] );
                }
                addrCount = 0;
        }
        ;

disallowed: DISALLOW OPEN_BRACE targets CLOSE_BRACE
        {
                int i;
                printf("Disallow these addresses:\n");
                for ( i = 0 ; i < addrCount ; i++ ) {
                        printf( "\t%s\n", address[i] );
```

```
63:                     }
64:                     addrCount = 0;
65:             }
66:             ;
67:
68:
69:     targets:
70:             |
71:             targets email_address
72:             ;
73:
74:
75:     email_address:
76:             WORD ATSYM WORD PERIODSYM WORD
77:             {
78:              if ( addrCount < 10 ) {
79:                        sprintf( address[addrCount++],
80:                                  "%s@%s.%s", $1, $3, $5 );
81:                        free($1);  free($3);  free($5);
82:                     }
83:             }
84:             ;
```

Our C code section of Listing 23.8 (lines 1–30) illustrates some of the key changes that make this parser useful in the configuration domain. Our `main` function, rather than simply calling `yyparse`, opens our desired configuration file and then calls the function `yyrestart` with the file pointer. This allows us to change the input from `stdin` to the file of our choice. We then call the `yyparse` function to perform the grammar parse. Upon return (the parse is either complete or an error was encountered), we close our input file pointer. We also declare a few resources (lines 27 and 28), which will be used to store the e-mail addresses and the number parsed so far. We'll see how these are actually used in the `bison` rules section.

In the `bison` declaration section (lines 31–33) we define the tokens that are expected from the lexer. The result of this line (when the parser generator is created) is a header file containing symbolic constants for each of these tokens.

The rules section (lines 34–84) defines the grammar for the configuration parser. We include a base rule to allow one or more configuration sections (at lines 36–38) and then a rule to match either an `allow` section or a `disallow` section (lines 40–44). The `allowed` and `disallowed` rules are similar in pattern, except for what they represent (permitted or rejected e-mail addresses). Each begins with the respective reserved word (`allow` or `disallow` and then an open brace (`{`) followed by zero or more e-mail addresses, terminated by a close brace (`}`). We see C code sections for each of the two rules; we'll return to these after we dive into the `targets` rule.

In the targets rule, we specify that zero or more e-mail addresses can be recognized (lines 69–72). Our email_address rule (lines 75–84) specifies a simple e-mail address recognizer. In this case, that is a word followed by a @ symbol, followed by another two words, separated by a ‘. Therefore, e-mail addresses such as:

```
mtj@mtjones.com
```

will be recognized, but:

```
mtj@mail.mtjones.com
```

will not. For this demonstration, the simple e-mail address spec will suffice.

The action portion of the email_address rule defines what we do when we recognize a valid e-mail address. In this case (as long as we haven't exhausted our storage resources), we populate an address entry with the aggregated e-mail address. Note here that the token list:

```
WORD ATSYM WORD PERIODSYM WORD
```

is accessible through a special set of references. The reference $1 represents the first WORD in the list. Recall in our lexer (Listing 23.7 line 10) that we duplicated the current token using strdup to yylval. The $1 here represents the yylval stored by the lexer. Reference $3 represents the third WORD and $5 the last WORD. Therefore, we use sprintf (to concatenate the strings together into our address array), resulting in a contiguous e-mail address from the individual WORD parts. Once we've stored the address recognized, we free each of the string references. Recall that we used strdup to copy the current token. This has the effect of mallocing memory for the new object, and therefore, we must free this resource.

Once the CLOSE_BRACE is recognized for the section (line 46 or 57), we emit the addresses that were found. We know the type of addresses these are (allowed or disallowed), given the context of the rule. If we're in the allowed rule, then we know that these addresses are within an allow section. Similarly, within the disallowed rule, we emit the addresses as disallowed. After we emit each section, we zero the addrCount to permit additional sections to be recognized and stored.

Let's look at an example of the parser. Listing 23.9 provides an example file (properly formed).

Listing 23.9 Sample Configuration File

```
allow {

  mtj@mtjones.com
  dan_5422@yahoo.com
```

```

    }

    disallow {

      you@yahoo.com
      them@excite.com

    }
```

Calling this file `config.file` (as required by the parser), we demonstrate its use as follows:

```
# ./parser
Allow these addresses:
        mtj@mtjones.com
        dan_4522@yahoo.com
Disallow these addresses:
        you@yahoo.com
        them@excite.com
#
```

If any sections were not properly formed, we'd not see any addresses (since the addresses aren't actually `printf`'d until the entire section is parsed).

While this was a simple example, it should illustrate some of the power of `flex` and `bison`. `flex` and `bison` could be used to build lexers and parsers for very complex languages (such as the C language) or very simple grammars as illustrated here.

THE BIG PICTURE

In Figure 23.3, we showed the phases involved in taking an input file, breaking it down into tokens (via the lexical analyzer), and then passing these tokens to the parser to generate a parse tree. Let's now amend this figure to illustrate what we've done here (see Figure 23.4).

As we've shown in this chapter, the lexical analyzer is generated from the `flex` utility, given an input file representing the regular expressions that recognize the tokens of the grammar. The parser is generated from the `bison` utility, again specified by a grammar file. Each of these phases is built together in a single image, with connectivity between the two specified by the `flex` and `bison` tools and also by the developer.

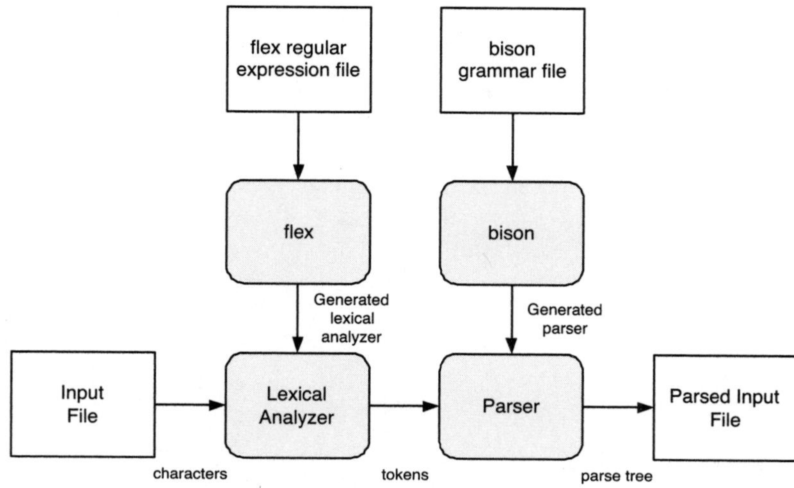

FIGURE 23.4 Parsing phases with `flex` and `bison` flows.

The flow of the lexer and parser is partly provided internally but is visible in the grammar definitions (see Figure 23.5). Our main function (provided in the `bison` grammar file) calls `yyparse` to perform the parsing function. This in turn calls `yylex` to retrieve the tokens as they're extracted from the input stream. The `yylex` function returns the type of token found, with any other data needed by the parser returned in other variables (such as `yylval`).

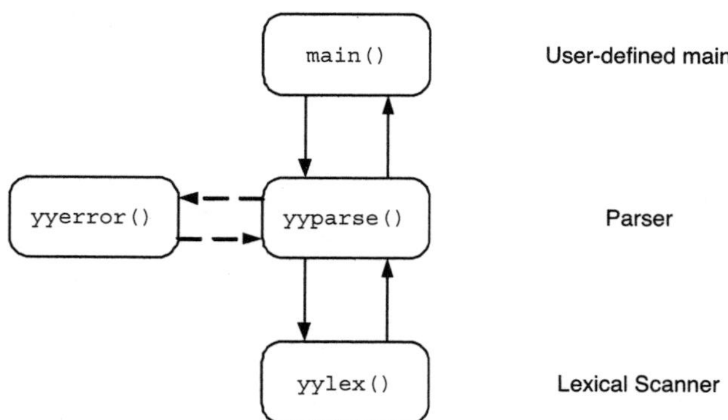

FIGURE 23.5 Grammar definitions in lexer and parser flow.

We've seen a few of the internal functions and variables provided in the scanner and parser. Table 23.1 provides a list of some of the others that you may encounter in your use of `flex` and `bison`.

TABLE 23.1 Useful Scanner and Parser Functions and Variables

Name	Type	Description
`yyparse`	Function	Parser function (called by `main`)
`yyerror`	Function	Error function (can be provided by user)
`yylex`	Function	Scanner functions (returns tokens, used by `yyparse`)
`yyterminate`	Function	Terminates the parsing process
`yylval`	`char*/union`	Token value
`yytext`	`char*`	Pattern string used by the lexer
`yydebug`	`int`	Set to 1 to enable debug mode

Designing and specifying parsers (and lexers) can be a difficult task, but in the end, the act of specifying how the grammar works is necessary even if it's to be done by hand. Once the specification is done, the generation of the parser with `bison` is trivial (compared to writing one by hand) and therefore `flex` and `bison` can be very useful tools in our development toolbox.

SUMMARY

The `flex` and `bison` tools can be two very important elements for software specification and generation of lexers and grammar parsers. The `flex` tool allows the specification and generation of lexical analyzers that provide the ability to tokenize the input. The `bison` tool allows the specification of a grammar, which when generated can take the tokens from the lexical analysis phase to recognize correctly formed input. We demonstrated these tools in two scenarios, looking at two different parsing examples. The build process was also discussed, including linkage between the lexer and parser phases.

Part V Debugging and Testing

In this final part of the book, we'll look at the topics of debugging and testing. This includes unit-testing frameworks, using the GNU source-level debugger, and finally code hardening techniques for creating higher quality and more reliable GNU/Linux applications.

Chapter 24, "Software Unit Testing Frameworks"

The topic of software testing is an important one with quite a bit of development in the open source community. After an introduction to unit and system testing, unit testing frameworks are explored, including a look at how to make your own and two open source distributions. The `expect` utility is also covered as a means to test applications at a high level.

Chapter 25, "Debugging with GDB"

The GNU Debugger is a source-level debugger that is a staple for GNU/Linux application development. GDB is integrated into the GNU toolchain and offers both command-line and GUI versions. This chapter presents GDB in a tutorial fashion and walks through the debugging of a simple application using breakpoints. Features such as data inspection, stack frame viewing, and the GDB commands used most are covered. More advanced features for multiprocess and multithreaded application debugging and core-dump file debugging are also discussed.

Chapter 26, "Code Hardening"

The topic of code hardening, otherwise known as defensive programming, is the final chapter of the book. Code hardening is an umbrella for a variety of techniques that have the goal of increasing the quality and reliability of software. In this chapter, we'll look at numerous coding methods as well as tools (such as static source checking tools) to help build better software.

24 Software Unit Testing Frameworks

In This Chapter

- Unit Testing versus System Testing
- Brew Your Own Frameworks
- Testing with the C Unit Test Framework
- Testing with the Embedded Unit Test Framework
- Testing with `expect`

INTRODUCTION

Writing software is difficult, even when building on an outstanding operating system such as GNU/Linux. The complexity of software grows as the sizes of the systems that we develop grow. But even when developing smaller systems, problems can still find there way in. This is where testing comes in. Even if we develop software that doesn't work the first time, performing tests on the software can identify the shortcomings, allowing us to fix them.

If our tests are repeatable, we can easily retest our software once we've made changes to it (otherwise known as *regressing*). This makes it much easier to update our software, since we know it's easy to verify that we haven't broken anything when we're done. If we do find something that's broken but wasn't tested before, we simply update the regression test to check it in the future.

In this chapter, we'll look at a number of available open source unit testing frameworks and how they can be used to improve the quality of our software.

Unit Testing

First, let's address what is meant by "unit" testing. For this discussion, let's divide testing into two unique categories. The first is end-to-end (or system) tests, which test specific user-level features of the software (see Figure 24.1).

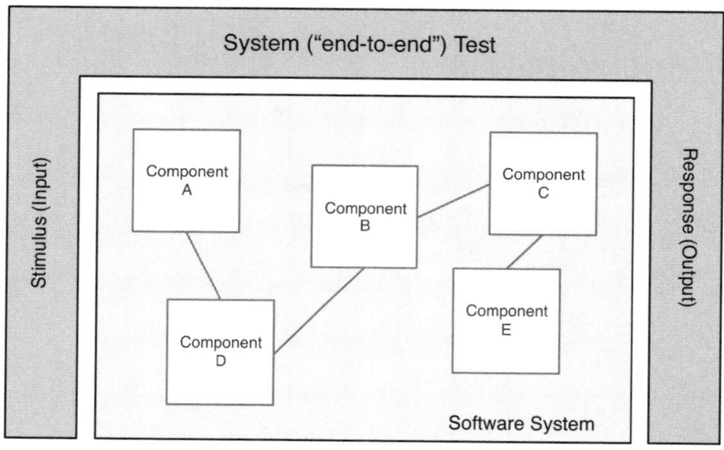

FIGURE 24.1 System testing (or end-to-end) perspective.

We can also think about this as testing based upon the requirements of the software. In this category, we typically don't consider how the software is constructed (black-box testing) but instead simply test for what the system should do. As illustrated in Figure 24.2, the components of our system can be further broken down into smaller modules called *units*.

Unit testing assumes more knowledge and insight into the software to be tested. Unit tests address units of a software system and therefore don't typically address system requirements but instead internal behavior of the system. Let's say that our software system included a queue unit that was to be used for internal task communication. We could separate this queue unit from the rest of the system and test it in isolation with a number of different tests. This is a unit test, since we're addressing a unit of the system (see Figure 24.3). We also commonly consider how the unit was constructed and therefore test with this knowledge (white-box testing).

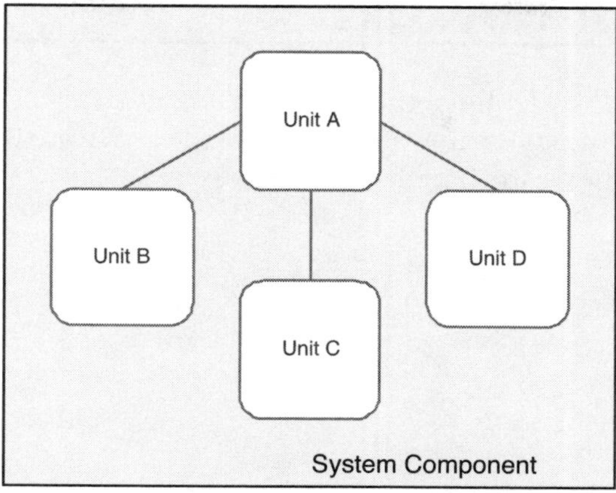

FIGURE 24.2 System components are made up of units.

This permits us to ensure that we've tested all of the elements of the unit (recall the use of the GNU `gcov` utility in Chapter 7, "Coverage Testing with GNU `gcov`").

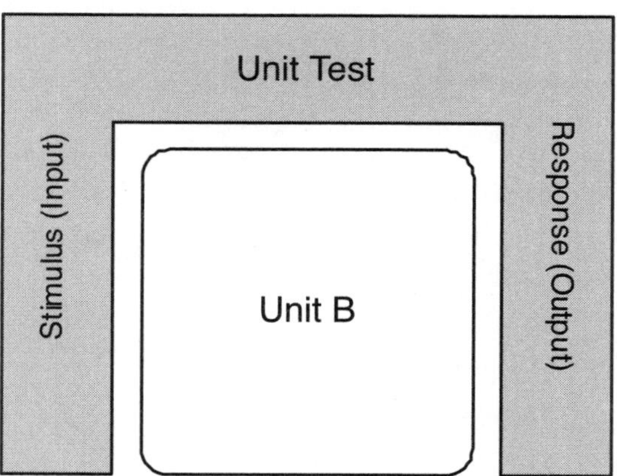

FIGURE 24.3 Unit testing perspective of an individual software unit.

Therefore, a unit test simply invokes the unit's APIs and verifies that a given stimulus produces an expected result.

UNIT TESTING FRAMEWORKS

Now that we've identified the scope of the unit test, let's look at some unit testing frameworks to explore how they can be used to increase the quality of our software. The following are the frameworks we'll review.

- Brew your own
- C unit test (cut) system
- Embedded unit test
- `expect`

Brew Your Own

Building your own simple unit test framework is not difficult to do. Even the simplest architecture can yield great benefits. Let's look at a simple architecture for testing a software unit.

Consider that we have a simple stack module with an API consisting of the following:

```
typedef struct { ... } stack_t;
int stackCreate( stack_t *stack, int stackSize );
int stackPush( stack_t *stack, int element );
int stackPop( stack_t *stack, int *element );
int stackDestroy( stack_t *stack );
```

This very simple LIFO stack API permits us to create a new stack, push an element on the stack, pop an element from the stack, and finally destroy the stack. Listing 24.1 shows the code (`stack.c`) for this simple module, and Listing 24.2 shows the header file (`stack.h`).

ON THE CD

Listing 24.1 Stack Module Source (on the CD-ROM at `./source/ch24/byo/stack.c`)

```
#include <stdlib.h>
#include "stack.h"


int stackCreate( stack_t *stack, int stackSize )
{
  if ((stackSize == 0) || (stackSize > 1024)) return -1;

  stack->storage = (int *)malloc( sizeof(int) * stackSize );

```

```
  if (stack->storage == (void *)0) return -1;

  stack->state = STACK_CREATED;
  stack->max = stackSize;
  stack->index = 0;

  return 0;
}


int stackPush( stack_t *stack, int element )
{
  if (stack == (stack_t *)NULL) return -1;
  if (stack->state != STACK_CREATED) return -1;
  if (stack->index >= stack->max) return -1;

  stack->storage[stack->index++] = element;

  return 0;
}


int stackPop( stack_t *stack, int *element )
{
  if (stack == (stack_t *)NULL) return -1;
  if (stack->state != STACK_CREATED) return -1;
  if (stack->index == 0) return -1;

  *element = stack->storage[—stack->index];

  return 0;
}


int stackDestroy( stack_t *stack )
{
  if (stack == (stack_t *)NULL) return -1;
  if (stack->state != STACK_CREATED) return -1;

  stack->state = 0;
  free( (void *)stack->storage );

  return 0;
}
```

Listing 24.2 Stack Module Header File (on the CD-ROM at `./source/ch24/byo/stack.h`)

```
#define STACK_CREATED   0xFAF32000

typedef struct {

  int state;
  int index;
  int max;
  int *storage;

} stack_t;


int stackCreate( stack_t *stack, int stackSize );

int stackCreate( stack_t *stack, int element );

int stackPop( stack_t *stack, int *element );

int stackDestroy( stack_t *stack );
```

Let's look at a simple regression that when built with this stack module can be used to verify that it works as expected. Since there are many individual tests that can be used to validate this module, we'll concentrate on just a few to illustrate the approach.

First, in this regression, we provide two infrastructure functions as wrappers for the regression. The first is a simple `main` function that invokes each of the tests, and the second is a result checking function. The result checking function simply tests the result input, and if zero, the test failed, otherwise the test passed. This function is shown in Listing 24.4 (we'll look at its use shortly).

We declare a failed integer, which will be used to keep track of the number of actual failures. This is used by our `main`, which allows it to determine if the regression passed or failed. The `checkResult` function (Listing 24.3) takes two inputs; a test number and the individual test result. If the test result is zero, then we mark the test as failed (and increment our failed count); otherwise the test passed (result is nonzero), and we indicate this.

Listing 24.3 Result Checking Function for a Simple Regression (on the CD-ROM at `./source/ch24/byo/regress.c`)

```
int failed;

```

```
void checkResult( int testnum, int result )
{
  if (result == 0) {
    printf( "*** Failed test number %d\n", testnum );
    failed++;
  } else {
    printf( "Test number %2d passed.\n", testnum );
  }
}
```

Our `main` program simply calls our regression tests in order, clearing the failed count (as shown in Listing 24.4).

Listing 24.4 Simple Regression `main` (on the CD-ROM at `./source/ch24/byo/regress.c`)

```
int main()
{

  failed = 0;
  test1();

  return 0;
}
```

Now let's look at a regression that focuses on creation and destruction of stacks. As we saw in the stack module source, there are numerous ways that a stack creation and destruction can fail. This test tries to address each of them so that we can convince ourselves that it's coded properly (see Listing 24.5).

In this regression, we call an API function with a set of input data and then check the result. In some cases, we pass good data, and in others we pass data that will cause the function to exit with a failure status. Consider lines 8–9, which test stack creation with a null stack element. This creation should fail and return -1. At line 9, we call `checkResult` with our test number (first argument) and then the test result as argument two. Note here that we test the `ret` variable with -1, since that's what we expect for this failure case. If `ret` wasn't -1, then the expression results in 0, indicating that the test failed. Otherwise, if `ret` is -1, the expression reduces to 1, and the result is a pass.

Our regression also explores the stack structure in order to ensure that the creation function has properly initialized the internal elements. At line 20, we check the internal state variable to ensure that it has been properly initialized with `STACK_CREATED`.

Listing 24.5 Stack Module Regression Focusing on Creation and Destruction (on the CD-ROM at `./source/ch24/byo/regress.c`)

```
void test1( void )
{
  stack_t myStack;
  int ret;

  failed = 0;

  ret = stackCreate( 0, 0 );
  checkResult( 0, (ret == -1) );

  ret = stackCreate( &myStack, 0 );
  checkResult( 1, (ret == -1) );

  ret = stackCreate( &myStack, 65536 );
  checkResult( 2, (ret == -1) );

  ret = stackCreate( &myStack, 1024 );
  checkResult( 3, (ret == 0) );

  checkResult( 4, (myStack.state == STACK_CREATED) );

  checkResult( 5, (myStack.index == 0) );

  checkResult( 6, (myStack.max == 1024) );

  checkResult( 7, (myStack.storage != (int *)0) );

  ret = stackDestroy( 0 );
  checkResult( 8, (ret == -1) );

  ret = stackDestroy( &myStack );
  checkResult( 9, (ret == 0) );

  checkResult( 10, (myStack.state != STACK_CREATED) );

  if (failed == 0) printf( "test1 passed.\n");
  else printf("test1 failed\n");
}
```

At the end of this simple regression, we indicate whether the entire test passed (all individual tests passed) or failed. A sample run of the regression is illustrated as:

```
# gcc -Wall -o test regress.c stack.c
# ./test
Test number  0 passed.
Test number  1 passed.
Test number  2 passed.
Test number  3 passed.
Test number  4 passed.
Test number  5 passed.
Test number  6 passed.
Test number  7 passed.
Test number  8 passed.
Test number  9 passed.
Test number 10 passed.
test1 passed.
#
```

While this method is quite simple, it's also very effective and permits the quick revalidation of a unit after changes are made. Once we've run our regression, we can deliver it safely with the understanding that we're less likely to break the software that uses it.

C UNIT TEST SYSTEM

The C Unit Test system, or *cut* for short, is a simple architecture for building unit test applications. Cut provides a simple environment for the integration of unit tests, with a set of tools that are used to pull together tests and then build a main framework to invoke and report them. Cut parses the unit test files and then builds them together into a single image, performing each of the provided unit tests in order.

Cut provides a utility, written in Python, called `cutgen.py`. This utility parses the unit test source files and then builds a main function around the tests that are to be run. The unit test files must be written in a certain way; we'll explore this in the sample source.

Let's walk through an example that verifies the push and pop functions of our previous stack example. Cut presents a simple interface that we build to, which is demonstrated in Listing 24.6.

We declare our locals at line 5 and 6 (our two stacks for which we'll perform our tests) and then four functions that serve as the interface to cut.

The first two functions to note are the first and last. These are the initialization function, `__CUT_BRINGUP__Explode` (at lines 8–20), and the post test execution function, `__CUT_TAKEDOWN__Explode` (at lines 74–85). The cut framework calls the bringup function prior to test start, and once all of the test functions have been performed, the takedown function is called. Note that we simply perform our necessary initialization (initialize our two stacks with stackCreate), but we also perform unit testing here to ensure that the required elements are available for us in the unit test. Similarly, in the takedown function, which is called once all unit tests have been called, we destroy our stacks, but we also check the return status of these calls.

The unit tests are encoded as function with a prefix __CUT__. This allows the cutgen utility to find them and call them within the test framework. As we perform the necessary elements of our unit tests, we call a special function called ASSERT, which is used to log errors. The ASSERT function has two pieces: a test expression and a string emitted if the test fails. The test expression identifies the success condition, and if false, then the test element fails.

Note that in some cases, we have multiple tests for each API element (such as shown for lines 64–65).

Listing 24.6 Unit Test Example Written for Cut (on the CD-ROM at ./source/ch24/cut/test_1.c)

```
1:      #include <stdio.h>
2:      #include "stack.h"
3:      #include "cut.h"
4:
5:      stack_t myStack_1;
6:      stack_t myStack_2;
7:
8:      void __CUT_BRINGUP__Explode( void )
9:      {
10:       int ret;
11:
12:       printf("Stack test bringup called\n");
13:
14:       ret = stackCreate( &myStack_1, 5 );
15:       ASSERT( (ret == 0), "Stack 1 Creation." );
16:
17:       ret = stackCreate( &myStack_2, 5 );
18:       ASSERT( (ret == 0), "Stack 2 Creation." );
19:
20:     }
21:
22:
23:     void __CUT__PushConsumptionTest( void )
24:     {
25:       int ret;
```

```

    /* Exhaust the stack */

    ret = stackPush( &myStack_1, 1 );
    ASSERT( (ret == 0), "Stack Push 1 failed." );

    ret = stackPush( &myStack_1, 2 );
    ASSERT( (ret == 0), "Stack Push 2 failed." );

    ret = stackPush( &myStack_1, 3 );
    ASSERT( (ret == 0), "Stack Push 3 failed." );

    ret = stackPush( &myStack_1, 4 );
    ASSERT( (ret == 0), "Stack Push 4 failed." );

    ret = stackPush( &myStack_1, 5 );
    ASSERT( (ret == 0), "Stack Push 5 failed." );

    ret = stackPush( &myStack_1, 6 );
    ASSERT( (ret == -1), "Stack exhaustion failed." );

  }


  void __CUT__PushPopTest( void )
  {
    int ret;
    int value;

            /* Test two pushes and then two pops */

    ret = stackPush( &myStack_2, 55 );
    ASSERT( (ret == 0), "Stack Push of 55 failed." );

    ret = stackPush( &myStack_2, 101 );
    ASSERT( (ret == 0), "Stack Push of 101 failed." );

    ret = stackPop( &myStack_2, &value );
    ASSERT( (ret == 0), "Stack Pop failed." );
            ASSERT( (value == 101), "Stack Popped Wrong Value." );

    ret = stackPop( &myStack_2, &value );
    ASSERT( (ret == 0), "Stack Pop failed." );
    ASSERT( (value == 55), "Stack Popped Wrong Value." );
```

```

}


void __CUT_TAKEDOWN__Explode( void )
{
  int ret;

  ret = stackDestroy( &myStack_1 );
  ASSERT( (ret == 0), "Stack 1 Destruction." );

  ret = stackDestroy( &myStack_2 );
  ASSERT( (ret == 0), "Stack 2 Destruction." );

  printf( "\n\nTest Complete\n");
}
```

Now that we have our unit test file (encoded for the cut environment), let's look at how we make this an executable test. While this can be easily encoded in a simple Makefile, we'll demonstrate command-line building for this example.

Three files are needed from the cut system (the URL from which these files can be obtained is provided in the "Resources" section of this chapter). The `cutgen.py` utility builds the unit test environment, given our set of unit test source files. This is a Python file, so a Python interpreter will be necessary on the target system. Two other files are `cut.h` and `libcut.inc`, which are ultimately linked with our unit test application.

The first step is creating the cut unit test environment. This creates C main and brings together the necessary APIs used by the unit tests. The `cutget.py` utility provides this for us, as demonstrated here:

```
cutgen.py test_1.c > cutcheck.c
```

Given our unit test file (Listing 24.6), we provide this as the single argument to `cutgen.py` and redirect the output into a source file called `cutcheck.c`. To further understand what `cutgen` has provided, let's now look at this file (shown in Listing 24.7).

The automatically generated file from `cutgen` simply brings together the unit tests present in our unit test files and calls them in order. We could provide numerous unit test files to `cutgen`, which would result in additional unit test functions being invoked within the generated file.

Listing 24.7 Unit Test Environment Source Created By `cutgen`

```
#include "libcut.inc"


extern void __CUT_BRINGUP__Explode( void );
extern void __CUT__PushConsumptionTest( void );
extern void __CUT__PushPopTest( void );
extern void __CUT_TAKEDOWN__Explode( void );


int main( int argc, char *argv[] )
{
  cut_init( -1 );

  cut_start( "Explode", __CUT_TAKEDOWN__Explode );
  __CUT_BRINGUP__Explode();
  __CUT__PushConsumptionTest();
  __CUT__PushPopTest();
  cut_end( "Explode" );
  __CUT_TAKEDOWN__Explode();


  cut_break_formatting();
  return 0;
}
```

Finally, we simply compile and link the files together to build a unit test image. This implies the automatically generated source file, unit test file, and the source to test (`stack.c`). This is illustrated here:

```
# gcc -o cutcheck stack.c test_1.c cutcheck.c
```

We can then execute the unit test by simply invoking `cutcheck`. This will emit numbers and `.` characters to indicate progress through the unit test process.

```
# ./cutcheck
Stack test bringup called

    0.........   10.....

Test Complete

#
```

The cut system provides some additional features that we've not addressed, but from this quick review, it's easy to see how powerful and useful this simple utility can be.

Embedded Unit Test

The Embedded Unit Test framework (called Embunit) is an interesting framework that's designed for embedded systems. The framework can operate without the need for standard C libraries and allocates objects from a `const` area. Embunit also provides a number of tools to generate test templates and also the `main` function for the test environment.

The Embunit framework is very similar to cut and provides a very useful API for testing. Some of the test functions that are provided in Embunit are shown in Table 24.1.

TABLE 24.1 Test Functions Provided By Embunit

Function	Purpose
`TEST_ASSERT_EQUAL_STRING(exp,actual)`	Assert on failed string compare
`TEST_ASSERT_EQUAL_INT(exp,actual)`	Assert on failed integer compare
`TEST_ASSERT_NULL(pointer)`	Assert if pointer is `NULL`
`TEST_ASSERT_NOT_NULL(pointer)`	Assert if pointer is not `NULL`
`TEST_ASSERT_MESSAGE(cond,message)`	Assert and emit message if the condition is false
`TEST_ASSERT(condition)`	Assert if the condition is false
`TEST_FAIL(message)`	Fail the test, emitting the message

Let's now look at a unit test coded for the Embunit framework, and then we'll look at the main program that sets up the environment. In Listing 24.8, a minimal unit test for Embunit is shown. At line 1, we include the `embUnit` header file (which makes the test interface available to our unit test). We then define two functions that are called before and after each unique unit test that's identified to `embUnit`: `setUp` at lines 7–10 and `tearDown` at lines 13–16.

We then define our individual unit tests (lines 19–54). Three tests are illustrated here; the first is called `testInit` (lines 19–25), the second is called `testPushPop` (lines 28–45), and the third is called `testStackDestroy` (lines 48–54). Like our earlier unit tests, we perform an action and then test the result. In this case, we use the Embunit-provided `TEST_ASSERT_EQUAL_INT` function to check the response, and if the assert fails, an error message is printed.

Finally, we specify the unit tests that are available within the `StackTest_tests` function (lines 57–68). This is done in the context of a test fixtures structure. We define `fixtures` as simply an array that's initialized with the provided functions and function names. Note that we provide our three unit tests here, providing a name for each to indicate the specific test in the event a failure occurs. We then create another structure at lines 64–65 that defines our test, `setup` function, `teardown` function, and `fixtures` (our list of unit tests). This new structure (called `StackTest`) is returned to the caller, which we'll explore shortly.

Listing 24.8 Unit Test Coded for the Embunit Framework (on the CD-ROM at `./source/ch24/emb/stackTest.c`)

```
 1:   #include <embUnit/embUnit.h>
 2:   #include <stdio.h>
 3:   #include "stack.h"
 4:
 5:   stack_t myStack;
 6:
 7:   static void setUp( void )
 8:   {
 9:     printf("setUp called.\n");
10:   }
11:
12:
13:   static void tearDown( void )
14:   {
15:     printf("tearDown called.\n");
16:   }
17:
18:
19:   static void testInit( void )
20:   {
21:     int ret;
22:
23:     ret = stackCreate( &myStack, 5 );
24:     TEST_ASSERT_EQUAL_INT( 0, ret );
25:   }
26:
27:
28:   static void testPushPop( void )
29:   {
30:     int ret, value;
```

```

  ret = stackPush( &myStack, 55 );
  TEST_ASSERT_EQUAL_INT( 0, ret );

  ret = stackPush( &myStack, 101 );
  TEST_ASSERT_EQUAL_INT( 0, ret );

  ret = stackPop( &myStack, &value );
  TEST_ASSERT_EQUAL_INT( 0, ret );
  TEST_ASSERT_EQUAL_INT( 101, value );

  ret = stackPop( &myStack, &value );
  TEST_ASSERT_EQUAL_INT( 0, ret );
  TEST_ASSERT_EQUAL_INT( 55, value );
}


static void testStackDestroy( void )
{
  int ret;

  ret = stackDestroy( &myStack );
  TEST_ASSERT_EQUAL_INT( 0, ret );
}


TestRef StackTest_tests( void )
{
  EMB_UNIT_TESTFIXTURES( fixtures ) {
    new_TestFixture("testInit", testInit ),
    new_TestFixture("testPushPop"", testPushPop ),
    new_TestFixture("testStackDestroy", testStackDestroy ),
  };
  EMB_UNIT_TESTCALLER( StackTest, "StackTest",
                       setUp, tearDown, fixtures );

  return( TestRef)&StackTest;
}
```

The `EMB_UNIT_TESTFIXTURES` *and* `EMB_UNIT_TESTCALLER` *are macros that create special arrays representing the individual unit tests (fixtures) as well as the unit test aggregate (*`StackTest`*).*

Our `main` program provides the means to invoke the unit tests (see Listing 24.9). We include the `embUnit` header file to gather the types and symbols. At line 3, we declare our previous function, which creates the text fixtures array (recall from Listing 24.8, lines 57–68). We call the `TestRunner_start` function to initialize the test environment and then invoke `TestRunner_runTest` with our unit test fixture `init` function (`StackTest_tests`). This invokes all of the unit tests that we discussed from Listing 24.8. When done, we call `TestRunner_end`, which emits statistics about the unit test, including the number of tests run and the number of failed tests.

ON THE CD

Listing 24.9 Embunit Main Program (on the CD-ROM at `./source/ch24/emb/main.c`)

```
 1:     #include <embUnit/embUnit.h>
 2:
 3:     TestRef StackTest_tests( void );
 4:
 5:     int main( int argc, const char *argv[] )
 6:     {
 7:       TestRunner_start();
 8:         TestRunner_runTest( StackTest_tests() );
 9:       TestRunner_end();
10:       return 0;
11:     }
```

Building the unit test within Embunit simply involves compiling and linking our source files together with the Embunit library. In order to find the Embunit library, we must specify its location (as well as the location of the header files). Building and running the unit test is illustrated as follows:

```
# gcc -Wall -I/usr/local/src/embunit/ \
      -L/usr/local/src/embunit/lib \
      -o stackTest main.c stack.c stackTest.c  -lembUnit
# ./stackTest
.setUp called.
testInit called
tearDown called.
.setUp called.
tearDown called.
.setUp called.
tearDown called.

OK (3 tests)
#
```

Using GCC, we build our image called `stackTest` and then invoke it. We see that `setUp` and `tearDown` are called three times each, before and after each of our unit tests. In the end, we see that the Embunit test environment reports that three tests were run and all were okay.

`expect` Utility

The `expect` utility has found wide use in the testing domain. `expect` is an application that scripts programmed dialogues with interactive programs. Using `expect`, we can spawn an application and then perform a script consisting of dialogue with the application. This dialogue consists of a series of statements and expected responses. In the test domain, the statements are the stimulus to the application, and the expected response is what we expect from a valid application under test.

The `expect` utility can even talk to numerous applications at once and has a very rich structure for application test.

Consider the following test in Listing 24.10. At line 3, we set an internal variable of `timeout` to 1. This tells expect that when the `timeout` keyword is used, the `timeout` will represent 1 second (instead of the default of 10 seconds). Next, we declare a procedure called `sendexpect`. This function provides both the `send` and `expect` behaviors in one function. It sends the `out` string (argument one) to the attached process. The `expect` function in this case uses the pattern match behavior. Two possibilities exist for what we expect as input from the attached process. If we receive the `in` string (argument 2 of the `sendexpect` procedure), then we've received what we expected and emit a `passed` message. Otherwise, we wait, and when the timeout occurs, we call it a failure and exit.

At line 14, we spawn our test process (in this case we're testing the `bc` calculator). We consume the input from `bc`'s startup by expecting the string warranty. We then emit an indicator to `stdout`, indicating that the test has started (line 17). At this point, we begin our test. Using the `sendexpect` procedure, we send a command to the application and then provide what we expect as a response. Since we're testing a calculator process, we provide some simple expressions (lines 20–22), followed by a slightly more complex function example. In the end, we emit the `bc quit` command, expecting `eof` (an indication that the `bc` application terminated).

ON THE CD

Listing 24.10 Testing the `bc` Calculator with `expect` (on the CD-ROM at `./source/ch24/expect/test_bc`)

```
1:     #!/usr/local/bin/expect -f
2:
3:     set timeout 1
4:
```

```
proc sendexpect { out in } {
  send $out
  expect {
    $in         { puts "passed\n" }
    timeout     { puts "***failed\n" ; exit }
  }
}

# Start the bc test
spawn bc
expect "warranty"

puts "Test Starting\n"

# Test some simple math
sendexpect "2+4\n" "6"
sendexpect "2*8\n" "16"
sendexpect "9-2\n" "7"

# Test a simple function
sendexpect \
  "define f (x) { if (x<=1) return(1); return(f(x-1) * x); }\n"
  "\r"
sendexpect "f(5)\n" "120"

# End the test session
sendexpect "quit\n" eof

puts "Test Complete\n"

exit
```

The `expect` method differs greatly from our unit test examples, but it is a very powerful mechanism not only for testing but also for automated tasks, even those on remote systems.

SUMMARY

The unit testing discipline has improved with open source tools. These tools (and more complex ones such as the DejaGnu framework) provide efficient and simple mechanisms for unit testing. In this chapter, we explored unit testing (as opposed to system testing) and investigated a number of methods to achieve it. We first reviewed a brew your own method for testing that was simple but got the job done. We then reviewed two open source tools for unit testing, the C unit test system (cut), and then the Embedded Unit Test system. Finally, we took a quick look at `expect` and its capabilities for process-based testing.

RESOURCES

The C Unit Test System at *http://sourceforge.net/projects/cut/*
Embedded Unit Test Framework at *http://sourceforge.net/projects/embunit/*
Expect Home Page at *http://expect.nist.gov/*

25 Debugging with GDB

In This Chapter

- Source Debugging with GDB
- Debugging Multiprocess Applications
- Debugging Multithreaded Applications
- Debugging Programs Already Running
- Post-mortem Debugging

INTRODUCTION

The GNU Debugger (also known as GDB) is a source-level debugger that provides the ability to debug applications at the source and machine levels. Additionally, GDB permits the debugging of already running applications (by attaching to the application's process ID) as well as debugging applications post-mortem. All of the traditional features one would expect from a source-level debugger are available with GDB, including multilanguage and multiarchitecture support.

This chapter will introduce GDB and explore its features and capabilities in a tutorial manner.

Compiling for GDB

Before we jump into GDB, we'll first need to know how to build our application so that it's debuggable by GDB. The `-g` flag tells the compiler to include debugging information in the image, which can be used by GDB to understand variable types (for data inspection) and machine instruction to source-line mappings. Compiling is illustrated as:

```
# gcc -g testapp.c -o testapp
```

The test image can now be successfully debugged via GDB. One very important point to note is that debugging with optimization enabled can yield odd results. The optimizer may move code around or remove code altogether. This can make an optimized debugging session confusing and hard to follow. Therefore, while GDB will still debug optimized code, it's much easier to debug an unoptimized image.

USING GDB

Let's now dive into GDB and look at its capabilities for debugging C applications. We'll look at some of the most common methods for debugging with GDB using breakpoints. We'll demonstrate using command-line GDB, though GUI versions exist.

One important point to note before we jump in is that the program being debugged uses the same terminal input and output as GDB. This is suitable for our purposes here.

It is possible to redirect `stdin` *and* `stdout` *for the program's I/O. We can redirect the output on the command line when we start the GDB session. We can also specify a new terminal for the program's* `stdin` *using the* `tty` *shell command.*

To demonstrate GDB, we'll use the source shown in Listing 25.1. This source represents a very simple stack implementation that provides math operators.

Listing 25.1 Example Source for the GDB Debugging Session (on the CD-ROM at `./source/ch25/testapp.c`)

```
#include <stdio.h>
#include <assert.h>

#define MAX_STACK_ELEMS        10

#define OP_ADD                 0
#define OP_SUBTRACT            1
#define OP_MULTIPLY            2
#define OP_DIVIDE              3
```

```
typedef struct {
  int stack[MAX_STACK_ELEMS];
  int index;
} STACK_T;


void initStack( STACK_T *stack )
{
  assert( stack );
  stack->index = 0;
}


void push( STACK_T *stack, int elem )
{
  assert( stack );
  assert( stack->index < MAX_STACK_ELEMS );

  stack->stack[stack->index++] = elem;
  return;
}


int pop( STACK_T *stack )
{
  assert( stack );
  assert( stack->index > 0 );

  return( stack->stack[—stack->index] );
}


void operator( STACK_T *stack, int op )
{
  int a, b;

  assert( stack );
  assert( stack->index > 0 );

  a = pop(stack); b = pop(stack);

  switch( op ) {
```

```
    case OP_ADD:
      push( stack, (a+b) ); break;

    case OP_SUBTRACT:
      push( stack, (a-b) ); break;

    case OP_MULTIPLY:
      push( stack, (a*b) ); break;

    case OP_DIVIDE:
      push( stack, (a/b) ); break;

    default:
      assert(0); break;

  }

}


int main()
{
  STACK_T stack;

  initStack(&stack);

  push( &stack, 2 );
  push( &stack, 5 );
  push( &stack, 2 );
  push( &stack, 3 );
  push( &stack, 5 );
  push( &stack, 3 );
  push( &stack, 6 );

  operator( &stack, OP_ADD );
  operator( &stack, OP_SUBTRACT );
  operator( &stack, OP_MULTIPLY );
  operator( &stack, OP_DIVIDE );
  operator( &stack, OP_ADD );
  operator( &stack, OP_SUBTRACT );

  printf( "Result is %d\n", pop( &stack ) );
  return 0;
}
```

We compile our source with the `-g` flag to include debugging information for GDB, as:

```
# gcc -g -Wall -o testapp testapp.c
#
```

Starting GDB

To debug a program with GDB, we simply execute GDB with our program name as the first argument. We can also start GDB and then load our program using the `load` command. Here we start GDB with our application:

```
# gdb testapp
GNU gdb Red Hat Linux (5.3post-0.20021129.18rh)
Copyright 2003 Free Software Foundation, Inc.
GDB is free software, covered by the GNU General Public
License, and you are welcome to change it and/or distribute
copies of it under certain conditions.
Type "show copying" to see the conditions.
There is absolutely no warranty for GDB.  Type "show warranty"
for details.
This GDB was configured as "i386-redhat-linux-gnu"...
(gdb)
```

The `(gdb)` is the regular prompt for GDB and indicates that it is available for commands. We could start our application using the `run` command, but we'll look at a few other commands first.

Looking at Source

Once we start GDB, our application is not yet running, but instead is just loaded into GDB. Using the `list` command, we can view the source of our application, as demonstrated below:

```
(gdb) list
      }

    }


    int main()
    {
      STACK_T stack;
```

```
78
79          initStack(&stack);
    (gdb)
```

The main is our entry point, so list here shows this entry. We can also specify the lines of interest with list as:

```
    (gdb) list 75,85
75        int main()
76        {
77          STACK_T stack;
78
79          initStack(&stack);
80
81          push( &stack, 2 );
82          push( &stack, 5 );
83          push( &stack, 2 );
84          push( &stack, 3 );
85          push( &stack, 5 );
    (gdb)
```

Using the list command with no arguments will always list the source with the current line centered in the list.

Using Breakpoints

The primary strategy for debugging with GDB is the use of breakpoints to stop the running program and allow inspection of the internal data. A breakpoint can be set in a variety of ways, but the most common is specifying a function name. Here, we tell GDB to break at our main program:

```
(gdb) break main
Breakpoint 1 at 0x804855b: file testapp.c, line 79.
(gdb) run
Starting program: /home/mtj/gnulinux/ch25/testapp

Breakpoint 1, main () at testapp.c:79
79        initStack(&stack);
(gdb)
```

Once we give the break command, GDB tells us our breakpoint number (since we may set multiple) and the address, filename, and line number of the breakpoint. We then start our application using the run command, which results in hitting our previously set breakpoint. Once the breakpoint is hit, GDB shows the line that will be executed next. Note that this statement, line 79, is the first executable statement of our application.

Recall that in all C applications, the `main` *function is the user entry point to the application, but various other work goes on behind the scenes to start and end the program. Therefore, when we break at the* `main` *function, we break at our user entry point, but not the true start of the application.*

We can view the available breakpoints using the `info` command:

```
(gdb) info breakpoints
Num Type           Disp Enb Address    What
1   breakpoint     keep y   0x0804855b in main at testapp.c:79
        breakpoint already hit 1 time
(gdb)
```

We see our single breakpoint and an indication from GDB that this breakpoint has been hit.

If our breakpoint is now of no use, we remove it using the `clear` command:

```
(gdb) clear 79
Deleted breakpoint 1
(gdb)
```

Other methods for setting breakpoints are shown in Table 26.1.

TABLE 25.1 Available Methods for Setting Breakpoints

Command	Breakpoint Method
`break function`	Set a breakpoint at a function
`break file:function`	Set a breakpoint at a function
`break line`	Set a breakpoint at a line number
`break file:line`	Set a breakpoint at a line number
`break address`	Set a breakpoint at a physical address

One final interesting breakpoint method is the conditional breakpoint. Consider the following command:

```
(gdb) break operator if op = 2
Breakpoint 2 at 0x8048445: file testapp.c, line 48.
```

This tells GDB to break at the `operator` function if the `op` argument is equal to two (`OP_MULTIPLY`). This can be very useful if you're looking for a given condition rather than having to break at each call and check the variable.

Stepping Through the Source

When we left our debugging session, we had hit a breakpoint on our `main` function. Let's now step forward through the source. We have a few different possibilities, depending upon what we want to achieve (Table 26.2 lists these). To execute the next line of code, we can use the `step` command. This will also step into a function (if a function call is the next line to execute). If we'd prefer to step over a function, we could use the `next` command, which executes the next line and, if a function, simply performs it and sets the next line to execute to the line after the function. The `cont` command (short for continue) simply starts the program running.

TABLE 25.2 Methods for Stepping Through the Source

Command (shortcut)	Operation
`next (n)`	Execute next line, step over functions
`step (s)`	Execute next line, step into functions
`cont (c)`	Continue execution

We can also provide a count after the `next` and `step` commands, which performs the command the number of times specified as the argument. For example, issuing the command `step 5` will perform the `step` command five times.

We illustrate the `next` and `step` commands within our debugging session as follows:

```
Breakpoint 1, main () at testapp.c:79
79        initStack(&stack);
(gdb) s
initStack (stack=0xbfffde60) at testapp.c:20
20        assert( stack );
(gdb) s
21        stack->index = 0;
```

```
(gdb) s
22      }
(gdb) s
main () at testapp.c:81
81          push( &stack, 2 );
(gdb) n
82          push( &stack, 5 );
(gdb)
```

In this last debugging fragment, we step into the `initStack` function. GDB then lets us know where we are (the function name and stack address). We step through the lines of `initStack` and upon returning, GDB let's us know again that we're back in the `main` function. We then use the `next` command to perform the `push` function with a value of 2.

Inspecting Data

GDB makes it easy to inspect the data within a running program. Continuing from our debugging session, let's now look at our stack structure. We do this with the `display` command:

```
(gdb) display stack
1: stack = {stack = {2, 0, 1107383313, 134513378,
    1108545272, 1108544020, -1073750392, 134513265,
    1108544020, 1073792624}, index = 1}
(gdb)
```

If we simply display the `stack` variable, we see the aggregate components of the structure (first the array itself, then the `index` variable). Note that many of the stack elements are unusually large numbers, but this is only because the structure was not initialized. We can inspect specific elements of the `stack` variable, also using the `display` command:

```
(gdb) display stack.index
2: stack.index = 1
(gdb)
```

If we were dealing with an object reference (a pointer to the structure), we could deal with it as we would in C. For example, in this next example, we step into the `push` function to illustrate dealing with an object reference:

```
(gdb) s
push (stack=0xbffffae0, elem=5) at testapp.c:27
```

```
27        assert( stack );
(gdb) display stack->index
3: stack->index = 1
(gdb) display stack->stack[0]
4: stack->stack[0] = 2
(gdb)
```

One important consideration is the issue of static data. Static data names may be used numerous times in an application (bad coding policy, but it happens). To display a specific instance of static data, we can reference both the variable and file, such as `display 'file2.c'::variable`.

The `print` command (or its shortcut, `p`) can also be used to display data.

Changing Data

It's also possible to change the data in an operating program. We use the `set` command to change data, illustrated as:

```
(gdb) set stack->stack[9] = 999
(gdb) p *stack
$11 = {stack = {2, 0, 1107383313, 134513378,
    1108545272, 1108544020, -1073743096, 134513265,
    1108544020, 999}, index = 1}
(gdb)
```

Here we see that we've modified the last element of our stack array and then printed it back out to monitor the change.

Examining the Stack

The `backtrace` command (or `bt` for short) can be used to inspect the stack. This can tell us the current active function trace and the parameters passed. We're currently in the `push` function in our debugging session; let's look at the stack backtrace:

```
(gdb) bt
#0  push (stack=0xbffffae0, elem=5) at testapp.c:27
#1  0x08048589 in main () at testapp.c:82
#2  0x42015504 in __libc_start_main () from /lib/tls/libc.so.6
(gdb)
```

At the top is the current stack frame. We're in the `push` function, with a stack reference and an element of 5. The second frame is the function that called `push`, in this case, the `main` function. Note here that `main` was called by a function `__libc_start_main`. This function provides the initialization for `glibc`.

Stopping the Program

It's also possible to stop a debugging session using Ctrl+C. If the program is stopped in a function for which no debugging information is available (it wasn't compiled with `-g`), then only assembly will be displayed (since source debugging information is not available).

OTHER GDB DEBUGGING TOPICS

In this section, we'll discuss some other topics of GDB, such as multiprocess application debugging and post-mortem debugging.

Multiprocess Application Debugging

One problem with the debugging of multiprocess applications is which process to follow when a new process is created. Recall from Chapter 12, "Introduction to Sockets Programming," that the `fork` function returns to both the parent and child processes. We can tell GDB which to returns to `follow` using the `follow-fork-mode` command. For example, if we wanted to debug the child process, we'd specify to `follow` the child process as:

```
set follow-fork-mode child
```

Or, if we instead wanted to follow the parent (the default mode), we'd specify this as:

```
set follow-fork-mode parent
```

In either case, when GDB follows one process, the other process (child or parent) continues to run unimpeded. We can also tell GDB to ask us which process to follow when a `fork` occurs, as:

```
set follow-fork-mode ask
```

When the `fork` occurs, GDB will ask which to follow. Whichever is not followed will execute normally.

Multithreaded Application Debugging

There's no other way to put it: Debugging multithreaded applications is difficult at best. GDB offers some capabilities that assist in multithreaded debugging, and we'll look at those here.

The breakpoint is one of the most important aspects of debugging, but its behavior is different in multithreaded applications. If a breakpoint is created at a source line used by multiple threads, then every thread is affected by the breakpoint. We can limit this by specifying the thread to be affected. For example:

```
(gdb) break pos.c:17 thread 5
```

This installs a breakpoint at line 20 in `myfile.c`, but only for thread number 5. We can further refine these breakpoints using thread qualifiers. For example:

```
11      void *posThread( void *arg )
12      {
13        int ret;
14
15        ret = checkPosition( arg );
16
17        if (ret == 0) {
18
19          ret = move( arg );
20        }
21    (gdb) b pos.c:17 thread 5 if ret > 0
      Breakpoint 1 at 0x8048550: file pos.c, line 19
      (gdb)
```

In this example, we specify to break at line 17 in file `pos.c` for thread 5, but here we qualify that thread 5 will be stopped only if the local `ret` variable is greater than 0.

We can identify the threads that are currently active in a multithreaded application using the `info threads` command. This command lists each of the active threads and its current state. For example:

```
(gdb) info threads
  5 Thread -161539152 (LWP 2819)  posThread (arg=0x0) at pos.c:17
...
* 1 Thread -151046720 (LWP 2808)  init at init.c:154
(gdb)
```

The `*` before thread 1 identifies that it is the current focus of the debugger. We could switch to any thread using the `thread` command, which allows us to change the focus of the debugger to the specified thread.

```
    (gdb) thread 1
    [Switching to thread 1 (Thread -161539152 (LWP 2819))]#0  posThread
17    if (ret == 0) {
    (gdb)
```

As we step through a multithreaded program, we'll find that the focus of the debugger can change at any step. This can be annoying, especially when the current thread is what we're interested in debugging. We can instruct GDB not to preempt the current thread by locking the scheduler. For example:

```
(gdb) set scheduler-locking on
```

This tells GDB not to preempt the current thread. When we want to allow other threads to preempt our current thread, we can set the mode to off:

```
(gdb) set scheduler-locking off
```

Finally, we can identify the current mode using the `show` command:

```
(gdb) show scheduler-locking
Mode for locking scheduler during execution is "on".
(gdb)
```

One final important command for thread debugging is the ability to apply a single command to all threads within an application. The `thread apply all` command is used for this purpose. For example, the following command will emit a stack backtrace for every active thread:

```
(gdb) thread apply all backtrace
```

The `thread apply` command can also apply to a list of threads instead of all threads, as illustrated below:

```
(gdb) thread apply 1 4 9 backtrace
```

This performs a stack backtrace on threads 1, 4, and 9.

Debugging an Existing Process

We can debug an application that is currently running by attaching GDB to the process. All that we need is the process identifier for the process to debug. In this example, we've started our application in one terminal and then started GDB in another. Once GDB has started, we issue the `attach` command to attach to the process. This suspends the process, allowing us to control it.

```
$ gdb
GNU gdb Red Hat Linux (5.3post-0.20021129.18rh)
...
```

```
This GDB was configured as "i386-redhat-linux-gnu".
(gdb) attach 23558
Attaching to process 23558
Reading symbols from /home/mtj/gnulinux/ch25/testapp...done.
Reading symbols from /lib/tls/libc.so.6...done.
Loaded symbols for /lib/tls/libc.so.6
Reading symbols from /lib/ld-linux.so.2...done.
Loaded symbols for /lib/ld-linux.so.2
0x08048468 in operator (stack=0xbfffe9e0, op=1) at testapp.c:51
51        a = pop(stack); b = pop(stack);
(gdb) bt
#0  0x08048468 in operator (stack=0xbfffe9e0, op=1) at testapp.c:51
#1  0x080485cc in main () at testapp.c:93
#2  0x42015504 in __libc_start_main () from /lib/tls/libc.so.6
(gdb)
```

This method is very useful for dealing with "hung" programs where the fault occurs only after some period of time, or for dealing with unexpected hangs in production environments.

GDB starts by loading the symbols for the process and then identifying where the process was suspended (in the `operator` function). We issue the `bt` command to list the backtrace, which tells us which particular invocation of `operator` we're in (in this case, an `OP_SUBTRACT` call). Finally, if we're done debugging, we can release the process to continue by detaching from it using the `detach` call:

```
(gdb) detach
Detaching from program: /home/mtj/gnulinux/ch25/testapp,
process 23558
(gdb) quit
$
```

Once the `detach` command has finished, our process continues normally.

Postmortem Debugging

When an application aborts and dumps a resulting core dump file, GDB can be used to identify what happened. Our application has been hardened, but we'll remove a couple of asserts in the `push` function in order to force a core dump.

To enable GNU/Linux to generate a core dump, the command `ulimit -c unlimited` *should be performed. Otherwise, with limits in place, core dump files will not be generated.*

We execute our application to get the core dump:

```
$ ./testapp
Segmentation fault (core dumped)
$ ls
core.23730  testapp  testapp.c
```

Now that we have our core dump, we can use GDB to identify where things went wrong. In the following example, we specify the executable application and the core dump image to GDB. It loads the app and uses the core dump file to identify what happened at the time of failure. After all the symbols are loaded, we see that the function failure occurred at `push` (but we already knew that). What's most important is that we see someone called `push` with a stack argument of 0 (null pointer). We would have caught this with our `assert` function, but it was conveniently removed for the sake of demonstration.

Further down, we see that the offending call was made at `testapp` line 30. This happens to be a call that we added to force the creation of this core file.

```
# gdb testapp core.23730
GNU gdb Red Hat Linux (5.3post-0.20021129.18rh)
...
Core was generated by `./testapp'.
Program terminated with signal 11, Segmentation fault.
Reading symbols from /lib/tls/libc.so.6...done.
Loaded symbols for /lib/tls/libc.so.6
Reading symbols from /lib/ld-linux.so.2...done.
Loaded symbols for /lib/ld-linux.so.2
#0  0x0804839c in push (stack=0x0, elem=2) at testapp.c:30

30        stack->stack[stack->index++] = elem;
(gdb) bt
#0  0x0804839c in push (stack=0x0, elem=2) at testapp.c:30
#1  0x08048536 in main () at testapp.c:81
#2  0x42015504 in __libc_start_main () from /lib/tls/libc.so.6
(gdb)
```

Although that was a quick review, it covers many of the necessary features that are needed for debugging with GDB.

SUMMARY

A source-level debugger such as GDB is a necessary tool for developing applications of any size. This quick review of GDB introduced compiling for GDB debugging and many of the most useful commands. Other topics such as multiprocess application debugging and postmortem debugging were also discussed.

RESOURCES

GDB: The GNU Project Debugger Web site at *http://www.gnu.org/source/gdb/documentation/*

26 Code Hardening

In This Chapter

- An Introduction to Code Hardening
- Code Hardening Techniques
- Tools Support for Code Hardening
- Tracing Binary Applications

INTRODUCTION

The practice of code hardening (or defensive programming) is a useful technique to increase the quality and reliability of software. The practice entails anticipating where errors can occur in our code and then writing that code in a way that either avoids them altogether or identifies them immediately so that their source can be more easily tracked. Since C is not a safe language, some methods have proven invaluable to help build more reliable programs, which we'll detail here.

This chapter will cover a number of techniques under the umbrella of code hardening, all of which can be applied immediately. Since the benefits are clear, let's jump right into this chapter. We'll look at a variety of code-hardening methods, as well as tool-based techniques such as using the compiler or open source tools to help build secure and reliable GNU/Linux applications.

CODE HARDENING TECHNIQUES

Code hardening can take a number of different forms, and entire books have been written on the topic. In this section, we'll look at a variety of techniques that can help build better code.

Return Values

The failure to check return values is one of the most common mistakes made in modern software. Many applications call user or system functions and are very optimistic about their successful operation. When building hardened software, all reasonable attempts should be made to check return values, and if failures are found, deal with them appropriately. *Reasonable attempts* is a key here; consider the following bogus example:

```
ret = printf( "Current mode is %d\n", mode );
if ( ret < 0 ) {
  ret = printf( "An error occured emitting mode.\n" );
}
```

The point is easily illustrated, but in most cases (of user and system calls) the return value is relevant and should be checked in every case.

Strongly Consider User/Network I/O

Whenever we develop applications that take input either from a user or from the network (such as a Sockets application), it's even more critical to scrutinize the incoming data. Errors such as insufficient data for a given operation or more data received than buffer space is available for are two of the most common.

Use Safe String Functions

A number of standard C library functions suffer from security problems. The problem they present is that there's no bounds checking, which means that they can be exploited (we'll discuss the buffer overflow issue shortly). The simple solution to this problem is to avoid unsafe functions and instead use the safe versions (as shown in Table 26.1).

Buffer Overflow

Buffer overruns cause unpredictable software behavior in the best case and security exploits in the worst. Buffer overruns can be avoided very simply. Consider the following erroneous example:

```
static char myArray[10];
...
int i;
for ( i = 0 ; i < 10 ; i++ ) {
  myArray[i] = (char)(0x30+i);
}
myArray[i] = 0;     // <—Overrun
```

TABLE 26.1 Safe Replacements for C Library Functions

Unsafe Function	Safe Replacement	Header
gets	fgets	stdio.h
sprintf	snprintf	stdio.h
strcat	strncat	string.h
strcpy	strncpy	string.h
strcmp	strncmp	string.h
strcasecmp	strncasecmp	strings.h
vsprintf	vsnprintf	stdio.h

In this example, we've overrun the bounds of our array by writing to the eleventh element. Whatever object follows this array is now corrupted. There's actually a very simple solution to this problem, and it involves a better programming practice using symbolic constants. In the next example, we create a constant defining the size of our array, but then we add one more element for the trailing NULL.

```
#define ARRAY_SIZE    10
static char myArray[ARRAY_SIZE+1];
...
int i;
for (i = 0 ; i < ARRAY_SIZE ; i++ ) {
  myArray[i] = (char)(0x30+i);
}
myArray[ARRAY_SIZE] = 0;
```

We've automatically protected our array by an extra element at the end, but also—in good programming practice—we've used a symbol to denote the size of the array, rather than relying on a number.

Provide Logical Alternatives at Decision Points

A very common mistake that can yield unpredictable results is the absence of a `default` section in a `switch` statement. Consider the following example:

```
switch( mode ) {
  case OPERATIONAL_MODE:
    /* switch to operational mode processing */
    break;
  case BUILT_IN_TEST_MODE:
    /* switch to test processing */
    break;
}
```

In the event another mode was added but this particular code segment was not updated, the result after this segment has executed is unpredictable. The solution is to *always* include a `default` section that either asserts (in debugging mode) or at a minimum notifies the caller that a problem has occurred. If we're really not expecting another mode, we can simply `assert` here to catch the condition during debugging:

```
switch( mode ) {
  case OPERATIONAL_MODE:
    /* switch to operational mode processing */
    break;
  case BUILT_IN_TEST_MODE:
    /* switch to test processing */
    break;
  default:
    assert(0);
    break;
}
```

A similar problem exists with `if/then/else` chains. The following example illustrates the problem:

```
float multiplier = 0.0;
if (state == FIRST_STAGE) multiplier = 0.75;
else if (state == SECOND_STAGE) multiplier = 1.25;
```

If our state is corrupted or takes on a value that we did not expect, then our multiplier takes on the value of 0.0, and the result is unpredictable at best and, depending upon the application, catastrophic at worst. An `else` should be provided to at a minimum catch the issue, such as seen here:

```
float multiplier = 0.0;
if (state == FIRST_STAGE) multiplier = 0.75;
else if (state == SECOND_STAGE) multiplier = 1.25;
else multiplier = SAFE_MULTIPLIER;
```

In many cases, the trailing `else` isn't necessary, but whenever one is seen, it should be given extra scrutiny to avoid erroneous results.

Self-Identifying Structures

A self-identifying structure is a method that mimics the concept of runtime type checking present in strongly typed languages. In a strongly typed language, the use of an invalid type results in a runtime error. Consider the passing of pointers in a weakly typed language such as C. With C typecasting, it's not difficult to confuse one structure for another.

With a simple policy change to C structures and a limited amount of checking, we can help ensure that functions are dealing with the right types. Consider the C source shown in Listing 26.1. At lines 6–12, we see our target structure, which contains a special header called a *signature* (sometimes called a *runtime type identifier*). The type is shown at line 4, in this case simply a signature that uniquely represents our structure. We then provide two macro functions that initialize (`INIT_TARGET_MARKER`) and then check (`CHECK_TARGET_MARKER`) the signature in the structure.

Skipping ahead a little, we look at the `main` function at lines 34–54. We allocate two objects (both of size `targetMarket_t`) and then initialize one of them as an actual target marker using the `INIT_TARGET_MARKER` macro. Finally, we try to display each of the objects by passing each to the `displayTarget` function.

In our `displayTarget` function (lines 22–31), our first task is to check the signature of the received object by calling `CHECK_TARGET_MARKER`. If the signature is not correct, we `assert` rather than risk providing bogus information. Granted, in a production system we could probably handle this better, but this illustrates the concept.

ON THE CD

Listing 26.1 Illustrating a Self-identifying Structure (on the CD-ROM at `./source/ch26/selfident.c`)

```
1:     #include <stdio.h>
2:     #include <assert.h>
3:
4:     #define TARGET_MARKER_SIG        0xFAF32000
5:
6:     typedef struct {
7:
```

```
    unsigned int signature;
    unsigned int targetType;
    double       x, y, z;

  } targetMarker_t;


  #define INIT_TARGET_MARKER(ptr) \
            (((targetMarker_t *)ptr)->signature =
               TARGET_MARKER_SIG)
  #define CHECK_TARGET_MARKER(ptr) \
            assert(((targetMarker_t *)ptr)->signature == \
                      TARGET_MARKER_SIG)


  void displayTarget( targetMarker_t *target )
  {

    /* Pre-check of the target structure */
    CHECK_TARGET_MARKER(target);

    printf( “Target type is %d\n”, target->targetType );

    return;
  }


  int main()
  {
    void *object1, *object2;

    /* Create two objects */
    object1 = (void *)malloc( sizeof(targetMarker_t) );
    assert(object1);
    object2 = (void *)malloc( sizeof(targetMarker_t) );
    assert(object2);
    /* Init object1 as a target marker struct */
    INIT_TARGET_MARKER(object1);

    /* Try to display object1 */
    displayTarget( (targetMarker_t *)object1 );

    /* Try to display object2 */
    displayTarget( (targetMarker_t *)object2 );
```

```
    return 0;
}
```

Reporting Errors

The reporting of errors is an interesting topic because the policy that's chosen can be very different, depending upon the type of application we're developing. For example, if we're writing a command-line utility, emitting error messages to `stderr` is a common method to communicate errors to the user. But what happens if we're building an application that has I/O capabilities, such as an embedded Linux application? There are a number of possibilities, including the generation of a specialized log or use of the standard system log (`syslog`). The `syslog` function has the prototype:

```
#include <syslog.h>
void syslog( int priority, char *format, ... );
```

To the `syslog` function, we provide a priority, a format string, and some arguments (similar to `printf`). The priority can be one of `LOG_EMERG`, `LOG_ALERT`, `LOG_CRIT`, `LOG_ERR`, `LOG_WARNING`, `LOG_NOTICE`, `LOG_INFO`, or `LOG_DEBUG`. An example of using `syslog` to generate a message to the system log is shown in Listing 26.2.

Listing 26.2 Simple Example of `syslog` Use (on the CD-ROM at `./source/ch26/simpsyslog.c`)

```
#include <syslog.h>

int main()
{

  syslog( LOG_ERR, "Unable to load configuration!" );

  return 0;
}
```

This results in our system log (stored within our filesystem at `/var/log/messages`) being updated as:

```
Jul 21 18:13:10 camus sltest: Unable to load configuration!
```

In this example, our application in Listing 26.2 was called `sltest`, with the hostname of `camus`. The system log can be especially useful because it's an aggregate of many error reports. This allows a developer to see where a message was generated in relation to others, which can be very useful in helping to understand problems.

The `syslog` *is very useful for communicating information for system applications and daemons.*

One final topic on error reporting is that of being specific about the error being reported. The error message must uniquely identify the error in order for the user to be able to deal with it reasonably.

Reducing Complexity Also Reduces Potential Bugs

Code that is of higher complexity potentially contains more bugs. It's a fact of life, but one that we can use to help reduce defects. In some disciplines this is called *refactoring*, but the general goal is to take a complex piece of software and break it up so that it's more easily understood. This very act can lead to higher quality software that is more easily maintained.

Self-Protective Functions

Writing self-protective functions can be a very useful debugging mechanism to ensure that your software is correct. The programming language Eiffel includes language features to provide this mechanism (known as *programming-by-contract*).

Being self-protective means that when you write a function, you scrutinize the input to the function and, upon completion of its processing, scrutinize the output to ensure that what you've done is correct.

Let's look at an example of a simple function that illustrates this behavior (see Listing 26.3).

If an expression results in false (0), the `assert` *function causes the application to fail and an error to be generated to* `stdout`. *To disable* `asserts` *within an application, the* `NDEBUG` *symbol can be defined, which causes the* `assert` *calls to be optimized away.*

Listing 26.3 Example of a Self-protective Function (on the CD-ROM at `./source/ch26/selfprot.c`)

```
STATUS_T checkAntennaStatus( ANTENNA_T antenna, MODE_T *mode )
{
  ANTENNA_STS_T retStatus;

  /* Validate the input */
  assert( validAntenna( antenna ) );
  assert( validMode( mode ) );

```

```

    /*-------------------------------------*/
    /* Internal checkAntennaStatus processing */
    /*-------------------------------------*/


    /* We may have changed modes, check it. */
    assert( validMode( mode ) );

    return retStatus;
}
```

In Listing 26.3 we see a function that first ensures that it's getting good data (validating input) and then that what it's providing is correct (checking output). We also could have returned errors upon finding these conditions, but for this example, we're mandating proper behavior at all levels. If all functions performed this activity, finding the real source of bugs would be a snap.

The use of `assert` isn't restricted just to ensuring that function inputs and outputs are correct. It can also be used for internal consistency. Any critical failure that should be identified during debugging is easily handled with `assert`.

Using the `assert` *call for internal consistency is often the only practical way to find timing (race condition) bugs in threaded code.*

Maximize Debug Output

Too much output can disguise errors; too little and an error could be missed. The right balance must be found when emitting debug and error output to ensure that only the necessary information is presented, to avoid overloading an already overloaded user.

Memory Debugging

There are many libraries available that support debugging dynamic memory management on GNU/Linux. One of the most popular is called Electric Fence, which programs the underlying processor's MMU (memory management unit) to catch memory errors via segment faults. Electric Fence can also detect exceeding array bounds. The Electric Fence library is very powerful and identifies memory errors immediately.

Compiler Support

The compiler itself can be an invaluable tool to identify issues in our code. When we build software, we should always enable warnings using the `-Wall` flag. To further ensure that warnings aren't missed in large applications, we can enable the `-Werror` flag, which treats warnings as errors and therefore halts further compilation of a source file. When building an application that has many source files, this combination can be beneficial. This is demonstrated as:

```
gcc -Wall -Werror test.c -o test
```

If we want our source to have ANSI compatibility, we can enable checking for ANSI compliance (with pedantic checking) as:

```
gcc -ansi -pedantic test.c -o test
```

Identifying uninitialized variables is a very useful test, but in addition to the warning option, optimization must also be enabled, because the data flow information is available only when the code is optimized:

```
gcc -Wall -O -Wuninitialized test.c -o test
```

Chapter 4, "The GNU Compiler Toolchain," provides additional warning information. The `gcc` main page also contains numerous warning options about those enabled via `-Wall`.

SOURCE CHECKING TOOLS

To identify security vulnerabilities as well as common programming mistakes, source-checking tools should be part of the development process. In addition to being simple to use, they can easily be automated as part of the build process. One important note when using source checking tools is that while they can identify flaws, they can also miss them. Therefore, use your best judgment when using the tools, and always know your source.

The `splint` tool (short for *secure programming lint*) is a static source checking tool built by the Inexpensive Program Analysis group at the University of Virginia. It provides strong and weak checking of source and, with annotation, can perform a very complete analysis of source.

With unannotated source, the `-weak` option can be used (with header files found in the `./inc` subdirectory):

```
splint -weak *.c -I./inc
```

Splint also supports modes for standard checking (`-standard`, the default mode), moderate checking (`-checks`), and extremely strict checking (`-strict`).

The `flawfinder` tool (developed by David Wheeler) is another useful tool that statically checks source in search of errors. `flawfinder` provides useful error messages that can be tutorial in nature. Consider the following example:

```
$ flawfinder test.c
test.c:11:  [2] (buffer) char:
  Statically-sized arrays can be overflowed. Perform bounds
  checking, use functions that limit length, or ensure that
  the size is larger than the maximum possible length.
$
```

In this case, an array was found that does not necessarily present a security issue, but a gentle reminder is provided of the potential for exploitation.

Many other source checking tools exist, such as `RATS` (Rough Auditing Tool for Security) and `ITS4` (static vulnerability scanner). URLs for these tools can be found in the "Resources" section of this chapter.

CODE TRACING

One final useful topic is that of system call tracing. While not specifically a source auditing tool, it can be a very useful tool for understanding the underlying operation of a GNU/Linux application. The `strace` utility provides the capability to trace the execution of an application from the perspective of system calls (such as `fopen` or `fwrite`, to name just two).

Consider the application shown in Listing 26.4. This application violates many of the code hardening principles already discussed, but we'll see how we can still debug it using `strace`.

Listing 26.4 Poorly Hardened Application (on the CD-ROM at `./source/ch26/badprog.c`)

```
#include <unistd.h>
#include <fcntl.h>

#define MAX_BUF         128

int main()
{
  int fd;
  char buf[MAX_BUF+1];
```

```
10:
11:        fd = open( "myfile.txt", O_RDONLY );
12:
13:        read( fd, buf, MAX_BUF );
14:
15:        printf( "read %s\n", buf );
16:
17:        close( fd );
18:    }
```

The first thing to note about this application is that at line 11, where we attempt to open the file called `myfile.txt`, there is no checking to ensure that the file actually exists. Executing this application will give an unpredictable result:

```
$ gcc -o bad bad.c
$ ./bad
read @êÿ¿8Z@
$
```

This is not what we expected, so now let's use `strace` to see what's going on. We'll shrink the output a bit, since we're not interested in the libraries that are loaded and such.

```
$ strace ./bad
execve("./bad", ["./bad"], [/* 20 vars */]) = 0
uname({sys="Linux", node="camus", ...}) = 0
...
open("myfile.txt", O_RDONLY)        = -1 ENOENT (No such file or
                                              directory)
read(-1, 0xbfffef20, 128)           = -1 EBADF (Bad file descriptor)
fstat64(1, {st_mode=S_IFCHR|0620, st_rdev=makedev(136, 0), ...}) = 0
mmap2(NULL, 4096, PROT_READ|PROT_WRITE, MAP_PRIVATE|MAP_ANONYMOUS,
-1, 0) = 0x40017000
write(1, "read \300\357\377\2778Z\1@\n", 14read Àïÿ¿8Z@
) = 14
close(-1)                           = -1 EBADF (Bad file descriptor)
munmap(0x40017000, 4096)            = 0
exit_group(-1)                      = ?
$
```

After executing our app, we see that the `execve` system call is used to actually start the program. We then see an `open` shortly after execution, which matches our source (line 11, Listing 26.4). We can see at the right that the `open` system call returned –1, with an error of `ENOENT` (the file doesn't exist). This tells us right away what's going on with our application. The attempted `read` also fails, with the error of a bad file descriptor (since the `open` call failed).

The `strace` tool can be useful not only to understand the operation of our programs, but also the operation of programs for which we may not have source. From the perspective of system calls, we can at some level understand what binary applications are up to.

SUMMARY

Code hardening can increase our development time, but it routinely reduces our debugging time. By anticipating faults while we design, we automatically increase the reliability and quality of our software, so this technique is one to be mastered. In this chapter we discussed a variety of code hardening techniques, as well as non-coding methods to help create better software and understand its operation.

RESOURCES

Secure Programming for Linux and UNIX HOWTO at *http://www.dwheeler.com/secure-programs/Secure-Programs-HOWTO/index.html*

Electric Fence `malloc()` Debugger at *http://perens.com/FreeSoftware/*

Splint Source Checking Tool at *http://www.splint.org/*

Flawfinder Source Checker at *http://www.dwheeler.com/flawfinder/*

RATS Source Checker at *http://www.securesoftware.com/security_tools_download.htm*

ITS4 Static Vulnerability Scanner at *http://www.cigital.com/its4/*

Appendix

A Acronyms and Partial Acronyms

AMD	Advanced Micro Devices
API	Application Programmer's Interface
ASCII	American Standard Code for Information Interchange
AT&T	American Telephone and Telegraph
AWK	Aho-Weinberger-Kernighan
BASH	Bourne-Again SHell
BB	Basic Block
BBG	Basic Block Graph
BNF	Backus-Naur Form
BSD	Berkeley Software Distribution
BSS	Block Started by Symbol
CMU	Carnegie Mellon University
COW	Copy-On-Write
CPU	Central Processing Unit
CSE	Common Subexpression Elimination
DEC	Digital Equipment Corporation
DMA	Direct Memory Access
DNS	Domain Name Server
DL	Dynamically Loaded
DWARF	Debugging with Attribute Record Format
EMACS	EMACS Makes a Computer Slow
ELF	Executable and Linking Format
EOF	End of File
EXT2	2nd Extended Filesystem
EXT3	3rd Extended Filesystem
FIFO	First-In-First-Out
FQDN	Fully Qualified Domain Name
FS	Field Separator
FSF	Free Software Foundation
GCC	GNU Compiler Collection

GCOV	GNU Coverage
GDB	GNU DeBugger
GID	Group ID
GLIBC	GNU C Library
GMT	Greenwich Mean Time
GNU	GNU's Not Unix
GOT	Global Offset Table
GPROF	GNU Profiler
HTML	Hypertext Markup Language
HTONL	Host TO Network Long
HTONS	Host TO Network Short
HUP	HangUP
IP	Internet Protocol
IBM	International Business Machines
IPC	Inter-Process Communication
IPv4	Internet Protocol Version 4
KB	KiloByte
LIFO	Last-In-First-Out
LISP	List Processor
MINIX	Miniature Unix
MIT	Massachusetts Institute of Technology
MMU	Memory Management Unit
MUTEX	Mutual Exclusion
NF	Number of Fields
NFS	Network File System
NPTL	Native POSIX Thread Library
NR	Number of Record
NTOHL	Network TO Host Long
NTOHS	Network TO Host Short
OFS	Output Field Separator
ORS	Output Record Separator
OSI	Open Source Initiative
PDP	Programmed Data Processor
PGRP	Process Group
PIC	Position Independent Code
PID	Process Identifier
POSIX	Portable Operating System Interface
PWD	Present Working Directory
QID	Queue Identifier
QPL	Qt Public License
RAM	Random Access Memory

REGEX	Regular Expression
RS	input Record Separator
SCSH	Scheme Shell
SED	Stream Editor
SPLINT	Secure Programming Lint
STDERR	Standard Error
STDIN	Standard Input
STDOUT	Standard Output
SYSV	System 5
TAR	Tape Archive
TCL	Tool Command Language
TCP	Transmission Control Protocol
TLB	Translation Lookaside Buffer
UDP	User Datagram Protocol
UID	User ID
UTC	Coordinated Universal Time
VFS	Virtual File System
VI	Visual Interface
VPATH	Virtual Path

Appendix

B About the CD-ROM

The CD-ROM included with GNU/Linux Application Programming includes all example applications found in the book.

CD-ROM FOLDERS

source: Contains all the code from examples in the book, by chapter.

figures: Contains all the figures in the book, by chapter.

OVERALL SYSTEM REQUIREMENTS

- Linux with a 2.4 or 2.6 Kernel (tested with Red Hat and Fedora)
- Pentium I Processor or greater
- CD-ROM drive
- Hard drive
- 256 MBs of RAM
- 1 MB of hard drive space for the code examples

Appendix

C Software License

GNU GENERAL PUBLIC LICENSE

Preamble

The licenses for most software are designed to take away your freedom to share and change it. By contrast, the GNU General Public License is intended to guarantee your freedom to share and change free software—to make sure the software is free for all its users. This General Public License applies to most of the Free Software Foundation's software and to any other program whose authors commit to using it. (Some other Free Software Foundation software is covered by the GNU Library General Public License instead.) You can apply it to your programs, too.

When we speak of free software, we are referring to freedom, not price. Our General Public Licenses are designed to make sure that you have the freedom to distribute copies of free software (and charge for this service if you wish), that you receive source code or can get it if you want it, that you can change the software or use pieces of it in new free programs; and that you know you can do these things.

To protect your rights, we need to make restrictions that forbid anyone to deny you these rights or to ask you to surrender the rights. These restrictions translate to certain responsibilities for you if you distribute copies of the software, or if you modify it.

For example, if you distribute copies of such a program, whether gratis or for a fee, you must give the recipients all the rights that you have. You must make sure that they, too, receive or can get the source code. And you must show them these terms so they know their rights.

We protect your rights with two steps: (1) copyright the software, and (2) offer you this license which gives you legal permission to copy, distribute and/or modify the software.

Also, for each author's protection and ours, we want to make certain that everyone understands that there is no warranty for this free software. If the software is modified by someone else and passed on, we want its recipients to know that what they have is not the original, so that any problems introduced by others will not reflect on the original authors' reputations.

Finally, any free program is threatened constantly by software patents. We wish to avoid the danger that redistributors of a free program will individually obtain patent licenses, in effect making the program proprietary. To prevent this, we have made it clear that any patent must be licensed for everyone's free use or not licensed at all.

The precise terms and conditions for copying, distribution and modification follow.

GNU GENERAL PUBLIC LICENSE
TERMS AND CONDITIONS FOR COPYING, DISTRIBUTION AND MODIFICATION

0. This License applies to any program or other work which contains a notice placed by the copyright holder saying it may be distributed under the terms of this General Public License. The "Program," below, refers to any such program or work, and a "work based on the Program" means either the Program or any derivative work under copyright law: that is to say, a work containing the Program or a portion of it, either verbatim or with modifications and/or translated into another language. (Hereinafter, translation is included without limitation in the term "modification.") Each licensee is addressed as "you."

Activities other than copying, distribution and modification are not covered by this License; they are outside its scope. The act of running the Program is not restricted, and the output from the Program is covered only if its contents constitute a work based on the Program (independent of having been made by running the Program). Whether that is true depends on what the Program does.

1. You may copy and distribute verbatim copies of the Program source code as you receive it, in any medium, provided that you conspicuously and appropriately publish on each copy an appropriate copyright notice and disclaimer of warranty; keep intact all the notices that refer to this License and to the absence of any warranty; and give any other recipients of the Program a copy of this License along with the Program.

You may charge a fee for the physical act of transferring a copy, and you may at your option offer warranty protection in exchange for a fee.

2. You may modify your copy or copies of the Program or any portion of it, thus forming a work based on the Program, and copy and distribute such modifications or work under the terms of Section 1 above, provided that you also meet all of these conditions:

a) You must cause the modified files to carry prominent notices stating that you changed the files and the date of any change.

b) You must cause any work that you distribute or publish, that in whole or in part contains or is derived from the Program or any part thereof, to be licensed as a whole at no charge to all third parties under the terms of this License.

c) If the modified program normally reads commands interactively when run, you must cause it, when started running for such interactive use in the most ordinary way, to print or display an announcement including an appropriate copyright notice and a notice that there is no warranty (or else, saying that you provide a warranty) and that users may redistribute the program under these conditions, and telling the user how to view a copy of this License. (Exception: if the Program itself is interactive but does not normally print such an announcement, your work based on the Program is not required to print an announcement.)

These requirements apply to the modified work as a whole. If identifiable sections of that work are not derived from the Program, and can be reasonably considered independent and separate works in themselves, then this License, and its terms, do not apply to those sections when you distribute them as separate works. But when you distribute the same sections as part of a whole which is a work based on the Program, the distribution of the whole must be on the terms of this License, whose permissions for other licensees extend to the entire whole, and thus to each and every part regardless of who wrote it.

Thus, it is not the intent of this section to claim rights or contest your rights to work written entirely by you; rather, the intent is to exercise the right to control the distribution of derivative or collective works based on the Program.

In addition, mere aggregation of another work not based on the Program with the Program (or with a work based on the Program) on a volume of a storage or distribution medium does not bring the other work under the scope of this License.

3. You may copy and distribute the Program (or a work based on it, under Section 2) in object code or executable form under the terms of Sections 1 and 2 above provided that you also do one of the following:

a) Accompany it with the complete corresponding machine-readable source code, which must be distributed under the terms of Sections 1 and 2 above on a medium customarily used for software interchange; or,

b) Accompany it with a written offer, valid for at least three years, to give any third party, for a charge no more than your cost of physically performing source distribution, a complete machine-readable copy of the corresponding source code, to be distributed under the terms of Sections 1 and 2 above on a medium customarily used for software interchange; or,

c) Accompany it with the information you received as to the offer to distribute corresponding source code. (This alternative is allowed only for noncommercial distribution and only if you received the program in object code or executable form with such an offer, in accord with Subsection b above.)

The source code for a work means the preferred form of the work for making modifications to it. For an executable work, complete source code means all the source code for all modules it contains, plus any associated interface definition files, plus the scripts used to control compilation and installation of the executable. However, as a special exception, the source code distributed need not include anything that is normally distributed (in either source or binary form) with the major components (compiler, kernel, and so on) of the operating system on which the executable runs, unless that component itself accompanies the executable.

If distribution of executable or object code is made by offering access to copy from a designated place, then offering equivalent access to copy the source code from the same place counts as distribution of the source code, even though third parties are not compelled to copy the source along with the object code.

4. You may not copy, modify, sublicense, or distribute the Program except as expressly provided under this License. Any attempt otherwise to copy, modify, sublicense or distribute the Program is void, and will automatically terminate your rights under this License. However, parties who have received copies, or rights, from you under this License will not have their licenses terminated so long as such parties remain in full compliance.

5. You are not required to accept this License, since you have not signed it. However, nothing else grants you permission to modify or distribute the Program or its derivative works. These actions are prohibited by law if you do not accept this License. Therefore, by modifying or distributing the Program (or any work based on the Program), you indicate your acceptance of this License to do so, and all its terms and conditions for copying, distributing or modifying the Program or works based on it.

6. Each time you redistribute the Program (or any work based on the Program), the recipient automatically receives a license from the original licensor to copy, distribute or modify the Program subject to these terms and conditions. You may not impose any further restrictions on the recipients' exercise of the rights granted herein. You are not responsible for enforcing compliance by third parties to this License.

7. If, as a consequence of a court judgment or allegation of patent infringement or for any other reason (not limited to patent issues), conditions are imposed on you (whether by court order, agreement or otherwise) that contradict the conditions of this License, they do not excuse you from the conditions of this License. If you cannot distribute so as to satisfy simultaneously your obligations under this License and any other pertinent obligations, then, as a consequence, you may not distribute the Program at all. For example, if a patent license would not permit royalty-free redistribution of the Program by all those who receive copies directly or indirectly through you, then the only way you could satisfy both it and this License would be to refrain entirely from distribution of the Program.

If any portion of this section is held invalid or unenforceable under any particular circumstance, the balance of the section is intended to apply and the section as a whole is intended to apply in other circumstances.

It is not the purpose of this section to induce you to infringe any patents or other property right claims or to contest validity of any such claims; this section has the sole purpose of protecting the integrity of the free software distribution system, which is implemented by public license practices. Many people have made generous contributions to the wide range of software distributed through that system in reliance on consistent application of that system; it is up to the author/donor to decide if he or she is willing to distribute software through any other system and a licensee cannot impose that choice.

This section is intended to make thoroughly clear what is believed to be a consequence of the rest of this License.

8. If the distribution and/or use of the Program is restricted in certain countries either by patents or by copyrighted interfaces, the original copyright holder who places the Program under this License may add an explicit geographical distribution limitation excluding those countries, so that distribution is permitted only in or among countries not thus excluded. In such case, this License incorporates the limitation as if written in the body of this License.

9. The Free Software Foundation may publish revised and/or new versions of the General Public License from time to time. Such new versions will be similar in spirit to the present version, but may differ in detail to address new problems or concerns.

Each version is given a distinguishing version number. If the Program specifies a version number of this License which applies to it and "any later version," you have the option of following the terms and conditions either of that version or of any later version published by the Free Software Foundation. If the Program does not specify a version number of this License, you may choose any version ever published by the Free Software Foundation.

10. If you wish to incorporate parts of the Program into other free programs whose distribution conditions are different, write to the author to ask for permission. For software which is copyrighted by the Free Software Foundation, write to the Free Software Foundation; we sometimes make exceptions for this. Our decision will be guided by the two goals of preserving the free status of all derivatives of our free software and of promoting the sharing and reuse of software generally.

NO WARRANTY

11. BECAUSE THE PROGRAM IS LICENSED FREE OF CHARGE, THERE IS NO WARRANTY FOR THE PROGRAM, TO THE EXTENT PERMITTED BY APPLICABLE LAW. EXCEPT WHEN OTHERWISE STATED IN WRITING THE COPYRIGHT HOLDERS AND/OR OTHER PARTIES PROVIDE THE PROGRAM "AS IS" WITHOUT WARRANTY OF ANY KIND, EITHER EXPRESSED OR IMPLIED, INCLUDING, BUT NOT LIMITED TO, THE IMPLIED WARRANTIES OF MERCHANTABILITY AND FITNESS FOR A PARTICULAR PURPOSE. THE ENTIRE RISK AS TO THE QUALITY AND PERFORMANCE OF THE PROGRAM IS WITH YOU. SHOULD THE PROGRAM PROVE DEFECTIVE, YOU ASSUME THE COST OF ALL NECESSARY SERVICING, REPAIR OR CORRECTION.

12. IN NO EVENT UNLESS REQUIRED BY APPLICABLE LAW OR AGREED TO IN WRITING WILL ANY COPYRIGHT HOLDER, OR ANY OTHER PARTY WHO MAY MODIFY AND/OR REDISTRIBUTE THE PROGRAM AS PERMITTED ABOVE, BE LIABLE TO YOU FOR DAMAGES, INCLUDING ANY GENERAL, SPECIAL, INCIDENTAL OR CONSEQUENTIAL DAMAGES ARISING OUT OF THE USE OR INABILITY TO USE THE PROGRAM (INCLUDING BUT NOT LIMITED TO LOSS OF DATA OR DATA BEING RENDERED INACCURATE OR LOSSES SUSTAINED BY YOU OR THIRD PARTIES OR A FAILURE OF THE PROGRAM TO OPERATE WITH ANY OTHER PROGRAMS), EVEN IF SUCH HOLDER OR OTHER PARTY HAS BEEN ADVISED OF THE POSSIBILITY OF SUCH DAMAGES.

END OF TERMS AND CONDITIONS

Appendix: How to Apply These Terms to Your New Programs

If you develop a new program, and you want it to be of the greatest possible use to the public, the best way to achieve this is to make it free software which everyone can redistribute and change under these terms.

To do so, attach the following notices to the program. It is safest to attach them to the start of each source file to most effectively convey the exclusion of warranty; and each file should have at least the "copyright" line and a pointer to where the full notice is found.

> one line to give the program's name and a brief idea of what it does.
> Copyright © 19yy name of author
>
> This program is free software; you can redistribute it and/or modify it under the terms of the GNU General Public License as published by the Free Software Foundation; either version 2 of the License, or (at your option) any later version.
>
> This program is distributed in the hope that it will be useful, but WITHOUT ANY WARRANTY; without even the implied warranty of MERCHANTABILITY

> or FITNESS FOR A PARTICULAR PURPOSE. See the GNU General Public License for more details.
>
> You should have received a copy of the GNU General Public License along with this program; if not, write to the Free Software Foundation, Inc., 675 Mass Ave, Cambridge, MA 02139, USA.

Also add information on how to contact you by electronic and paper mail.

If the program is interactive, make it output a short notice like this when it starts in an interactive mode:

> Gnomovision version 69, Copyright © 19yy name of author
> Gnomovision comes with ABSOLUTELY NO WARRANTY; for details type 'show w'.
>
> This is free software, and you are welcome to redistribute it under certain conditions; type 'show c' for details.

The hypothetical commands 'show w' and 'show c' should show the appropriate parts of the General Public License. Of course, the commands you use may be called something other than 'show w' and 'show c'; they could even be mouse-clicks or menu items—whatever suits your program.

You should also get your employer (if you work as a programmer) or your school, if any, to sign a "copyright disclaimer" for the program, if necessary. Here is a sample; alter the names:

> Yoyodyne, Inc., hereby disclaims all copyright interest in the program 'Gnomovision' (which makes passes at compilers) written by James Hacker.
>
> signature of Ty Coon, 1 April 1989
> Ty Coon, President of Vice

This General Public License does not permit incorporating your program into proprietary programs. If your program is a subroutine library, you may consider it more useful to permit linking proprietary applications with the library. If this is what you want to do, use the GNU Library General Public License instead of this License.

Index

T